Lecture Notes on Data Engineering and Communications Technologies

Volume 258

Series Editor

Fatos Xhafa, Technical University of Catalonia, Barcelona, Spain

The aim of the book series is to present cutting edge engineering approaches to data technologies and communications. It will publish latest advances on the engineering task of building and deploying distributed, scalable and reliable data infrastructures and communication systems.

The series will have a prominent applied focus on data technologies and communications with aim to promote the bridging from fundamental research on data science and networking to data engineering and communications that lead to industry products, business knowledge and standardisation.

Indexed by SCOPUS, INSPEC, EI Compendex.

All books published in the series are submitted for consideration in Web of Science.

Ghislaine Pellat · Žaneta Lacová · Marian Šuplata
Editors

Data-Centric Business and Applications

How a Business and Territorial Approach Can Challenge the Crises in Europe?

Springer

Editors
Ghislaine Pellat
Université Grenoble Alpes
Grenoble, France

Marian Šuplata
Faculty of Management
Comenius University
Bratislava, Slovakia

Žaneta Lacová
Faculty of Economics
Matej Bel University
Banská Bystrica, Slovakia

ISSN 2367-4512 ISSN 2367-4520 (electronic)
Lecture Notes on Data Engineering and Communications Technologies
ISBN 978-3-031-95067-4 ISBN 978-3-031-95068-1 (eBook)
https://doi.org/10.1007/978-3-031-95068-1

This Springer imprint is published by the registered company Springer Nature Switzerland AG
The registered company address is: Gewerbestrasse 11, 6330 Cham, Switzerland

Introduction

Europe in Times of Crisis: A Shared Reflection

The contributions in this volume explore the multiple transformations shaping contemporary Europe. Through a wide range of case studies and analyses, they address the ways individuals, businesses, institutions, and territories are navigating a time marked by uncertainty, disruption, and change. The backdrop to these reflections is a growing recognition that Europe is not facing one crisis, but many as health, environmental, economic, political and technological crises. These crises often overlap and reinforce each other. This condition, often described as a "polycrisis," demands new orientations of thinking and acting.

Throughout the chapters, a shared question emerges: how can Europe respond, adapt, and move forward in a world defined by complexity and volatility? The answers offered are diverse. They converge around several key ideas.

First one is that there is a strong focus on the changing role of individuals, especially young people. Whether through civic engagement, new forms of education, or digital inclusion, many chapters highlight a generational shift toward more collaborative, values-based participation in European society. This shift reflects both a desire for meaning and a reaction to recent shocks such as the COVID-19 pandemic and geopolitical tensions.

The second idea is the transformation of work and employment which seems central. Digital tools, remote work, and artificial intelligence are reshaping how people work, what they expect from employers, and what kinds of skills are now essential. The contributions underline the challenges of managing human capital in a fast-changing world, while also pointing to opportunities for innovation and resilience.

At the same time, the role of firms is being redefined. No longer seen as purely economic actors, companies are now expected to take on responsibilities linked to sustainability, ethics, and social impact. New regulations and reporting standards reflect this evolution. Businesses that embrace transparency, environmental care, and inclusive governance are shown to be more trusted and more future-ready.

The territorial dimension is also present across the chapters. Local and regional experiences reveal both vulnerability and innovation. Some contributions emphasize how territories, through public policies and local cooperations, can act as laboratories for adaptation and collective problem-solving.

Finally, the volume raises essential questions about the future of European integration. While crises expose tensions between different economic models, political visions, or levels of development, they also call for stronger cooperation. The European Union's capacity to act as a cohesive and strategic force depends on its ability to support flexible, plural, and coordinated responses.

Rather than presenting crises as breaks in progress, the volume treats them as catalysts for necessary transformation. It argues for moving beyond temporary solutions and toward a more inclusive, sustainable, and intelligent model of development. This model would be one that is equipped for uncertainty, responsive to social needs, and grounded in shared values.

Contents

European Crises as Catalysts

The chapters gathered in this volume address a wide range of transformations currently shaping the European space. From political reconfigurations and youth engagement to technological adaptation and institutional resilience, these contributions form a cohesive reflection on how crises, both recent and ongoing, have tested and in some cases redefined, the European project.

Three major thematic axes emerge.

The first concerns the evolving relationship between citizens, particularly youth, and the European Union. Educational initiatives and cross-border learning projects reveal a generational shift toward more collaborative, value-driven European identities. Civic engagement is increasingly shaped by transnational experiences, intercultural teamwork, and exposure to EU-level norms.

The second concerns, the recurring crises of the last decade, namely the COVID-19 pandemic and the war in Ukraine which were acting as accelerators of institutional stress and behavioral change. These events serve as analytical lenses for understanding shifts in political attitudes, consumer behavior, and cross-border cooperation. The dual impact of health and geopolitical emergencies has highlighted not only the fragility of interconnected systems but also the capacity of communities and institutions to adapt under pressure.

Thirdly, digital transformation features prominently across sectors. Whether in rural development, food retail, or monetary innovation, digital tools are reframing access to services, modes of participation, and economic structures. The deployment of artificial intelligence and digital currencies illustrates both opportunities and risks: improved efficiency on the one hand, but also ethical, social, and institutional concerns on the other.

Beneath these common threads lie several tensions. The promise of deeper integration is often counterbalanced by structural divergences as economic models, governance preferences, or digital divides. In similarity, while innovation is presented as a solution, its uneven distribution risks which reinforce the pre-existing inequalities.

Overall, the contributions suggest that Europe's capacity to navigate uncertainty depends less on uniformity than on its ability to cultivate flexible, plural, and cooperative frameworks. Crises, rather than halting progress, have exposed the emergency of rethinking foundational models toward more resilient, inclusive, and strategically coordinated responses at both national and supranational levels.

The Commitment of EU Youths to the European Project

Andreia Gabriela Andrei, Ghislaine Pellat, Teodora Roman, Adriana Zait, Irina Manolescu, Katarina Chomova, Zuzana Kittová, Paulína Krnáčová, Žaneta Lacová, Jana Marasova, Anna Vallusova, Juliana Vassileva, and Evelina Staykova

Abstract The European Union (EU) has faced significant challenges for many years, starting with the 2008 global financial crisis that led to the recession of the Eurozone, continuing with migration issues, Brexit in 2016, and the recent rise of populist and nationalist parties across Europe. In this context, the paper advances CitEuroPass questionaire and examines European youth's opinions and attitudes towards civic engagement and social values promoted by EU (i.e. commitment to democracy, human rights, equality, diversity, inclusion)—as increasingly important aspects of the EU's resilience. EU citizenship and European identity shape how individuals view, relate to and participate in the EU project at cultural, social and political levels. While multicultural exchange and emotional connections can strengthen active, engaged EU citizenship, strong social values provide a common foundation for member states to work together, and help to ensure that all citizens are treated fairly and equally under the law. Together, the social values and civic engagement are important aspects of the European Union (EU) as they ensure that social concerns are addressed effectively, that citizens are actively involved in decision-making processes and they participate in shaping the future of their communities.

Keywords Civic engagement · Social values · EU citizenship · EU youths · Resilience

A. G. Andrei (✉) · T. Roman · A. Zait · I. Manolescu
Faculty of Economics and Business Administration, Alexandru Ioan Cuza University of Iasi, Iasi, Romania
e-mail: andrei.andreia@gmail.com; andreia.andrei@uaic.ro

G. Pellat
CERAG, University Grenoble Alpes, Grenoble, France

K. Chomova · Z. Kittová · P. Krnáčová
Faculty of Commerce, University of Economics in Bratislava, Bratislava, Slovakia

Ž. Lacová · J. Marasova · A. Vallusova
Faculty of Economics, Matej Bel University in Banská Bystrica, Banská Bystrica, Slovakia

J. Vassileva · E. Staykova
Department of Administration and Management, New Bulgarian University, Sofia, Bulgaria

G. Pellat et al. (eds.), *Data-Centric Business and Applications*, Lecture Notes on Data Engineering and Communications Technologies 258,
https://doi.org/10.1007/978-3-031-95068-1_1

1 Introduction

The paper reports the prelimiary findings of a study applied in the spring of 2023 with students enroled in universities from France, Slovakia, Bulgaria and Romania.

The study focused on developing and testing the beta version of CitEuroPass questionaire, an instrument measuring respondents opinions, attitudes, and behavioral intentions related to the topics adressed by CitEuroPass project. The study was developed within the project *Erasmus + "CitEuroPass—Learning, Experiencing and Validating European citizenship through collaborative and innovative projects"*, as a collaborative work of five universities: Université Grenoble Alpes, France; University of Economics in Bratislava, Slovakia; Matej Bel University in Banská Bystrica, Slovakia; New Bulgarian University, Sofia, Bulgaria; Alexandru Ioan Cuza University of Iasi, Romania.

Therefore, the CitEuroPass project represents the context of this study. So, the paper begins with a brief presentation of the aims of CitEuroPass project—as the starting point for the development and the composition of CitEuroPass questionaire. Then the paper focusses on discussing study's approach, the development of CitEuroPass questionaire and its first application: the context, the literature, data collection, the sample, the prelininary results and measurement reliability. The results are presented in the last part, followed by final discussions.

2 CitEuro Pass Project—A Brief Description of Approach and Aims

Aiming to cultivate European citizenship, the spirit of cooperation among youth, and the feeling of a European identity, CitEuroPass project embraces the approach of Moccia (2018). According to the author, European citizenship is rooted in the concept of civitas, which refers to a community of people who recognize each other as fellow citizens within a shared public space. This notion of citizenship is not defined by membership in a pre-existing state, but rather emerges organically from the relationships and bonds formed between citizens. At its core, European citizenship is about people coming together voluntarily around a common purpose and shared values (Moccia 2018).

Therefore, the CitEuroPass project brings together students from 5 European universities to participate in training sessions and collaborative projects with peers from other EU countries. Through collaborative projects based on frugal innovation, it gives participants the opportunity to work and learn together, to experiment and create solutions that address environmental problems. In this way, it allows students to understand citizenship not as an abstract concept but as an enacted, cooperative practice. By working together, students participating in the project discover what unites Europeans and how collaboration can solve shared problems. This real-world training equips students with practical skills and an expanded, inclusive outlook needed to thrive in today's Europe. Furthermore, the relationships built during this program are expected to last beyond the project itself, contributing to a long-lasting spirit of togethernes, to a more integrated European community.

3 CitEuroPass Questionaire—Instrument Development

CitEuroPass questionnaire was developed between November 2022 and March 2023 as a pilot version of an anonymous survey, and a first data collection was organized between March and mid-June 2023 on a convenience sample of participants recruited in France, Slovakia, Bulgaria and Romania from students enrolled in the universities involved in the CitEuroPass project. The questionnaire was developed to measure respondents attitudes, opinions and behavioral intentions related to the topics adressed by CitEuroPass project—i.e. teamwork with fellows from other EU countries, aspects related to civic engagenment, European citizenship and identity, European values, social responsibility, innovation and entrepreneurial spirit directed to bring solutions to environmental issues.

The questionaire aimed to assess students' attitudes and behavioral intentions related to the topics of CitEuroPass project, and find out whether participants from different EU countries have rather similar or divergent opinions. The reliability of the measurement instrument was also an objective of evaluation.

The Citeuropass questionnaire comprised 65 items for the topics of interest plus socio-demographic characteristics. Most of the items (62 items) used 1 to 5 Likert scales (ratings from 1 = *not at all* to 5 = *very much*) to collect responses to closed-ended questions allowing statistical analysis, one closed-ended item used a Yes/No scale to asses the willingness to participate in collaborative projects, while 2 items were open-ended questions meant to get deeper insights regarding perceptions of European Citizenship. These two open-ended questions asked participants to input words that they think would best describe the European spirit and to enter comments about what European citizenship means to them, respectively.

The questionnaire was developed from the literature (Spears et al. 1997; Francois-Lecompte, and Roberts 2006; Doolittle and Faul 2013; La Barbera and Capone 2016; Moccia 2018, European Commission—Directorate-General for Education, Youth, Sport and Culture, Publications Office (2019); Schimmelfennig and Sedelmeier 2020; Becuwe and Baneth 2021; Dąbrowski et al. 2021; Bogoslov et al. 2022; Arslan and Wong 2022; European Commission 2023; Remr 2023; Hadjitchoneva et al. 2024), through the inclusion and adaptation of multi-item constructs needed to measure the dimensions of interest, such as: openness towards intercultural collaboration and teamwork, collaborative work between peers from different EU countries, European citizenship and identity, European values, citizen engagement - civic attitude and civic behavior, social responsibility and environment, entrepreneurial orientation, and entrepreneurship for sustainability).

A brief review of the literature and official websites of European Union on the themes covering aforementioned dimensions of interest indicates that EU is actively working to foster an integrated European community with shared identity and values (European Union. Official Journal of the European Union 2012; European Commission 2023), where citizen engagement, rule of law, visions of unity, purposeful cooperation and innovation building offer paths forward (Sedelmeier 2017; Sedelmeier and Avery 2018; Schimmelfennig and Sedelmeier 2020; Gastil 2021). The following lines presents few words about each dimension.

European Citizenship and Identity. European citizenship grants important rights and a sense of belonging to EU residents. Most of EU citizens feel attached to Europe, indicating a nascent community identity (European Commission 2023). EU citizenship links closely to 'European identity' (Mäkinen 2019), whereas European identity is seen as a supranational social identity, shaping individuals' emotional connection and self-concept tied to EU membership (Tajfel 1978). Trying to capture the nuances of identification, scholars typically examine both cognitive and affective dimensions - i.e. self-identifying as European and emotional attachment to Europe (Mitchell 2015; Prieto-Arranz et al. 2023). Education plays a key role in fostering identity rooted in shared values, though stronger emotional ties are still developing (Delanty 1997; Lehning 2001; Bruter 2003; European Commission, Directorate-General for Education, Youth, Sport and Culture, Publications Office, 2019). Active citizenship is seen as attitudinal foundation for constructive engagement from local to EU levels (European Commission 2019). Work is underway to increase civic participation in EU policymaking and strengthen democratic legitimacy across member states (European Commision, 2023a, b, c, d).

Civic engagement. The literature (Jacoby 2009; Doolittle and Faul 2013; Adler and Goggin 2005; Barrett and Zani 2014; Andrei et al. 2020; Zait and Andrei 2019; Remr, 2023) indicates that civic engagement is associated with involvement in a multitude of community, environmental and cultural initiatives. Measurement scales assess civic attitudes or quantify civic behavior through specific activities, such as volunteering, voting, etc., or combine both components—attitudes and corresponding behaviors (i.e., Doolittle and Faul's Civic Engagement Scale 2013). That is: both civic attitudes and civic behaviors are important. For example, if citizens participated more consistently through voting and civil society channels, governance accountability would improve (Heinelt 2018; Bussu 2022). As higher turnout in European Parliament elections is needed to strengthen democratic representation (European Parliament 2023), limited awareness of its relevance to everyday life undermines engagement. More outreach programs explaining the EU's roles coupled with online tools facilitating consultation input could motivate higher turnouts (Gastil 2021).

European Values. The European Union is founded on the values of respect for human dignity, freedom, democracy, equality, the rule of law and respect for human rights, values which are mentioned in Article 2 TEU of the Treaty on European Union (TEU) and supported by the Charter of Fundamental Rights of the European Union which puts EU core values into EU law to uphold principles such as human dignity, freedom, equality, solidarity, citizens' rights, justice through its laws and policies (European Union. Official Journal of the European Union, 2012). While most citizens consider the EU as important for peace, stability and the economy (Directorate General Communication 2022), an increased awareness of people about fundamental European values is very important, as civil society pressure might act as a tool of application to reinforce the core values and the shared ethical practices.

Social Responsibility. Responsible business practices that respect workers, communities and the environment are critical for sustainability (European Commission 2023; Arslan and Wong, 2022; Pérez-Barea et al. 2018; Webb et al. 2008; Wong 2008; Francois-Lecompte and Roberts 2006). The EU's strong ethics, transparency and non-financial reporting framework leads globally on spurring accountable conduct (European Commission, Directorate-General for Financial Stability, Financial Services and Capital Markets Union, Groen, Alcidi, Simonelli et al. 2021). However, cases of ethics violations, greenwashing and inequality persist, indicating a need for stricter oversight and standards (Transparency International EU et al. 2023).

Environmental Responsibility and Entrepreneurship for Sustainability. The EU spurs eco-innovation and startups through significant funding programs like Horizon Europe and the European Innovation Council (European Commission 2023). These schemes award grants for research, development and commercialization to provide solutions for pressing issues like climate challenges and pollution (European Commission 2023; European Innovation Council 2023). Still, faster scaling is required for sustainability entrepreneurship as well as building collaborative platforms to accelerate progress by connecting researchers, entrepreneurs and policymakers (Fellnhofer 2022; Bogoslov et al. 2022; Blomsma et al. 2023).

4 Data Collection and Sample

The CitEuroPass questionnaire was applied online between the end of March and middle of June 2023 with students enrolled in one of the five universities involved in the study: Université Grenoble Alpes, France; University of Economics in Bratislava, Slovakia; Matej Bel University in Banská Bystrica, Slovakia; New Bulgarian University, Sofia, Bulgaria; Alexandru Ioan Cuza University of Iasi, Romania.

Invitation to participate in the study and the link to the online questionnaire was sent to the students, resulting in a convenience sample based on student availability and willingness to spend around 10 min to fill-in the form. The anonymity of the respondents was granted. The survey was conducted anonymously, in compliance with GDPR, so participants were assured of their confidentiality and anonymity in the study reporting.

The research sample (N = 679) comprised European students (75.1% undergraduates and 29.9% graduate students), most of them (90.9%) aged between 18–25 years old. Only 9.1% of student participants were 25 years old or older, while 38% were aged between 18–21, and 52.9% were between 21–25 years old. Study participants were enrolled in fields of study such as: Social Sciences, Economics and Business (75.7%), Mathematics, Informatics and Natural Sciences (13.6%), or others (10.7%), at the universities from France, Slovakia, Bulgaria or Romania involved in the project. The responses regarding the permanent residence of student participants indicated the following distribution of the country of provenience in the sample: France (22.2%), Slovakia (28.4%), Bulgaria (12.7%), Romania (32.4%), and other countries (4.3%) such as: Belgium, Germany, Italy, Spain, Switzerland, Norway, Poland, Estonia, Moldavian Republic, Ukraine (between 1 to 3 students from each of the mentioned countries).

5 Responses Evaluation and Discussion

The Citeuropass questionnaire was applyed anonimously and collected data on 65 items covering the topics of interest (63 closed-ended items and 2 open-ended questions) plus socio-demographic characteristics.

Most of the items (62 items) used 1 to 5 Likert scales ratings (from 1 = *not at all* to 5 = *very much*) to collect participants' responses. One item used a Yes/No closed-ended question to assess respondents' willingness to participate in collaborative projects, and 2 items used open-ended questions.

The two items that used open-ended questions were meant to get deeper insights regarding perceptions of European citizenship by asking participants to input words they think would best describe European spirit and also to add comments on what European citizenship means for themselves.

According to paticipants' responses to the open questions, the top 10 words that best describe European spirit are: *freedom, diversity, unity, collaboration, culture, democracy, union, equality, solidarity, friendship, exchange, sharing, community, frienship, opprotunities, happiness, innovation, and peace*—as it is shown in Table 1 and the word-cloud of the most frequent words (Fig. 1).

Moreover, we can observe a high degree of similarity between the lists of the top 10 words provided by the respondents in the four countries (see Table 1), despite the fact that the words were generated by the participants in response to the open question placed at the beginning of the questionnaire: *Please name 3 words that best describe European spirit in your opinion.*

The second word-cloud presented below resulted from the 2nd open-ended question included at the end of the questionaire: *Please add here any comments you have on what European citizenship means to you.*

Participants' responses indicated an increased level of willingness to join collaborative projects mixing students from different EU countries (85,5% France, 77,1% Slovakia, 85% Romania, 85,1% Bulgaria) according to the YES answers to the Yes/No closed-ended item *Are you willing to participate in collaborative projects within teams mixing students from different EU countries?* Also we can observe a high degree of similarity between the four countries' frequency of YES responses (85,5% France, 77,1% Slovakia, 85% Romania, 85,1% Bulgaria).

Table 1 Words mentioned by respondents as best descriptors of European spirit

No	All respondents	French respondents	Bulgarian respondents	Slovack respondents	Romanian respondents
1	Freedom	Culture	Freedom	Freedom	Freedom
2	Diversity	Freedom	Diversity	Unity	Diversity
3	Unity	Diversity	Culture	Collaboration	Democracy
4	Collaboration	Exchange	Unity	Democracy	Culture
5	Culture	Collaboration	Opportunities	Friendship	Union
6	Democracy	Union	Democracy	Solidarity	Equality
7	Union	Unity	Equality	Diversity	Collaboration
8	Equality	Sharing	Community	Culture	Unity
9	Solidarity	Community	Innovation	Union	Innovation
10	Friendship	Solidarity	Solidarity	Opportunities	Peace

Fig. 1 Word-cloud that best describe European spirit in respondents opinion

The results presented in detail below, indicates increased similarities between the answers of participants from the four EU countries on all the 63 closed-ended items of the questionnaire. Before presenting the assessed dimensions one by one, we have to mention that the results also indicated a very good overall reliability of the measurement instrument at the level of CitEuroPass questionnaire, with Cronbach's Alpha = 0.95.

Commitment to EU citizenship, feeling good, and proud to be EU citizen. The results (see Table 2 and Fig. 3) also indicated that respondents feel good, and proud to be EU citizens. Mean score at sample level was 3.94. The highest scores were reported by Bulgarian students. Higher than average scores resulted in each country

Table 2 Commitment to EU citizenship, feeling good and proud as EU citizen

Country	Mean	N
Bulgaria	4.20	86
France	3.70	151
Romania	4.04	220
Slovakia	3.90	193
Total	3.94	679

(Table 2 and Fig. 2) for the 3rd topic approached through a multi-item reflective construct "Commitment to EU citizenship, feeling good and proud to be EU citizen" (4 items adapted from the literature (La Barbera 2015; La Barbera and Capone 2016): *q3.1. I am glad to be an EU citizen; q3.2. I think that EU citizens have a lot to be proud of; q3.3. Being an EU citizen gives me a good feeling; q3.8. I feel committed to the EU citizenship;* Cronbach's Alpha = 0.883).

Fig. 2 Word-cloud of respondents' comments on what European citizenship means for themselves

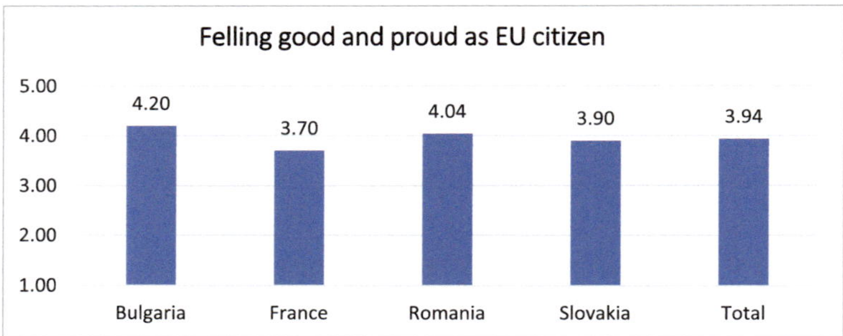

Fig. 3 Commitment to EU citizenship, felling good and proud as EU citizen

Table 3 EU Identity

Country	Mean	N
Bulgaria	3.40	86
France	2.98	151
Romania	3.63	220
Slovakia	3.22	193
Total	3.33	679

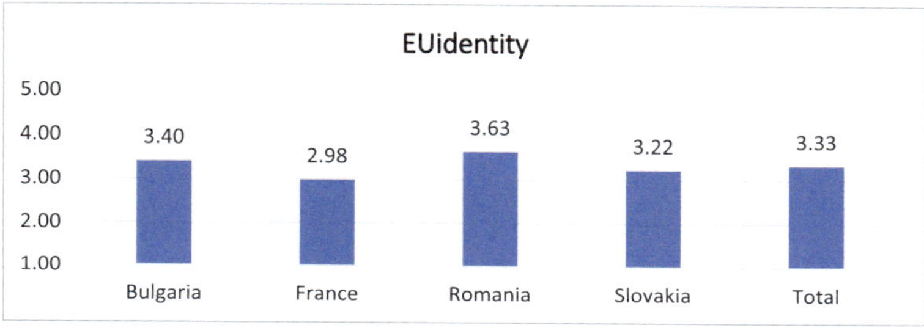

Fig. 4 EU Identity

European identity. Regarding the feeling of an EU identity (Table 3 and Fig. 4) the results indicated a mean score at sample level of 3.33 (higher than average).

"EU identity" was measured through a multi-item reflective construct (4 items adapted from the literature (Spears et al. 1997; La Barbera and Capone 2016): *q3.4. The fact that I am an EU citizen is an important part of my identity; q3.5. I feel I am deeply tied to other EU citizens; q3.6. EU citizens have a lot in common with each other; q3.7. I am similar to the average EU citizen;* Cronbach's Alpha = 0.817*)*.

As Table 3 shows, all countries scored close to the average value of 3 points, with higher ratings in eastern countries, indicating that EU citizenship starts to be seen as an important part of personal identity.

Supportive attitudes towards immediate action in the application of EU norms. Respondents' ratings were higher in eastern countries, but above the average value of 3 (M = 3.46 at sample level) in all countries for the single-item "*q3.9. I see positively a more immediate action in applying European norms"* (see Table 4 and Fig. 5). The item was adapted from the literature (La Barbera and Capone 2016).

Attitude towards EU enlargement. The results (Table 5 and Fig. 6) indicated significantly higher than average ratings for the EU enlargement dimension measured through a multi-item reflective construct" EU enlargement" (3 items adapted from La Barbera and Capone (2016): *q3.10. I see positively the enlargement of the EU beyond the countries that have joined from the beginning; q3.11. I think that EU enlargement represents an opportunity for all of us; Reversed item q3.12. EU enlargement scares me;* Cronbach's Alpha = 0.608*)*. Ratings ranged from an average of 3.74 in France

Table 4 Supportive attitudes towards immediate action in the application of EU norms

Country	Mean	N
Bulgaria	3.86	86
France	3.17	151
Romania	3.74	220
Slovakia	3.31	193
Total	3.49	679

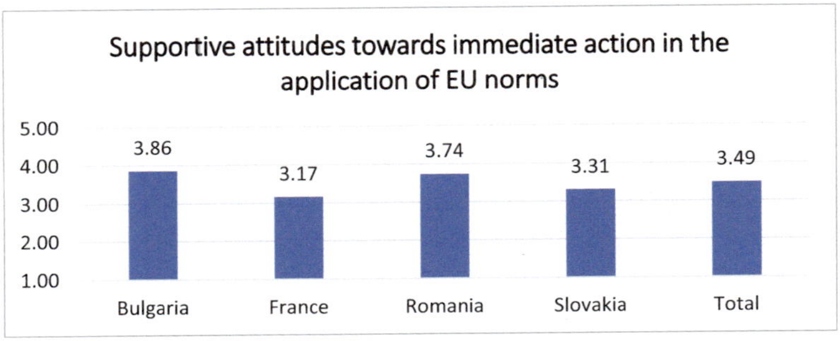

Fig. 5 Supportive attitudes towards immediate action in the application of EU norms

to 4.03 in Bulgaria. The ratings of Romania and Slovakia were closer to France, so the sample mean was M = 3.82.

The interest of increasing own knowledge about the 6 fundamental values of EU. The responses to the question *q4. Would you like to know more about the 6 EU*

Table 5 EU enlargement

Country	Mean	N
Bulgaria	4.03	86
France	3.74	151
Romania	3.79	220
Slovakia	3.82	193
Total	3.82	679

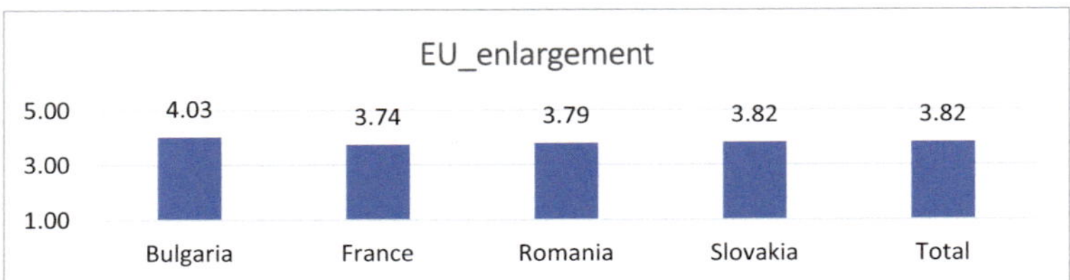

Fig. 6 EU enlargement

fundamental values?(6 items: q4.1. Human dignity; q4.2. Freedom; q4.3 Democracy; q4.4. Equality; q4.5. Rule of law; q4.6. Human rights?) a Cronbach's Alpha = 0.925 for the 6 items, and a higher-than-average interest of respondents to increase their knowledge about each of the 6 fundamental values of EU (see Table 6).

The results reported in Table 7 and Fig. 7 indicate the aggregate interest of the students of each country to increase their knowledge about the 6 fundamental values of EU. It ranged from an average of 3.33 in Slovakia to 3.86 in Romania. The overall interest for all of the 6 fundamental EU values at the level of entire sample (M = 3.56) is closed to the overall ratings of each country, and those specific: Human dignity (M = 3.54); Freedom of movement (M = 3.47); Democracy (M = 3.44); Equality (M = 3.56); Rule of law (M = 3.65); Human rights (M = 3.70).

Therefore, study participants perceive they don't know enough about the 6 fundamental values of EU and they are interested to learn about.

Table 6 Interest to know about each of the 6 fundamental values of EU

Value	N	Min	Max	Mean
q4.1. Human dignity	679	1	5	3,54
q4.2. Freedom	679	1	5	3,47
q4.3. Democracy	679	1	5	3,44
q4.4. Equality	679	1	5	3,56
q4.5. Rule of law	679	1	5	3,65
q4.6. Human rights	679	1	5	3,70

Table 7 Aggregate interest to know more about the 6 fundamental values of EU

Country	Mean	N
Bulgaria	3.69	86
France	3.39	151
Romania	3.86	220
Slovakia	3.33	193
Total	3.56	679

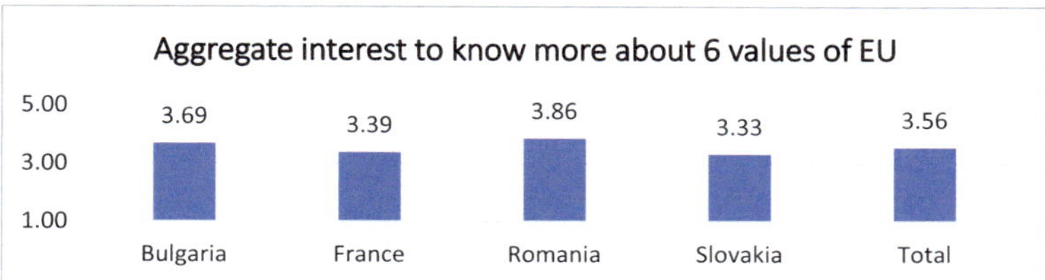

Fig. 7 Aggregate interest to know more about the 6 fundamental values of EU

Social Responsibility. The results (Table 8 and Fig. 8) indicated significantly higher than average ratings for social responsibility dimension measured through the multi-item reflective construct "social responsibility" (11 items: *q5.1. I am accountable for all my decisions and actions; q5.2. I have responsibility for my family, society, Europe, the world and planet earth to the extent of my ability and influence; 5.3. I am morally accountable for how I treat others; q5.4. I am responsible for doing my part to make the world a better place; q5.5. I try not to buy products from companies that harm the environment; q5.6. I buy fair trade products; q5.7. I try to reduce my consumption to what I really need; q5.8. I avoid doing all my shopping in big businesses /large retailers; q5.9. I go to small markets to support small producers; q5.10. I worry about environmental problems; q5.11. A better environment starts with myself;* Cronbach's Alpha = 0.862*)*.

The items were adapted from literature (Arslan and Wong 2022; Pérez-Barea et al. 2018; Webb et al. 2008; Wong 2008; Francois-Lecompte and Roberts 2006). Ratings ranged from an average of 3.65 to 4.02. The sample mean was M = 3.82.

Environmental preoccupation and responsibility. The results (Table 9 and Fig. 9) indicated significantly higher than average ratings for Environmental preoccupation and responsibility dimension measured through a multi-item reflective construct with 4 items: *q5.5. I try not to buy products from companies that harm the environment; q5.7. I try to reduce my consumption to what I really need; q5.10. I worry about environmental problems; q5.11. A better environment starts with myself;* Cronbach's Alpha = 0.738*)*. The four items were selected from the eleven items comprised in the Social Responsibility construct, which was adapted from literature (Arslan and

Table 8 Social responsibility

Country	Mean	N
Bulgaria	4.02	86
France	3.92	151
Romania	3.82	220
Slovakia	3.65	193
Total	3.82	679

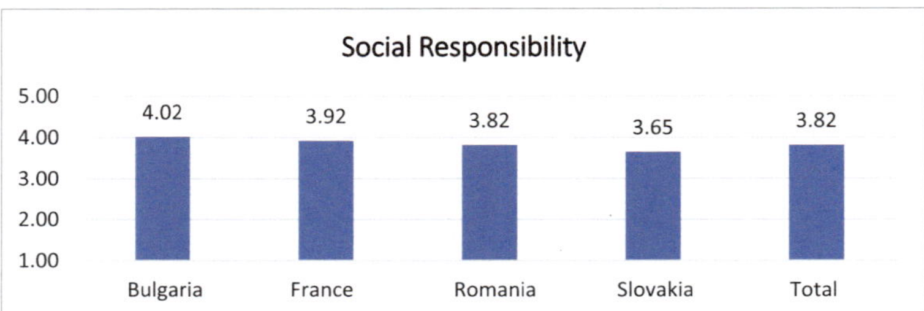

Fig. 8 Social responsibility

Table 9 Environmental preoccupation and responsibility

Country	Mean	N
Bulgaria	4.06	86
France	4.01	151
Romania	3.81	220
Slovakia	3.74	193
Total	3.87	679

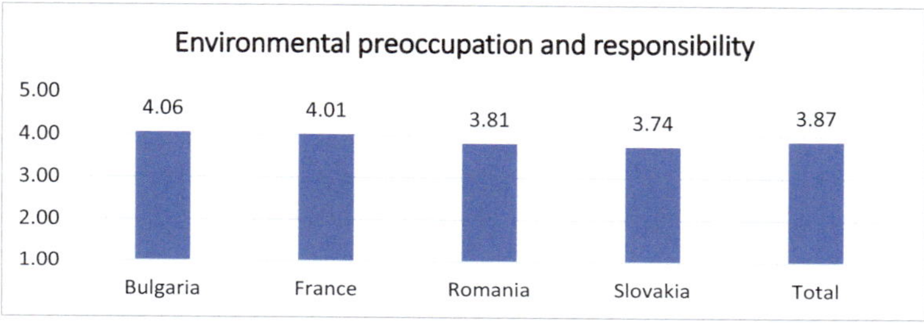

Fig. 9 Environmental preoccupation and responsibility

Wong, 2022; Pérez-Barea et al. 2018; Webb et al. 2008; Francois-Lecompte and Roberts 2006).

Ratings ranged from an average of 3.74 to 4.06. The sample mean was M = 3.82.

Citizen involvement at national and European level. The results (Table 10 and Fig. 10) indicated higher than average ratings for the dimension measured through a multi-item reflective construct „EU citizen involvement at national and EU level" (6 items: *q6.1. I pay close attention to the local news and events; q6.2. I pay a lot of attention to international events and news; q6.3. I am interested in reading the news related to the EU and European countries as much as I am interested in those related to my own country; q6.4. I always vote in local elections; q6.5. I always vote in national elections; q6.6. I always vote in the elections of our national representatives in the EU bodies;* Cronbach's Alpha = 0.821*)*. The items were adapted from literature (Becuwe and Baneth 2021; Dąbrowski et al. 2021) to comprise both national and European level of involvement. Ratings ranged from an average of 3.34 to 3.73. The sample mean was M = 3.57, with a slightly higher involvement at national level (M = 3.72) compared to international level (M = 3.42)—see Fig. 11.

Civic engagement. Civic engagement was measured through its 2 components - civic attitude and civic behavior - indicated in the literature (Jacoby 2009; Doolittle and Faul 2013; Adler and Goggin 2005; Barrett and Zani 2014; Andrei et al. 2020; Zait and Andrei 2019) portraying it in relation with involvement in environmental, cultural and community initiatives, and assessing it through specific attitudes and behaviors (i.e. Civic Engagement Scale of Doolittle and Faul 2013).

Table 10 EU citizen involvement at national and EU level

Country	Mean	N
Bulgaria	3.73	86
France	3.60	151
Romania	3.72	220
Slovakia	3.34	193
Total	3.57	679

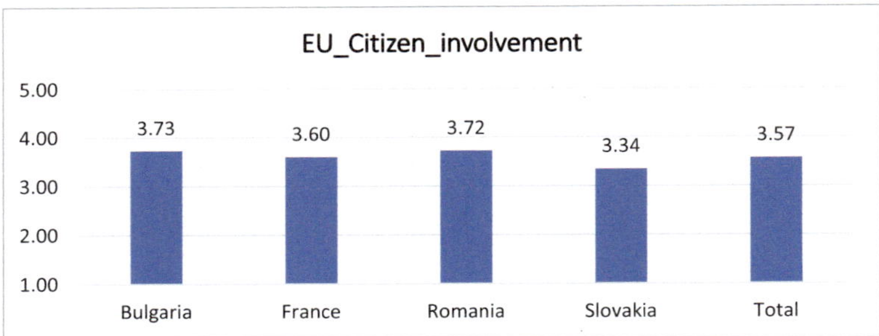

Fig. 10 EU citizen involvement at national and EU level

Fig. 11 Comparison between citizen involvement at European level and national level

Therefore, we adopted the civic engagement scale of Doolittle and Faul (2013) and used two multi-item reflective constructs: the construct Civic Attitude (7 items from Doolittle and Faul 2013: *q8.1. I feel responsible for my community; q8.2. I believe that I have a responsibility to help the poor and the needy; q8.3. I am committed to serve in my community; q8.4. I believe that all EU citizens have a responsibility to our enlarged community; q8.5. I believe that it is important to be informed of community issues; q8.6. I believe that it is important to volunteer; q8.7. I believe that it is important to financially support non-profit organizations;* Cronbach's Alpha = 0.876) and the construct Civic Behavior (6 items from Doolittle and Faul (2013): q8.*8. I am involved in structured volunteer position(s) in the community; q8.9. When working with others, I make positive changes in the community; q8.10. I help members of*

Table 11 Civic engagement: attitude

Country	N	Civic attitude	Civic behavior
Bulgaria	86	4.63	3.50
France	151	4.14	3.03
Romania	220	4.36	3.54
Slovakia	193	4.03	2.98
Total	679	4.25	3.26

my community; q8.11. I stay informed of events in my community; q8.12. I participate in discussions that raise issues of social responsibility; q8.13. I contribute to humanitarian organizations within the community; Cronbach's Alpha = 0.872).

The results (Table 11 and Fig. 11) indicated very high ratings for the construct Civic Attitude (M = 4.25), but significantly lower ratings (closed to neutral level of 3) for the construct Civic Behavior (ratings ranged from 2.98 to 3.50, with M = 3.26 at sample level).

Entrepreneurial spirit towards sustainability. The results (Table 12 and Fig. 12) indicated higher than neutral ratings for the dimension measured through a multi-item reflective construct "Entrepreneurial spirit" (6 items adapted from Covin et al. (2020): *q7.1. I have very little problems with renewal and change; q7.2. I quickly master new routines, procedures, and new ways of working; q7.3. I always search for creative solutions instead of familiar ones; q7.4. I am constantly looking for new ways to improve my performance at work; q7.5. I value new plans and ideas, even if I feel that they could fail in practice; q7.6. In order to be more productive, I sometimes act without the permission of my supervisor /boss;* Cronbach's Alpha = 0.785). Ratings ranged from an average of 3.34 to 3.85. The sample mean was M = 3.56.

Preference for cultural diversity (versus homogeneity) at work. The results (Table 13 and Fig. 13) indicated respondents' preference for cultural diversity (versus homogeneity) at work /in teamwork. A single-item question was used in this case: *Do you feel more comfortable to work within teams which are rather homogeneous, or do you prefer an increased variety of cultures among team members?* Ratings ranged from an average of 7.07 to 7.56. The sample mean was M = 7.38 on the *1 to 10 scale* (*from 1 = I like cultural homogeneity to 10 = I like cultural variety*)

Table 12 Entrepreneurial spirit

Country	Mean	N
Bulgaria	3.85	86
France	3.34	151
Romania	3.70	220
Slovakia	3.46	193
Total	3.56	679

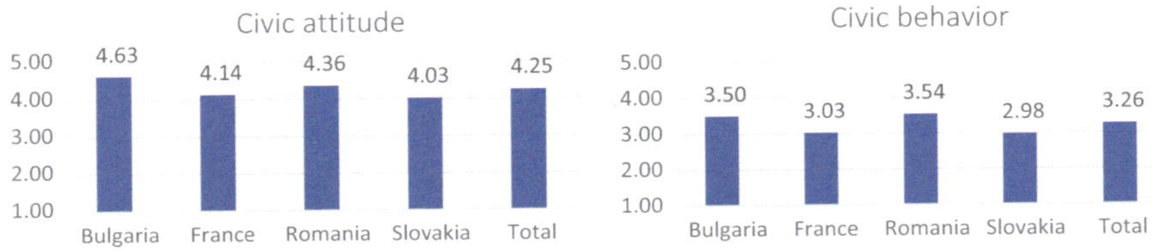

Fig. 12 Civic engagement: civic attitude and civic behavior

which indicates that participants from each country prefer diversity at work, but the preference for diversity over homogeneity in not very high.

What European citizenship means for EU citizens. The meaning of EU citizenship was adressed at the end of the survey through a question question requesting respondents to rate 7 closed-ended items: "q 10. *We would like to know what European citizenship means to you. Please rate the following propositions according to your beliefs, from 1 = not important to 5 = very important)"*.

The following set of 7 items adapted from the literature (Hadjitchoneva et al. 2024; Pellat et al. 2024), were listed to be rated: *q10.1. Free movement within the EU; q10.2. The right to vote democratically; q10.3. Democracy; q10.4. The greener and more sustainable European future; q10.5. The common values of the European Union, such as respect for human dignity, freedom, democracy, equality and the rule of law; q10.6. Possibilities of entrepreneurial initiative at European level; q10.7. Access to the European market.*

Table 13 Preference to cultural variety (vs. homogeneity) in teamwork

Country	Mean	N
Bulgaria	7.45	86
France	7.56	151
Romania	7.47	220
Slovakia	7.07	193
Total	7.38	679

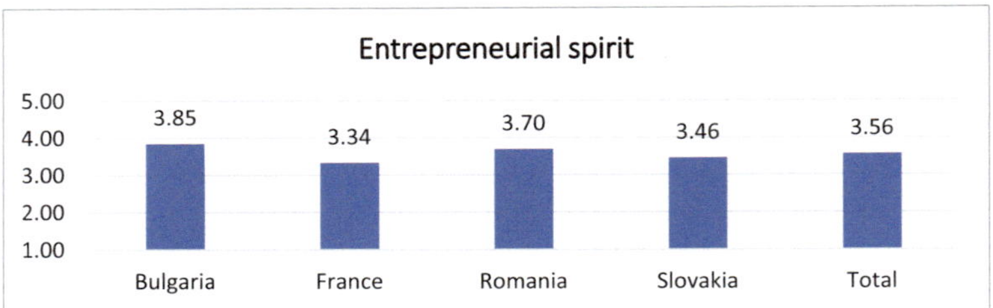

Fig. 13 Entrepreneurial spirit

Scale reliability computed on these 7 items indicated a Cronbach's Alpha = 0.901. The 7 items representing 7 meanings associated with EU citizenship received high ratings from our participants (see Table 14). Students' ratings indicated a high level level of importance of these 7 meanings associated with EU citizenship.

A high degree of similarity between the ratings provided by the participants of the 4 countries was found for the first 5 meanings associated with EU citizenship (*q10.1; q10.2; q10.3; q10.4; q10.5*) but a significant difference between France and the rest of the countries was also found for the last 2 meanings associated with EU citizenship (q10.6; q10.7) - see detailed results in Table 14 and Fig. 14, 15, 16, 17, 18, 19, 20, 21.

Table 14 What European citizenship means to EU citizens

EU citizenship means:*	Sample (N = 679)	Bulgaria (N = 86)	France (N = 151)	Romania (N = 220)	Slovakia (N = 193)
q10.1. Free movement within the EU	4,46	4.65	4.56	4.24	4.55
q10.2. The right to vote democratically	4,35	4.62	4.28	4.27	4.45
q10.3. Democracy	4,33	4.49	4.19	4.27	4.46
q10.4. A greener and more sustainable European future	4,14	4.30	4.08	4.15	4.09
q10.5. The common values of the European Union, such as respect for human dignity, freedom, democracy, equality and the rule of law	4,30	4.52	4.24	4.20	4.38
q10.6. The possibilities of entrepreneurial initiative at European level	3,93	4.33	3.13	4.13	4.15
q10.7. Access to the European market	4,09	4.50	3.42	4.20	4.34

* 10. We would like to know what European citizenship means to you. Please rate the following propositions according to your beliefs (from 1 = not important to 5 = very important)

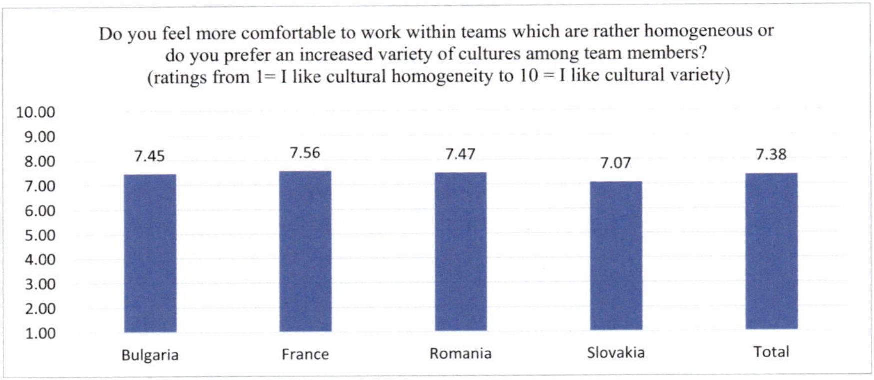

Fig. 14 Preference to cultural variety (vs. homogeneity) in teamwork

Fig. 15 EU citizenship means: Free movement within EU

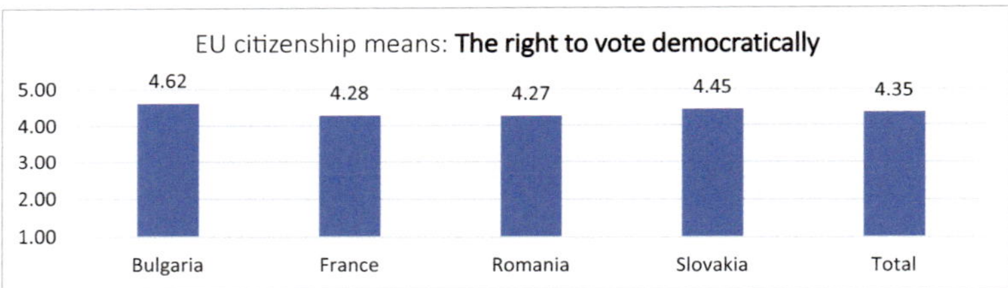

Fig. 16 EU citizenship means: The right to vote democratically

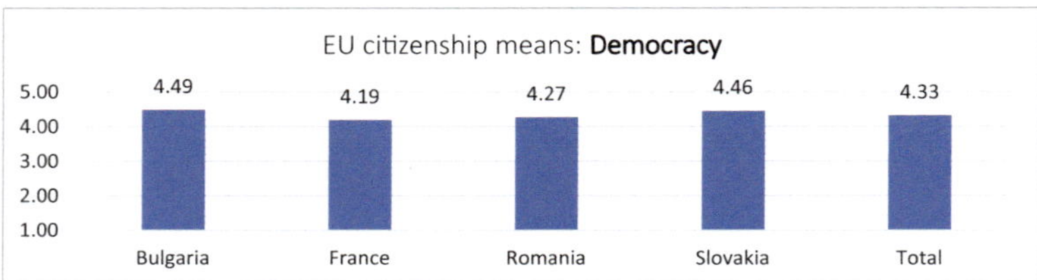

Fig. 17 EU citizenship means: Democracy

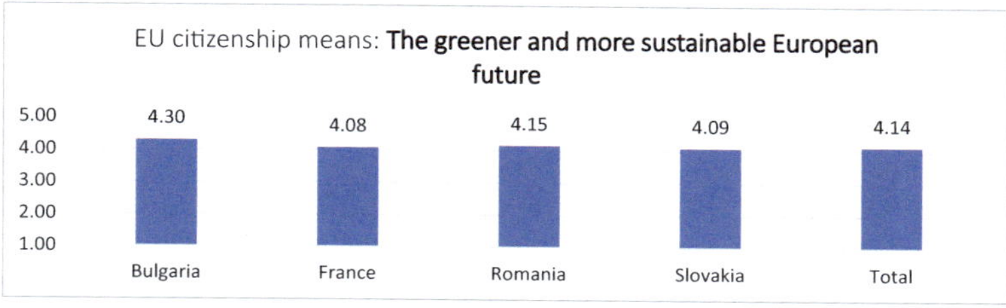

Fig. 18 EU citizenship means: The greener and more sustainable European future

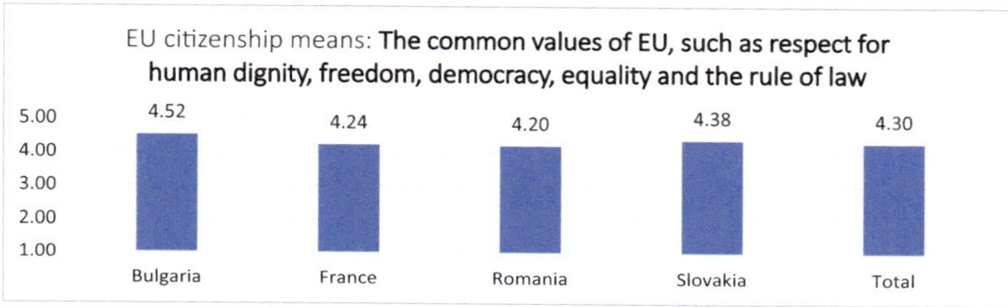

Fig. 19 EU citizenship the common values of the European union, such as respect for human dignity, freedom, democracy, equality, and the rule of law

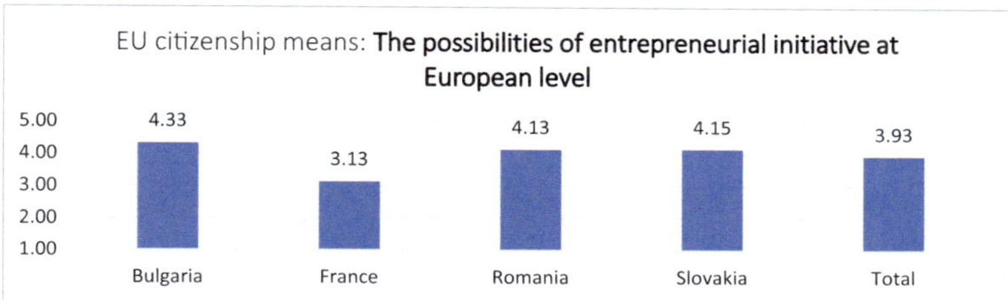

Fig. 20 EU citizenship means: Possibilities of entrepreneurial initiative at European level

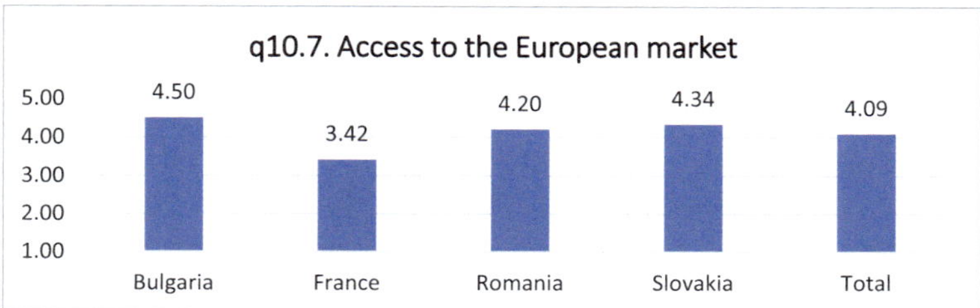

Fig. 21 EU citizenship means: Access to the European market

6 Conclusion

In conclusion, the CitEuro Pass questionnaire appears to be a reliable instrument according to the results of its first application detailed above.

Cronbach's Alpha was 0.95 at the overall level of CitEuroPass questionnaire, and high reliability coefficients were found for the sub-scales adapted from the literature and used to measure multi-item dimensions included in the questionnaire as follows:

- Commitment to EU citizenship, feeling good, and proud to be EU citizen: 4 items adapted from La Barbera and Capone 2016; Cronbach's Alpha = 0.883
- EU identity: 4 items adapted from Spears et al. 1997; La Barbera and Capone 2016; Cronbach's Alpha = 0.817.
- Supportive attitudes towards immediate action in the application of EU norms: single-item construct adapted from La Barbera and Capone (2016).
- Attitude towards EU enlargement: 3 items adapted from La Barbera and Capone 2016: Cronbach's Alpha = 0.608. Further improvement of the items is needed here to increase the reliability (Cronbach's Alpha should be higher than 0.7).
- The interest of increasing own knowledge about the core values of EU: 6 items listing the 6 fundamental values of EU; Cronbach's Alpha = 0.925.
- Social responsibility: 11 items adapted from literature (Arslan and Wong 2022; Pérez-Barea et al. 2018; Webb et al. 2008; Wong 2008; Francois-Lecompte and Roberts 2006); Cronbach's Alpha = 0.862.
- Environmental responsibility (4 items extracted from the 11 items of the more comprehensive construct Social responsibility presented above; Cronbach's Alpha = 0.738).
- EU citizen involvement at national and EU level: 6 items adapted from literature (Becuwe and Baneth 2021; Dąbrowski et al. 2021); Cronbach's Alpha = 0.821).
- Civic engagement (Attitude): 7 items from Doolittle and Faul 2013; Cronbach's Alpha = 0.876)
- Civic engagement (Behavior): 6 items from Doolittle and Faul 2013; Cronbach's Alpha = 0.872)
- Entrepreneurial spirit: 6 items adapted from Covin et al. (2020); Cronbach's Alpha = 0.785.
- Preference for cultural diversity (vs. homogeneity) at work /in teamwork: a single-item construct
- What European citizenship means for EU citizens (7 items derived from Hadjitchoneva et al. 2024; Pellat et al. 2024); Cronbach's Alpha = 0.901.

The CitEuroPass questionnaire comprised 65 items for the topics of interest plus socio-demographic characteristics. Most of the items (62 items) used 1 to 5 Likert scales (ratings from 1 = not at all to 5 = very much) to collect responses to closed-ended questions allowing statistical analysis, one closed-ended item used a Yes/No scale to assess the willingness to participate in collaborative projects, while 2 items were open-ended questions meant to get deeper insights regarding perceptions of European Citizenship and European spirit, respectively.

Results indicated an increased similarity between the ratings of participants from the four EU countries on the items of the questionnaire and a high level of willingness to join collaborative projects mixing students from different EU countries.

Acknowledgements This work was supported by the Citeuropass project 2022-1-FR01-KA220-HED-000088786, financed by the Erasmus+ Program of the European Union.

References

Adler RP, Goggin J (2005) What do we mean by "civic engagement"? J Transform Educ 2005(3):236–253

Andrei AG, Zait A, Stoian C, Tugulea O, Manolica A (2020) Citizen engagement in the "post-truth era": a knowledge management inquiry into the online spread of information. Kybernetes 49(5):1429–1443. https://doi.org/10.1108/K-03-2019-0178

Arslan G, Wong PT (2022) Measuring personal and social responsibility: an existential positive psychology approach. J Happiness Health 2(1):1–11

Becuwe N, Baneth O (2021) Special Eurobarometer 508 on Values and Identities of EU citizens

Barrett M, Zani B (2014) (2014) Political and Civic Engagement: Multidisciplinary perspectives. Routledge, London, UK

Bruter M (2003) Winning hearts and minds for Europe. Comp Pol Stud 36(10):1148–1179. https://doi.org/10.1177/0010414003257609

La Barbera F (2015) Framing the EU as common project vs. common heritage: effects on attitudes towards the EU deepening and widening. J Soc Psychol 155:617–635. https://doi.org/10.1080/00224545.2015.1041446

La Barbera F, Capone V (2016) Five dimensions of European identity: a contribution to the Italian adaptation and validation of the in-group identification scale. Eur J Psychol 12(2):288–303. https://doi.org/10.5964/ejop.v12i2.1058

Blomsma F, Bauwens T, Weissbrod I, Kirchherr J (2023) The 'need for speed': towards circular disruption—What it is, how to make it happen and how to know it's happening. Bus Strateg Environ 32(3):1010–1031

Bussu S, Bua A, Dean R, Smith G (2022) Introduction: embedding participatory governance. Critical Policy Studies 16(2):133–145

Bogoslov IA, Lungu AE, Stoica EA, Georgescu MR (2022) European green deal impact on entrepreneurship and competition: a free market approach. Sustainability 14(19):12335. https://doi.org/10.3390/su141912335

Consolidated Version of the Treaty on European Union. (2007). Article 2. Official Journal of the European Union, C326, pp 13–45. https://eur-lex.europa.eu/legal-content/EN/TXT/?uri=celex%3A12012M%2FTXT

Delanty G (1997) Models of citizenship: defining European identity and citizenship. Citizsh Stud 1(3):285–303

Directorate-General for Communication (2022) Eurobarometer: Public Opinion in the European Union. https://europa.eu/eurobarometer

Dąbrowski M, Spaans M, Fernandez-Maldonado AM, Rocco R (2021) Cohesion policy and the citizens' perceptions of the EU: the role of communication and implementation. Eur Plan Stud 29(5):827–843. https://doi.org/10.1080/09654313.2020.1805413

Doolittle A, Faul AC (2013) Civic engagement scale: a validation study. SAGE Open 3(3). https://doi.org/10.1177/2158244013495542

European Commision (2023). EU citizenship and democracy. https://commission.europa.eu/strategy-and-policy/policies/justice-and-fundamental-rights/eu-citizenship-and-democracy_en

European Commision (2023) EU priorities. https://european-union.europa.eu/priorities-and-actions/eu-priorities_en
European Commision (2023). EU at COP28 Climate Change Conference. https://commission.europa.eu/strategy-and-policy/priorities-2019-2024/european-green-deal/climate-action-and-green-deal/eu-un-climate-change-conference/eu-cop28-climate-change-conference_en
European Commission. (2023). Horizon Europe. https://ec.europa.eu/info/research-and-innovation/funding/funding-opportunities/funding-programmes-and-open-calls/horizon-europe_en
European Commision (2023). The European Green Deal. https://commission.europa.eu/strategy-and-policy/priorities-2019-2024/european-green-deal_en
European Commission, Directorate-General for Education, Youth, Sport and Culture, Key competences for lifelong learning, Publications Office (2019). https://doi.org/10.2766/569540
European Innovation Council (2023). EIC Horizon Prizes. https://eic.ec.europa.eu/eic-prizes/eic-horizon-prizes_en
European Union. Official Journal of the European Union (2012). Charter of Fundamental Rights of the European Union. https://eur-lex.europa.eu/eli/treaty/char_2012/oj
Fellnhofer K (2022) Entrepreneurial alertness toward responsible research and innovation: digital technology makes the psychological heart of entrepreneurship pound. Technovation 118:102384. https://doi.org/10.1016/j.technovation.2021.102384
Francois-Lecompte A, Roberts JA (2006) Developing a measure of socially responsible consumption in France. Mark Manag J 16(2)
Groen W, Alcidi C, Simonelli F et al (2021) European Commission, Directorate-General for Financial Stability, Financial Services and Capital Markets Union, Study on the non-financial reporting directive – Final report, Publications Office. https://doi.org/10.2874/229601
Gastil J (2021) A theoretical model of how digital platforms for public consultation can leverage deliberation to boost democratic legitimacy. J Deliberative Democr 17(1):78–89. https://doi.org/10.16997/10.16997/jdd.963
Hadjitchoneva J, Pellat G, Marasova J (2024) Reinvesting in European citizenship through entrepreneurship and frugal innovation. In: Pellat G, Zafiroski J, Šuplata M (eds) Cooperation and enlargement: two challenges to be addressed in the European Projects—2022. Studies in Systems, Decision and Control, vol 500. Springer. https://doi.org/10.1007/978-3-031-42253-9_3
Heinelt H (2018) Introduction. Handbook on Participatory Governance. Edward Elgar, pp 1–16
Jacoby B (2009) Civic engagement in today's higher education: an overview. In Civic Engagement in Higher Education: Concepts and Practices; Jacoby, B., Ed.; Jossey-Bass: San Francisco. CA, USA 2009:5–30
Lehning PB (2001) European citizenship: towards a European identity? Law Philos 20(3):239–282
Mazzoni D, Albanesi C, Ferreira PD, Opermann S, Pavlopoulos V, Cicognani E (2018) Cross-border mobility, European identity and participation among European adolescents and young adults. Eur J Dev Psychol 15(3):324–339. https://doi.org/10.1080/17405629.2017.1378089
Mäkinen K (2019) Interconceptualizing Europe and Peace: Identity Building Under the European Heritage Label. In Lähdesmäki T, Passerini L, Kaasik-Krogerus S, van Huis I (Eds.) Dissonant Heritages and Memories in Contemporary Europe (pp 51–78). Palgrave Macmillan. Palgrave Studies in Cultural Heritage and Conflict. https://doi.org/10.1007/978-3-030-11464-0_3
Mitchell K (2015) Rethinking the 'Erasmus E ffect'on European Identity. JCMS: J Common Mark Stud 53(2):330–348. https://doi.org/10.1111/jcms.12152
Moccia L (2018) European civic citizenship and EU integration policies. Civitas Europa 1:107–125. https://doi.org/10.3917/civit.040.0107
Pellat G, Zafiroski J, Šuplata M (eds) (2024) Cooperation and enlargement: two challenges to be addressed in the European Projects—2022. Studies in Systems, Decision and Control, vol 500. Springer
Pérez-Barea JJ, Fernández-Navarro F, Montero-Simó MJ, Araque-Padilla R (2018) A socially responsible consumption index based on non-linear dimensionality reduction and global sensitivity analysis. Appl Soft Comput 69:599–609. https://doi.org/10.1016/j.asoc.2018.04.059

Prieto-Arranz JI, Juan-Garau M, Mesquida-Mesquida F (2023) "Open your mind, sharpen your wits": a narrative approach to the benefits of study abroad as perceived by erasmus+ students. J Lang, Identity Educ 22(3):216–231. https://doi.org/10.1080/15348458.2020.1871352
Remr J (2023) Usefulness of a civic engagement scale for research on smart cities: measuring attitudes and behavior. Smart Cities. 6(6):3251–3265. https://doi.org/10.3390/smartcities6060144
Sedelmeier U (2017) Political safeguards against democratic backsliding in the EU: the limits of material sanctions and the scope of social pressure. J Eur Publ Policy 24(3):337–351
Sedelmeier U, Avery G (2018) EU Enlargement and Wider Europe. The European Union: How Does it Work, pp 167–192
Schimmelfennig F, Sedelmeier U (2020) The Europeanization of Eastern Europe: the external incentives model revisited. J Eur Publ Policy 27(6):814–833
Spears R, Doosje B, Ellemers N (1997) Self-stereotyping in the face of threats to group status and distinctiveness: The role of group identification. Pers Soc Psychol Bull 23(5):538–553. https://doi.org/10.1177/0146167297235009
Tajfel H (1978) Social categorization, social identity, and social comparison. In: Tajfel H (ed) Differentiation between social groups: Studies in the social psychology of intergroup relations. Academic Press, London, pp 61–76
Transparency International EU (2023). News. https://transparency.eu/
Verhaegen S (2018) What to expect from European identity? explaining support for solidarity in times of crisis. Comp Eur Polit 16:871–904
Webb DJ, Mohr LA, Harris KE (2008) A re-examination of socially responsible consumption and its measurement. J Bus Res 61(2):91–98
Wong PTP (2008) Freedom and responsibility. In: Shelton K, Bolz DL (eds) Responsibility 911: With great liberty comes great responsibility. Executive Excellence, pp 31–34
Zait A, Andrei AG (2019) Civic Engagement at the Crossroads of Online and Offline Spaces: A PLS-SEM Assessment. Sci Ann Econ Bus 66(4):559–572. https://doi.org/10.47743/saeb-2019-0045

The Consumer Purchasing Behavior During the Health and Economic Crisis Caused by the SARS-CoV-2 Pandemic and Its Impact on the Use of New Digital Distribution and Communication Channels of Food Stores in the Area of Value Delivery

Pavol Kita and Marta Žambochová

Abstract The SARS-CoV-2 pandemic has led to changes that have disrupted the way consumers consider and purchase food. At the same time, the corona crisis and its consequences have highlighted the necessity to take into account new consumer practices and the increased role of digital technologies. The mentioned facts were reflected in the structure of the paper. In the first part of the paper, the purchasing behavior of consumers during the health and economic crisis caused by the SARS-CoV-2 pandemic is analyzed, with an emphasis on food purchases and its impact on the investigated customer segments. The second part of the paper analyzes the impact of the crisis on the use of new digital distribution and communication channels of food stores in the area of value delivery. The paper is based on two researches . The results of the research carried out in 2021 provide information about consumer attitudes towards different sources of food purchases and their behavior in food stores during the SARS-CoV-2 pandemic. The results of the research carried out in 2022 show the extent to which new digital innovations related to the purchase of food are gradually diverting consumers from existing purchasing technologies. The conclusion of the paper provides a brief summary of the study, limitations of the study, and future research directions . The paper was supported by the project VEGA 1/0012/22 of the Ministry of Education of the Slovak Republic "The innovative business models of retail unit formats based on geo-marketing data and their influence on the creation of the value base and retail network of food stores in the digital period ".

P. Kita (✉)
Comenius University in Bratislava, Bratislava, Slovak Republic
e-mail: pavol.kita@uniba.sk

M. Žambochová
J. E. Purkyně University, Ústí Nad Labem, Czech Republic
e-mail: marta.zambochova@ujep.cz

G. Pellat et al. (eds.), *Data-Centric Business and Applications*, Lecture Notes on Data Engineering and Communications Technologies 258,
https://doi.org/10.1007/978-3-031-95068-1_2

Keywords Consumers purchasing behaviour · Digitalization · Food market · Retailing · Retail store · SARS-CoV-2 pandemic

1 Introduction

The SARS-CoV-2 pandemic has led to changes that have disrupted the way consumers consider and purchase food. At the same time, the corona crisis and its consequences have highlighted the necessity to take into account new consumer practices and the increased role of digital technologies. The mentioned facts were reflected in the structure of the paper. In the first part of the paper, the purchasing behavior of consumers during the health and economic crisis caused by the SARS-CoV-2 pandemic is analyzed, with an emphasis on food purchases and its impact on the investigated customer segments. The second part of the paper analyzes the impact of the crisis on the use of new digital distribution and communication channels of food stores in the area of value delivery. The paper is based on two researches. The results of the research carried out in 2021 provide information about consumer attitudes towards different sources of food purchases and their behavior in food stores during the SARS-CoV-2 pandemic. The results of the research carried out in 2022 show the extent to which new digital innovations related to the purchase of food are gradually diverting consumers from existing purchasing technologies. The conclusion of the paper provides a brief summary of the study, limitations of the study, and future research directions.

2 Literature Review

Pandemic crisis is an unprecedented experience of our time (Bogalska-Martin 2021). The pandemic forced national governments in Europe in the first months of 2020 to introduce restrictions on freedom of movement and some measures to prevent the spread of the disease (Masotti et al. 2023; Vidal-Mones 2021)), which had a significant impact on consumer decision-making (Ibn-Mohammed, 2021). Prevention of SARS-CoV-2 infection has become a critical issue. Self-protective behavior (Witte et al. 1998) of consumers related to health depended mainly on their individual attitude to the perception of health problems in connection with the ongoing pandemic (Huang 2023) and to the perception of the regulations of the Office of Public Health of the Slovak Republic as the competent state authority in the field of public of health (ÚVZ 2020) so that they do not become infected with the disease.

In the conditions of the health and economic crisis, the traditional model of food supply has shown its weaknesses. This model requires a pragmatic view, based on which consumers sought the provision of value depending on the dates of real deliveries of goods to the physical store through their visits and inquiries to the addressed store personnel. Given the hygiene measures taken and the often panic buying of food, contactless shopping (Verhoef et al. 2021) offered consumers safety in view of the fear of spreading disease and the possibility of contagion when shopping in a physical store. The health and economic crisis showed that not only food stores have many weaknesses in the area of digitalization, but also from the point of view of digital skills of consumers, it can be concluded that despite the overall positive development in recent years, the level of digital skills in 2020 was still highly differentiated. A comparison of the digital skills of individual socio-demographic groups of residents confirmed that Slovakia is still struggling with the phenomenon of informatization of society, which is usually referred to as the digital divide. It is a division of society into those who have access to information and communication technologies and use them, and those who do not have such access or digital literacy (according to data from the Statistical Office, at the end of 2020, up to 14.2% of Slovak households were without an internet connection). A significant reduction in regional disparities in the use of information and communication technologies can be considered positive (Čuláková et al. 2022). In 2022, Slovakia ranked 23rd in the digital economy and society index from the European Commission. Slovakia's e-commerce score has also decreased, as only 13% of small and medium-sized businesses sell online, compared to 17% in 2020. Only 16% of Slovak companies used electronic invoices this year, while the EU average is 32%.

In order for the crisis to become a springboard for food retail stores in the introduction of new digital technologies, it is necessary for managers to demonstrate a great ability to adapt and deal agilely with the incorporation of new digital elements into the business models of retail stores. In this way, they can guarantee the continuity of their activity and allow consumers access to several media channels that increase the complexity of customer choice. Therefore, it is important to map and have an overview of where and how digital technology (Rittera, Pedersena 2020; Stephens et al. 2022) is used in food stores, as a starting point for their ability to respond to European challenges, which can also be the basis to explore new opportunities for innovation (Bocken et al. 2014). The study uses quantitative analysis, but also qualitative methods, which made it possible to understand the extent to which retail operating units apply digitalization-related activities.

3 Methodology

The aim of this study is to examine some aspects of consumer purchasing behavior during the health and economic crisis caused by the SARS-CoV-2 pandemic, with an emphasis on the purchase of food and its impact on the investigated customer segments, as well as describe its impact on the use of some new digital distribution and communication channels. by which retail units provide value. The following research questions were formulated on the basis of the stated objective and the professional scientific literature:

RQ1: What has changed the most in consumer shopping behavior due to the covid 19 pandemic?

RQ2: Are there links between the observed changes in purchasing behavior?

RQ3: What are the main characteristics of respondents having similar changes in purchasing behavior as a result of the pandemic?

RQ4: What new digital distribution and communication channels are used by food stores?

The approach of the article is based on quantitative data from conducted researchs, which allows presenting facts about changes in consumer behavior in the period under review, about the main characteristics of respondents who have similar changes in purchasing behavior as a result of the pandemic, as well as about the use of new digital distribution and communication channels. Based on this, recommendations can be formulated for distribution professionals considering that businesses will need to understand consumer response and develop appropriate marketing strategies.

The main sources of knowledge about consumer purchasing behavior were primary data from empirical research conducted during the second wave of the SARS-CoV-2 pandemic in Slovakia in 2021 and secondary sources from scientific literature. The respondents were asked to express their attitude towards the phenomenon under investigation by recording a certain position on a five-point semantic scale. Empirical research was conducted in the form of a pre-prepared and agreed upon structured online interview with each respondent. The respondents were selected by contacting them based on their e-mail addresses. Respondents were asked to express their attitude towards the phenomenon under investigation by recording a certain position on a five-point semantic scale. This scale determines not only the content of the attitude, but also its approximate strength. For this reason, it is considered a very useful tool for attitude research. The questionnaire also included questions about the general characteristics of the household and the respondents. 533 respondents participated in the research on the basis of voluntariness and availability in order to obtain the widest and most accurate information possible. 57.12% of the core group were women and 42.88% were men. The largest group were respondents living in cities with up to 50,000 inhabitants. In terms of the number of household members, households with one member (9.18%), two members (23.22%), three members (24.91%), four members (21.91%), five members (14 0.42%), six members (4.49%), seven members (1.31%), eight members (0.56%). From the point of view of the subjective assessment of the financial situation, 2.06% of households were in a very bad financial situation, 11.98% in a bad financial situation, 43.63% in a sufficient financial situation, 38.20% in a good financial situation, 4.11% in a very good financial situation.

4 Results and Discussion

The conducted research answered the following research questions.

RQ1: What Has Changed the Most in the Purchasing Behavior of Consumers as a Result of the SARS-CoV-2 Pandemic?

To evaluate the research question, the non-parametric Friedman test was used to compare several dependent samples and subsequent post hoc analysis. The resulting p-value (< < 0.01) showed that there were statistically significant differences. The final result rankings from the Friedman test together with the average values of the respondents' answers and also with the color marking of the factors with a comparable level are shown in Table 1.

It is obvious that the most significant change in shopping behavior was compliance with hygiene regulations during the purchase, further online shopping, non-food items and maintaining a two-meter distance. Respondents preferred shopping in large retail stores, but complained that long queues at checkouts prevented them from maintaining safe distances. As an alternative to shopping in large retail stores, purchasing from farmers and local sellers, using a delivery service, became evident.

On the contrary, shopping behavior has changed the least in terms of online grocery shopping. A positive finding is the fact that consumers do not think that there is a shortage of disinfectants in stores and that poorly labeled shelves extend the time of purchase.

A more detailed look at the research results shows that during the pandemic, customers preferred shopping for basic necessities in large retail stores (45.50% of households) over shopping in small stores near their homes (33.89% of households). 76.60% of respondents moved the purchase of industrial goods to the sphere of electronic commerce, due to the closure of retail units with industrial goods and the exclusion of industrial goods from the offer of large retail stores. A total of 46.07% of households used contactless shopping, mainly home delivery of food and, to a lesser extent, online shopping. If this trend were to continue, it could have a long-term effect on the balance between contactless sales and physical sales. Retailers therefore have to deal with the question of whether customers will sporadically return to retail units after the crisis due to their new strategic procedures, which are based on the convenience of shopping directly from home based on digital technologies.

The sale of basic necessities in physical retail units required compliance with hygiene regulations from buyers during the pandemic (hand disinfection, wearing a mask)–95.50% of respondents comply with these measures. According to 61.42% of respondents, there were enough disinfectants in retail units, for 32.21% there was a lack of disposable gloves in food stores. 48.13% of respondents are convinced that one standard should apply to everyone, that is, also to retail units, in terms of compliance with hygiene regulations. According to 48, 50% of the respondents, the advice to the cash registers prevents the observance of a safe distance between customers in the store. 69.29% of respondents observe the mandated two-meter distance between

Table 1 Resulting order from the friedman test

Answers to questions		Mean Rank	Mean Value
I mainly buy groceries online.	Q.4.3	3,94	1,9532
There are few disinfectants in grocery stores.	Q.4.7	5,64	2,4438
The shelves and aisles in the stores are poorly marked, which increases the shopping time.	Q.4.10	5,66	2,4719
Queues before entering the store have a long waiting time.	Q.4.9	6,17	2,5918
I feel bad after shopping when I know others are in need.	Q.4.12	6,32	2,6536
There is a lack of disposable gloves in grocery stores.	Q.4.8	7,21	2,8652
Traders prevented the spread of panic buying by customers.	Q.4.11	7,45	2,9700
I only shop in small stores near my home.	Q.4.5	7,92	3,0787
I buy from farmers, local and national producers.	Q.4.15	8,44	3,2341
I use the delivery service.	Q.4.14	8,49	3,2434
Advice to cash registers prevents keeping a safe distance.	Q.4.6	8,86	3,3296
I prefer shopping in large retail stores.	Q.4.13	8,98	3,3876
I always keep a two-meter distance.	Q.4.2	10,56	3,8558
I mainly buy non-food goods online.	Q.4.4	11,00	3,9569
While shopping, I follow hygiene regulations (hand disinfection, mask).	Q.4.1	13,35	4,6704

Source Results of own processing in the scientific project VEGA 1/0012/22 of the Ministry of Education of the Slovak Republic The innovative business models of retail unit formats based on geo-marketing data and their influence on the creation of the value base and retail network of food stores in the digital period

individuals. According to 32.21% of the respondents, they have a long waiting time before entering the shops.

The crisis also affected the emotional state of consumers in relation to the need of some groups of individuals. 17.42% of respondents have a bad feeling when shopping when they know that others are in need. 40.82% of respondents have a bad feeling from time to time. By comparison, far more American citizens "feel embarrassed to shop when they know others don't have anything.“ This criterion changes from 20% before the crisis to 80% during the crisis in California, while it only changes from 14% before the crisis to 24% during the crisis in France (Pellat et al. 2020). By arranging their store, retailers provide consumers with appropriate information and orientation during the purchase. According to 16.48% of respondents in retail units, the shelves and zones for the movement of customers are poorly placed, which increases the time of purchase. 27.71% of respondents state that merchants prevented the spread of panic buying by customers and 45.69 took a neutral position.

RQ2: Are There Links Between the Observed Changes in Purchasing Behavior?

Rank correlation using Spearman's coefficient was used to evaluate the research question. The resulting coefficients are shown in Table 2. As can be seen from the table, during the pandemic, respondents always preferred either large retail stores or, on the contrary, small stores near their residence. It can be seen that both factors related to compliance with safety regulations, such as the use of masks, disinfectants and social distancing, are significantly related to each other. Furthermore, it is obvious that people buying food contactless buy mostly from farmers and local sellers and use delivery service and online shopping to a lesser extent. It can also be seen that individuals who were not satisfied with any of the monitored safety measures provided by the stores (provision of protective gloves, disinfection and proper labeling of shelves) are not satisfied with the other monitored measures either, and during the pandemic did not follow personal safety measures too much (mask, disinfection hands, keeping distance).

Table 2 Resulting coefficients

Correlations

Spearman's rho Correlation Coefficient		Ot.4.1	Ot.4.2	Ot.4.3	Ot.4.4	Ot.4.5	Ot.4.6	Ot.4.7	Ot.4.8	Ot.4.9	Ot.4.10	Ot.4.11	Ot.4.12	Ot.4.13	Ot.4.14	Ot.4.15
	Ot.4.1	1,000	,349**	,022	,222**	,004	,010	-,133**	-,005	-,084	-,199**	,162**	,022	,081	,039	,038
	Ot.4.2	,349**	1,000	,148**	,094*	,161**	,072	-,106*	-,112**	-,012	-,050	,138**	,095*	-,012	-,048	,194**
	Ot.4.3	,022	,148**	1,000	-,063	,087*	-,067	,020	,002	,026	,009	,080	,242**	,181**	,280**	,253**
	Ot.4.4	,222**	,094*	-,063	1,000	-,019	,111*	,002	,061	,052	-,029	,007	-,043	,010	,167**	,020
	Ot.4.5	,004	,161**	,087*	-,019	1,000	,126**	,048	,024	,065	,028	-,099*	-,034	-,372**	-,102*	,002
	Ot.4.6	,010	,072	-,067	,111*	,126**	1,000	,238**	,246**	,097*	,172**	-,114**	-,006	-,035	-,023	-,058
	Ot.4.7	-,133**	-,106*	,020	,002	,048	,238**	1,000	,526**	,268**	,442**	-,155**	,062	,024	,029	-,038
	Ot.4.8	-,005	-,112**	,002	,061	,024	,246**	,526**	1,000	,310**	,431**	-,055	,108*	,068	,054	,040
	Ot.4.9	-,084	-,012	,026	,052	,065	,097*	,268**	,310**	1,000	,379**	,002	,100*	-,001	,033	-,006
	Ot.4.10	-,199**	-,050	,009	-,029	,028	,172**	,442**	,431**	,379**	1,000	-,171**	,128**	,002	,006	,083
	Ot.4.11	,162**	,138**	,080	,007	-,099*	-,114**	-,155**	-,055	,002	-,171**	1,000	,110*	,251**	,059	,239**
	Ot.4.12	,022	,095*	,242**	-,043	-,034	-,006	,062	,108*	,100*	,128**	,110*	1,000	,073	,063	,218**
	Ot.4.13	,081	-,012	,181**	,010	-,372**	-,035	,024	,068	-,001	,002	,251**	,073	1,000	,135**	,115**
	Ot.4.14	,039	-,048	,280**	,167**	-,102*	-,023	,029	,054	,033	,006	,059	,063	,135**	1,000	,199**
	Ot.4.15	,038	,194**	,253**	,020	,002	-,058	-,038	,040	-,006	,083	,239**	,218**	,115**	,199**	1,000

Source Results of own processing in the scientific project VEGA 1/0012/22 of the Ministry of Education of the Slovak Republic The innovative business models of retail unit formats based on geo-marketing data and their influence on the creation of the value base and retail network of food stores in the digital period

RQ3: What Are the Main Characteristics of Respondents Having Similar Changes in Purchasing Behavior as a Result of the Pandemic?

To evaluate the research question, tests were first used to compare independent samples (Kruskal Wallis and Mann-Whit (ney test). The resulting p-values of the performed tests are shown in the following Table 3.

The tests showed that women, compared to men, followed hygiene rules more, recommended distances and were more grateful to traders for their approach to avoid panic. They also preferred shopping in small stores near their residence and online grocery shopping.

People between the ages of 40–59 followed the recommended distance the most and bought food online the most, while young people under 24 years of age did the least. Non-food goods were purchased most by young people under the age of 24, least by people over 60. In addition, old people preferred shopping in small stores closer to their place of residence and, on the contrary, complained least about the lack of disinfectants and gloves in stores. The older the respondents, the more they bought from local farmers and, conversely, the less they used the delivery service.

More educated respondents observed a safe distance more, did not lack protective gloves in stores, used delivery service more and bought more from local farmers and, on the contrary, shopped less in small stores near their residence.

People from small villages with up to 5,000 inhabitants followed the hygiene regulations the most, who at the same time shop more in small shops near their residence. In smaller cities, more people were bothered by queues at checkouts, incorrectly marked shelves and a lack of disinfectants and protective gloves.

The more the respondents felt their financial situation was good, the more they shopped online and used the delivery service, the more they felt the lack of disposable gloves in stores.

Respondents from large families observed the recommended distances more, bought non-food goods online and, conversely, bought food online less. Individuals were least bothered by long queues, poorly marked shelves, lack of disinfectants and gloves in stores.

Segmentation was further used to evaluate the research question. Due to the significant correlation between the individual monitored factors, factor analysis was first used to reduce the dimension of the task. KMO and Bartlett's Test (p-value $< < 0.01$) showed that the correlation matrix is indeed not an identity matrix and therefore it is appropriate to use factor analysis. Figure 1 shows a scree graph, from which it is evident that five factors were optimally selected (five eigenvalues were greater than 1).

The Varimax method was used to distinguish the affiliation of individual variables entering the factor analysis to the resulting factors.

Table 4 shows the individual variables entering the factor analysis and their affiliation to the resulting factors. It can be seen that the first factor contains items that could be summarized as "measures on the store side", the second factor refers to "online purchase of food primarily from local farmers", the third "preferred type of

Table 3 Resulting p-values of the performed tests

Answers to questions		**Gender**	**Age**	**Education**	**Commune size**	**Financial situation**	**Number of members**
While shopping, I follow hygiene regulations (hand disinfection, mask).	Q.4.1	,000	,428	,490	,023	,486	,286
I always keep a two-meter distance.	Q.4.2	,000	,000	,001	,355	,671	,042
I mainly buy groceries online.	Q.4.3	,079	,009	,195	,318	,016	,045
I mainly buy non-food goods online.	Q.4.4	,508	,000	,059	,298	,002	,048
I only shop in small stores near my home.	Q.4.5	,083	,001	,006	,000	,121	,289
Advice to cash registers prevents keeping a safe distance.	Q.4.6	,555	,284	,096	,002	,299	,015
There are few disinfectants in grocery stores.	Q.4.7	,877	,007	,335	,022	,301	,006
There is a lack of disposable gloves in grocery stores.	Q.4.8	,978	,022	,039	,020	,028	,006
Queues before entering the store have a long waiting time.	Q.4.9	,687	,107	,176	,399	,361	,162
The shelves and aisles in the stores are poorly marked, which increases the shopping time.	Q.4.10	,607	,369	,157	,014	,347	,016
Traders prevented the spread of panic buying by customers.	Q.4.11	,021	,205	,190	,824	,005	,550
I feel bad after shopping when I know others are in need.	Q.4.12	,174	,088	,710	,611	,249	,765
I prefer shopping in large retail stores.	Q.4.13	,538	,720	,393	,411	,371	,324
I use the delivery service.	Q.4.14	,177	,005	,000	,075	,000	,854
I buy from farmers, local and national producers.	Q.4.15	,819	,000	,029	,822	,128	,706

Source Results of own processing in the scientific project VEGA 1/0012/22 of the Ministry of Education of the Slovak Republic "The innovative business models of retail unit formats based on geo-marketing data and their influence on the creation of the value base and retail network of food stores in the digital period

store", the fourth "personal compliance measures" and the fifth "online purchase of non-food goods".

Each of the original variables can be written as a linear combination of newly formed factors. Based on this, for each measurement (each respondent), five new

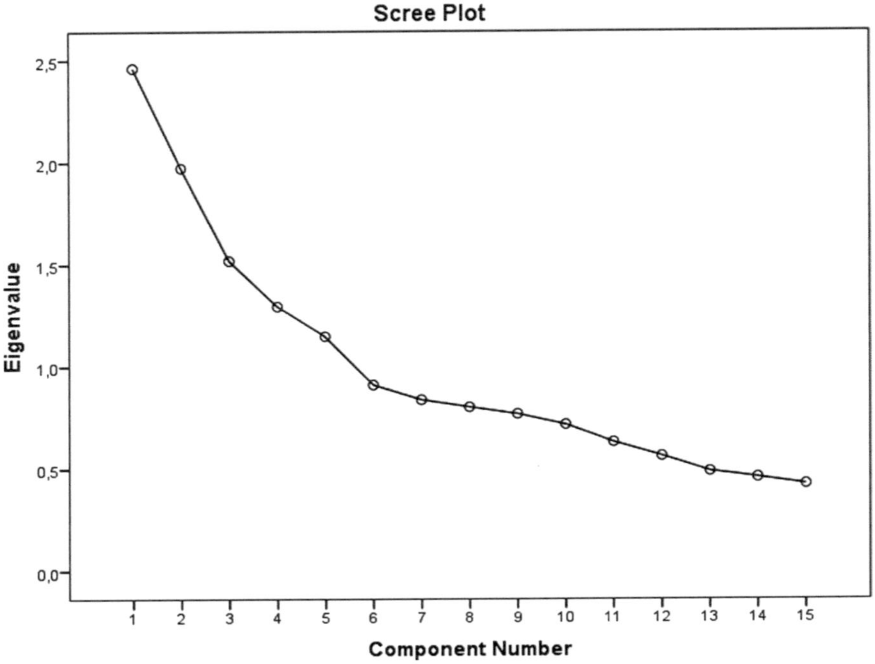

Fig. 1 Scree graph. *Source* Results of own processing in the scientific project VEGA 1/0012/22 of the Ministry of Education of the Slovak Republic The innovative business models of retail unit formats based on geo-marketing data and their influence on the creation of the value base and retail network of food stores in the digital period

values corresponding to belonging to the newly created variables, which, unlike the original ones, are only minimally correlated.

These newly created variables were used as input variables for segmentation using cluster analysis. The Two Step clustering method was specially used. This method showed the division into three clusters as optimal.

Subsequently, the classification of the belonging of objects (i.e. individual respondents) to individual clusters was carried out using decision trees, from four points of view. On the one hand, from an opinion point of view, in two ways, i. e. on the basis of artificial variables - factors created as part of the factor analysis, on the other hand, the explanatory variables entering the classification were the original variables entering the factor analysis, i.e. the answers to individual questions to what extent the respondent's families changed their behavior due to the pandemic. The third aspect was the demographic information of the respondents. The fourth aspect was moral characteristics, presented by subjective opinions from the point of view of the observed aspects.

From the structure of the decision trees, the characteristics of the representatives of individual clusters can be obtained from three points of view, which are:

Table 4 Variables included in the factor analysis

Rotated Component Matrixa						
Answers to questions		Components				
		1	2	3	4	5
While shopping, I follow hygiene regulations (hand disinfection, mask).	Q.4.1	-,130	-,128	,065	,713	,185
I always keep a two-meter distance.	Q.4.2	-,031	,270	-,222	,724	-,069
I mainly buy groceries online.	Q.4.3	-,032	,751	-,111	-,084	,190
I mainly buy non-food goods online.	Q.4.4	,078	-,119	,043	,273	,754
I only shop in small stores near my home.	Q.4.5	,062	,155	-,802	,142	-,080
Advice to cash registers prevents keeping a safe distance.	Q.4.6	,474	-,167	-,255	,190	,184
There are few disinfectants in grocery stores.	Q.4.7	,759	-,007	-,017	-,126	,018
There is a lack of disposable gloves in grocery stores.	Q.4.8	,785	,023	,077	-,016	,079
Queues before entering the store have a long waiting time.	Q.4.9	,583	,105	,019	,001	-,024
The shelves and aisles in the stores are poorly marked, which increases the shopping time.	Q.4.10	,723	,128	-,012	-,192	-,095
Traders prevented the spread of panic buying by customers.	Q.4.11	-,118	,270	,420	,444	-,101
I feel bad after shopping when I know others are in need.	Q.4.12	,169	,568	,114	,035	-,116

Table 4 (continued)

I prefer shopping in large retail stores.	Q.4.13	,084	,201	,728	,058	,056
I use the delivery service.	Q.4.14	-,025	,364	,084	-,218	,731
I buy from farmers, local and national producers.	Q.4.15	,028	,655	,100	,187	,048
Extraction Method: Principal Component Analysis. Rotation Method: Varimax with Kaiser Normalization.						
a. Rotation converged in 7 iterations.						

Source Results of own processing in the scientific project VEGA 1/0012/22 of the Ministry of Education of the Slovak Republic „The innovative business models of retail unit formats based on geo-marketing data and their influence on the creation of the value base and retail network of food stores in the digital period

1st Cluster

- respondents without a high school diploma who feel bad about their financial situation;
- respondents who do not like to be the center of attention and are concerned about the opinion of others;
- respondents who do not use the delivery service, who do not think that the shelves in stores are poorly marked;

*2nd Cluste*r

- women living either in smaller or, on the contrary, in large cities in families with more than three members, who do not feel their financial situation is bad;
- respondents who like to be the center of attention and like established things;
- respondents who think that the shelves in stores are incorrectly marked and there was a lack of disinfectants;

3rd cluster

- respondents with at least a high school diploma who feel bad about their financial situation or respondents who do not feel bad about their situation, especially individuals living in municipalities with a population of 50,000 to 200,000, or men living either in smaller or, on the contrary, in large families with more as three members;
- respondents who like to be the center of attention and like new things, and further individuals who do not like to be the center of attention, but are not worried about the opinion of others;

- respondents using a delivery service who do not think the stores have poorly labeled shelves, but also individuals who think the stores have poorly labeled shelves but do not think the stores have a shortage of disinfectants.

It can be stated that:

- low-educated and low-earning respondents with little self-confidence did not allow themselves to criticize stores and went there to shop during the pandemic because they did not have money for a delivery service;
- mothers of families like their orders and are not afraid to criticize stores;
- more educated respondents with a good income and especially men are self-confident, they do not complain too much about stores, but they like to use the delivery service.

4.1 The Impact of the Health and Economic Crisis Caused by the SARS-CoV 2 Pandemic on the Digital Transformation of Food Retail Stores in the Area of Value Delivery

Digital transformation is based on the use of digital technology in order to make changes in the technological structure of a retail store. This structure is not only about products and services, processes, but also about providing a personalized and seamless experience to existing and new customers across multiple channels (web, mobile, social media, etc.). The aim of the research was to characterize some of the new digital components of the distribution and communication channels through which retail units provide value. Questionnaire research was carried out in five districts of Bratislava, the capital of Slovakia. 244 retail stores in the five size categories shown in Tables 5 and 6 participated in the research.

Table 5 Districts of bratislava and the number of food stores of the examined sample N

Territorial districts of bratislava	Stores up to 40 m^2	Stores up to 100 m^2	Stores up to 400 m^2	Stores up to700 m^2	Stores over 700 m^2	Total
Bratislava 1	12	14	9	6	3	44
Bratislava 2	14	6	10	12	12	54
Bratislava 3	9	15	7	10	7	48
Bratislava 4	4	14	12	8	15	53
Bratislava 5	13	2	8	10	12	45
Total	52	51	46	46	49	244

Source Results of own processing in the scientific project VEGA 1/0012/22 of the Ministry of Education of the Slovak Republic The innovative business models of retail unit formats based on geo-marketing data and their influence on the creation of the value base and retail network of food stores in the digital period

Table 6 Districts of bratislava and share of food stores in %

Territorial districts of bratislava	Stores up to 40 m²	Stores up to 100 m²	Stores up to 400 m²	Stores up to 700 m²	Stores over 700 m²	Total
Bratislava 1	23,07	27,45	19,56	13,04	6.12	18,03
Bratislava 2	26,92	11,76	21.73	26,08	24,48	22.13
Bratislava 3	17,30	29,41	15,21	21,73	14,28	19,67
Bratislava 4	7,69	27,45	26,08	17,39	30,61	21,72
Bratislava 5	25,00	3,92	17,39	21,73	24,48	18,44

The research included questions about the distribution and communication channels through which stores deliver value to customers. In addition to these questions, the questionnaire also included questions about the basic characteristics of the store format, including its address. The questionnaire survey was conducted in the period March - June 2022 in the form of a structured interview with managers of retail stores.

Results of own processing in the scientific project VEGA 1/0012/22 of the Ministry of Education of the Slovak Republic The innovative business models of retail unit formats based on geo-marketing data and their influence on the creation of the value base and retail network of food stores in the digital period.

Data Analysis

Various statistical techniques were used to analyze the data, including higher level statistical techniques and components of descriptive statistics. Cluster analysis is among the most important. Among the methods of cluster analysis, the TwoStep method implemented in the statistical system SPSS was chosen due to the nature of the data. The division of retail unit formats into four clusters proved to be optimal. The resulting Silhouette coefficient with a value of almost 0.5 indicates a good quality of the resulting clustering. The quality is also underlined by the Ratio of Size value of 1.78, which shows the comparability of the clusters and thus meets all the necessary recommendations (Fig. 2).

The most significant factor in determining cluster assignment is the size of the grocery store:

- the 1st cluster is characterized by grocery stores over 700 m²
- the 2nd cluster is characterized by grocery stores of up to 100 m², which have a distance from the center of up to 3,244 m² (at the same time, they have a greater distance to the nearest retail unit than the grocery stores of the third cluster)
- the 3rd cluster is characterized by grocery stores up to 100 m², which have a distance from the center of more than 3244 m² (at the same time, they have a smaller distance to the nearest retail unit than the grocery stores of the second cluster)
- the 4th cluster is characterized by grocery stores from 100 to 700 m².

The conducted research answered the research question:

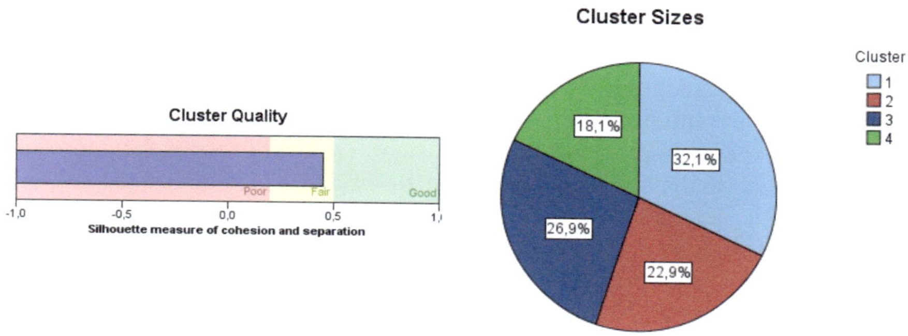

Fig. 2 Cluster quality and cluster sizes. *Source* Results of own processing in the scientific project VEGA 1/0012/22 of the Ministry of Education of the Slovak Republic The innovative business models of retail unit formats based on geo-marketing data and their influence on the creation of the value base and retail network of food stores in the digital period

RQ4: What New Digital Distribution and Communication Channels Are Used by Food Stores?

New innovations related to grocery shopping are gradually moving consumers away from existing shopping technologies. As a result, even some grocery stores are gradually revising their business model, thereby helping the digital transformation. Out of the total number of 244 investigated food stores, 84 food stores (34.42%), which are included in the first, second and third clusters, offer e-delivery of goods to the home. 160 (65.57%) stores do not offer e-delivery of goods to the home. Table 7 illustrates the number of stores by territorial districts of Bratislava that offer e-delivery of goods to the home.

With access to the Internet, consumers can shop through an online marketplace website whenever it suits them. Instead of home delivery, they can pick up the goods in

Table 7 Number of grocery stores by territorial districts of bratislava that offer e-delivery of goods to the home

Territorial districts of bratislava	Cluster 1	Cluster 2	Cluster 3	Total	Share of the total number of stores (%)
Bratislava 1	3	14	5	22	48,88
Bratislava 2	6	5	2	13	24,52
Bratislava 3	4	10	5	19	39,58
Bratislava 4	7	6	4	17	32,07
Bratislava 5	8	2	3	13	28,88
Total	28	37	19	84	34,42

Source Results of own processing in the scientific project VEGA 1/0012/22 of the Ministry of Education of the Slovak Republic The innovative business models of retail unit formats based on geo-marketing data and their influence on the creation of the value base and retail network of food stores in the digital period

person. Developing an easy-to-use online mapping portal helps customers visualize their pickup options. Out of the total number of 244 investigated food stores, 27 (11.06%) stores, which are included in the second and third clusters, realize the ordering of goods online and their collection in the store. 217 (88.93%) stores do not order goods online and pick them up in the store. Table 8 illustrates the number of stores according to the territorial districts of Bratislava, which realize the ordering of goods online and their collection in the store.

For a growing group of consumers, online shopping is the main sales and communication channel, especially in the non-food trade. Out of the total number of 244 investigated food stores, 79 stores (34.42%), which are included in the first, second, third and fourth clusters, improve the current online offer and sales. 165 (67.62%) stores do not improve the current online offer and sales. Table 9 illustrates the number of stores by territorial districts of Bratislava that are improving the current online offer and sales.

Instead of browsing paper flyers and searching for deals directly in stores, modern solutions such as smartphone apps are also available. Out of the total number of 244 surveyed grocery stores, 26 (10.65%) stores that are included in the second and third clusters use shopping through mobile applications and have their own applications that offer a number of useful functions. 218 (89, 34%) stores do not use shopping via mobile applications. Table 10 illustrates the number of stores by territorial districts of Bratislava that use shopping via mobile applications.

Social networks are currently one of the most popular advertising channels in the world for the sale of new products, which requires the store to deal with unique content and an interesting form of presentation. Out of the total number of 244 investigated food stores, 24 (9.83%) stores, which are included in the second and third clusters, move the sale of new products to social networks and media. 220 (90, 16%) food stores do not move the sale of new products to the social network and

Table 8 Number of grocery stores by territorial districts of bratislava which realize ordering of goods online and picking them up in the store

Territorial districts of bratislava	Cluster 2	Cluster 3	Total	Share of the total number of stores (%)
Bratislava 1	1	6	7	15,55
Bratislava 2	1	4	5	9,43
Bratislava 3	3	3	6	12,5O
Bratislava 4	1	4	5	9,43
Bratislava 5		4	4	8,88
Total	6	21	27	11,06

Source Results of own processing in the scientific project VEGA 1/0012/22 of the Ministry of Education of the Slovak Republic The innovative business models of retail unit formats based on geo-marketing data and their influence on the creation of the value base and retail network of food stores in the digital period

Table 9 Number of grocery stores by territorial districts of Bratislava, which improve the current online offer and sales

Territorial districts of bratislava	Clustre 1	Cluster 2	Cluster 3	Cluster 4	Total	Share of the total number of stores (%)
Bratislava 1	8		4		12	26,66
Bratislava 2	5		9	2	16	30,18
Bratislava 3	3	1	9	4	17	35,41
Bratislava 4	9		11		20	37,73
Bratislava 5	7		4	3	14	31,11
Total	32		37	9	79	32,37

Source Results of own processing in the scientific project VEGA 1/0012/22 of the Ministry of Education of the Slovak Republic The innovative business models of retail unit formats based on geo-marketing data and their influence on the creation of the value base and retail network of food stores in the digital period

Table 10 Number of grocery stores by territorial districts of Bratislava that use shopping via mobile applications

Territorial districts of Bratislava	Cluster 2	Cluster 3	Total	Share of the total number of stores (%)
Bratislava 1	9	2	11	24,44
Bratislava 2	2		2	3,77
Bratislava 3	7	1	8	16,66
Bratislava 4	3		3	5,66
Bratislava 5	1	1	2	4,44
Total	22	4	26	10,65

Source Results of own processing in the scientific project VEGA 1/0012/22 of the Ministry of Education of the Slovak Republic The innovative business models of retail unit formats based on geo-marketing data and their influence on the creation of the value base and retail network of food stores in the digital period

media. Table 11 illustrates the number of stores by territorial districts of Bratislava that move the sale of new products to the social network and media.

Geolocation applications represent an intelligent option for business and its visibility on the map. The use of geolocation applications also expands personalized advertising. Out of the total number of 244 investigated food stores, 102 (41.80%) stores (34.42%) that are included in the first, second and fourth clusters use the geolocation system. 142 (58.19%) food stores do not use the geolocation system. Table 12 illustrates the number of stores by territorial districts of Bratislava that use the geolocation system.

Stores can use contact data for advertising via SMS from external databases owned by mobile operators, who can define the selection criteria according to the client's requirements in such a way that micro-segments of users are filtered out. Out of the

Table 11 Number of grocery stores by territorial districts of Bratislava, which move the sale of new products to social networks and media

Territorial districts of bratislava	Cluster 2	Cluster 3	Total	Share of the total number of stores (%)
Bratislava 1	4	1	5	11,11
Bratislava 2	1	2	3	5,66
Bratislava 3	4	3	7	14,58
Bratislava 4	2	3	5	9,43
Bratislava 5	2	2	4	8,88
Total	13	11	24	9,83

Source Results of own processing in the scientific project VEGA 1/0012/22 of the Ministry of Education of the Slovak Republic The innovative business models of retail unit formats based on geo-marketing data and their influence on the creation of the value base and retail network of food stores in the digital period

Table 12 Number of grocery stores by territorial districts of Bratislava that use the geolocation system

Territorial districts of bratislava	Cluster 1	Cluster 2	Cluster 3	Cluster 4	Total	Share of the total number of stores (%)
Bratislava 1	1	10	6	6	23	51,11
Bratislava 2		2	5	16	23	43,39
Bratislava 3		6	2	11	19	39,58
Bratislava 4		3	3	11	17	32,07
Bratislava 5		3	6	11	20	44,44
Total	1	24	22	55	102	41,80

Source Results of own processing in the scientific project VEGA 1/0012/22 of the Ministry of Education of the Slovak Republic The innovative business models of retail unit formats based on geo-marketing data and their influence on the creation of the value base and retail network of food stores in the digital period

total number of 244 investigated food stores, 29 (11.88%) stores (34.42%), which are included in the second, third and fourth clusters, send advertising in the form of SMS to customers. 215 (88.11%) food stores do not send advertising in the form of SMS to customers. Table 13 illustrates the number of stores by territorial districts of Bratislava that send advertising in the form of SMS to customers.

The research carried out showed that all five types of food stores classified into four clusters use differentiated approaches to the use of new digital distribution and communication channels in ways of delivering value. The benefits of digitization, which is associated with better performance, are not equally enjoyed by stores, although better interaction with customers can lead to higher sales and performance. Stores will need to constantly adapt to the changing characteristics of consumers and digital technology in order to face increasingly fierce competition and stay in line

Table 13 Number of grocery stores by territorial districts of Bratislava that send advertising in the form of SMS to customers

Territorial districts of bratislava	Cluster 2	Cluster 3	Cluster 4	Total	Share of the total number of stores (%)
Bratislava 1		1	3	4	8,88
Bratislava 2	1		10	11	20,75
Bratislava 3	3	1	6	10	20,83
Bratislava 4			4	4	7,54
Bratislava 5		1	6	7	15,55
Total	4	3	29	29	11,88

Source Results of own processing in the scientific project VEGA 1/0012/22 of the Ministry of Education of the Slovak Republic The innovative business models of retail unit formats based on geo-marketing data and their influence on the creation of the value base and retail network of food stores in the digital period

with technological changes. Those retailers who have adopted the advanced digital components of the management model are creating a clearer gap between the leaders and the laggards by increasing their market value. Those food stores that cannot change are destined to disappear in quick or slow agony (Pagliaci 2021).

The differences within the clusters into which the investigated food stores were included indicate significant differences between consumers, which can be useful for sustainable marketing communication and defining the offer. High store density implies a shorter inter-store distance, promoting customer overlapping (Stassen et al. 1999). Of all supermarket shoppers, 75% visit multiple stores regularly every week (Stassen et al. 1999), and most (83%) maintain such a shopping pattern (Gijsbrechts et al. 2008). This typical consumer behavior tends to motivate conventional retail formats to introduce new added value and differentiate themselves from other retail businesses. Knowing the characteristics of customers and the context in which they develop will make it easier to influence their behavior and decision making. Therefore, these factors should be taken into account when determining policies to promote access to food through online shopping channels.

5 Conclusion

The digital age is associated with complex, turbulent and large-scale changes that are linked to digital transformation. To survive and thrive in this new market environment, organizations must be agile with dynamic capabilities. In the face of unexpected crises, they have to rely on their innovative abilities in relation to the purchasing behavior of consumers. During the SARS-CoV-2 pandemic, consumers overwhelmingly bought food in large-scale units, in stores near their homes, from farmers and local sellers. Compared to the online purchase of industrial goods, the online purchase of food was used to a small extent. However, consumers used a contactless form of purchase, namely home delivery of food. This situation suggests that there is great potential to improve the use of online food channels among consumers. In general, consumer attitudes were significantly influenced by education, age, household financial situation, place of residence and whether the decision-maker is a man or a woman.

To support the digital connection of food stores and consumers, it is necessary to use digital technologies more intensively on the part of retail units. It can be concluded that successful grocery stores will be those that can adapt to and respond to the dynamic changes caused by the SARS-CoV-2 pandemic by:

a. take into account new consumer practices and the increased role of digital technologies.
b. they will formulate new marketing strategies with the aim of building permanent customer loyalty.
c. due to the diversity of distribution and communication channels, they will develop individualized marketing strategies focused on the customer experience,
d. redefine their business models towards innovative digital models based on data and the use of new digital and communication channels,
e. by digitizing their activities, they will create new forms of interaction between consumers and the food store, which allow consumers access to several media channels and which increase the complexity of customer choice,
f. strengthening and supporting the use of digital and communication channels to deliver value can be an important tool for the commercial policy of food stores.

It is also a challenge to enable the Internet user to understand how the collected personal data is used. The task of food stores will be to create an atmosphere of trust in relation to the consumer (Pellat, Hoareau 2019). In this regard, food stores still have a difficult path to digitization ahead of them, but this process is in full swing and will not stop.

Limitations of the Study and Future Research

This study only evaluates and examines retail grocery operations. Obtaining a broader view of activities related to digitization would require comparative studies not only from the Bratislava area, but also from other regions of Slovakia. The research on shopping behavior during the health and economic crisis caused by the SARS-CoV-2 pandemic was conducted online, so additional investigation into the reasons for some of the answers provided by respondents was not possible, such as trust in online shopping channels, food prices and quality, and technical barriers to online shopping when investigating the determinants of people's online shopping behavior during the pandemic. Also, comparing the activities of food stores in the field of digitization on a larger scale would allow researchers and business policy makers to gain important insights. Future research should explore the factors and barriers that prevent retail operations from implementing digital technologies as well as business model elements.

References

Bogalska-Martin E (2021) Experiences and learnings from the time of Covid-19. Manag Gouv 26:11–12

Bocken N, Short W, Samuel R, Padmakshi E, Steve. (2014) A literature and practice review to develop sustainable business model archetypes. J Clean Prod 65:42–56. https://doi.org/10.1016/j.jclepro.2013.11.039

Editorial. Viete, koľko Slovákov nemá internet? Pozrite si, ktorý kraj je na tom najlepšie a ktorý najhoršie. Topky.sk. November 26

Gijsbrechts E, Campo K, Nisol P (2008) Beyond promotion-based store switching: Antecedents and patterns of systematic multiple-store shopping. Int J Res Mark 25:5–21. https://doi.org/10.1016/j.ijresmar.2007.07.001

Hsien-Long H (2023) Challenges for contactless online food delivery services during the COVID-19 pandemic in Taiwan: moderating effects of perceived government response. Eval Program Plann 97:102249. https://doi.org/10.1016/j.evalprogplan.2023.102249

Ibn-Mohammed T, Mustapha KB, Godsell J, Adamu Z, Babatunde KA, Akintade DD, Acquaye A, Fujii H, Ndiaye MM, Yamoah FA, Koh SC (2021) A critical review of the impacts of COVID-19 on the global economy and ecosystems and opportunities for circular economy strategies Resour, Conserv Recycl 164:105169. https://pubmed.ncbi.nlm.nih.gov/32982059/

Kritan K C K B F, Hencelová KD (2022) Dosah pandémie COVID-19 na zmeny v nákupnom správaní spotrebiteľov na Slovensku. Sociológia - Slovak Sociological Review, 54:385–411. https://doi.org/10.31577/sociologia.2022.54.5.14

Kita P, Žambochova M, Mazalánová K, Veronika K, Jaroslav. (2020) Model marketingovej komunikacie na zdravie orientovane nákupné správanie spotrebiteľov so zreteľom na postoje k spotrebe zdravých potravín. GUPRESS, Bratislava

Maciejewski G, Mokrysz S, Wróblewski Ł (2020) Consumers towards marketing straegies of coffee producers. Wageningen academic Publishers, Wageningen

Masotti M, van der Haar S, Janssen A, Iori E, Zeinstra G, Bos-Brouwers H, Vittuari M (2023) Food waste in time of COVID-19: The heterogeneous effects on consumer groups in Italy and the Netherlands. Appetite 180. https://doi.org/10.1016/j.appet.2022.106313

Pagliacci GR, Mario. (2021) Les business angels à l'époque du Covid. Management & Gouvernance 26:45–53

Pellat G, Hoareau E (2019) Comment l´utilisation de la RGPD effecte le comportement des consommateurs lors d´action internet? Manag Gouv 22:59–68

Pella G, Rauchwerger S, Peyroux C (2020) La stratégie d´achat en période perturbée induite par le Covid -19. Etude comparé entre deux pays: France – USA/Californie. In: Acte colloque 10 – 11 septembre 2020. Grenoble: PGV, l'IUT2 et l'UGA

Rittera T, Pedersena C (2020) Digitization capability and the digitalization of business models in business-to-business firms: Past, present, and future. Ind Mark Manage 86:180–190. https://doi.org/10.1016/j.indmarman.2019.11.019

Stassen RE, Mittelstaedt JD, Mittelstaedt RA (1999) Assortment overlap: its effect on shopping patterns in a retail market when the distributions of prices and goods are known. J Retail 75:371–386. https://doi.org/10.1016/S0022-4359(99)00013-5

Stephens HM, Mack EA, Mann J (2022) Broadband and entrepreneurship: An empirical assessment of the connection between broadband availability and new business activity across the United States. Telematics Inform 74:101873. https://doi.org/10.1016/j.tele.2022.101873

Úrad verejného zdravotníctva Slovenskej republiky. 2020. Opatrenie Úradu verejného zdravotníctva (UVZ) Slovenskej republiky pri ohrození verejného zdravia. ÚVZ, June 5

Verhoef PC, Broekhuizen T, Bart Y, Bhattacharya A, Dong JQ, Fabian N, Haenlein M (2021) Digital transformation: a multidisciplinary reflection and research agenda J Bus Res122: 889–901. https://doi.org/10.1016/j.jbusres.2019.09.022

Vidal-Mones B, Cobalea HB, Diaz-Ruiz R, Fernández-Zamudio M-A (2021) Citizens food habit behavior and food waste consequences during the first COVID-19 lockdown in Spain. Sustainability 13:3381. https://doi.org/10.3390/su13063381

Witte K, Berkowitz JM, Cameron J, McKeon JK (1998) Preventing the spread of genital warts: Using fear appeals to promote self-protective behaviors. Health Educ Behav 25:571–585. https://doi.org/10.1177/109019819802500505

Zmušková B (2022) Slovensko je v digitalizácii piate najhoršie v Únii Euroaktiv.sk. Jul 29

The European Project Confronted with the War in Ukraine and Its Impact on the French and Slovak Attitudes and Involvement in Their National Legislative Elections

Mária Huraj Rošteková and Katarína Chovancová

Abstract This study examines the mobilization of citizens for the European project in times of a crisis of values. We focus on the changes in attitudes that underlie the feeling of engagement among citizens of various EU member countries since the start of the military conflict in 2022 at the eastern border of the European Union, the largest and most brutal in the heart of the European continent since the Second World War. This crisis, which revealed the fragility of the international and European security order, is marked in many EU countries by a rise in populism and extremist positions, and the dissemination of propaganda, particularly in Central and Eastern European countries. We observe, compare, and analyze the political discourse as it is conveyed, reflected, and filtered by the media during electoral campaigns leading to

---

Mária Huraj Rošteková (works at the Faculty of Political Sciences and International Relations of Matej Bel University in Banska Bystrica, Slovakia. In her research work, she deals with European values and citizenship, the European Neighborhood Policy, the Eastern Partnership, and the relations of the countries of the southern Mediterranean with the EU. Her works are also devoted to the analysis of political and media discourse and the tools of EU soft power. She is a member of the ERECO-PGV Network.

Katarína Chovancová is professor in general linguistics at the Faculty of Arts, Matej Bel University in Banska Bystrica (UMB), Slovakia. In her research, she deals with general and contrastive linguistics, compared and contrastive grammar, pragmatics of communication and multilingual terminology. Her scientific work focuses on the research of communication events, discourse analysis and linguistic pragmatics. The main area of her interest is mediated communication in a broad sense, including didactic mediation of knowledge of linguistic systems through oriented description of grammar structures.

M. H. Rošteková (✉)
Faculty of Political Sciences and International Relations, Matej Bel University, Banska Bystrica, Slovakia
e-mail: maria.rostekova@umb.sk

K. Chovancová
Faculty of Arts, Matej Bel University, Banska Bystrica, Slovakia
e-mail: katarina.chovancova@umb.sk

G. Pellat et al. (eds.), *Data-Centric Business and Applications*, Lecture Notes on Data Engineering and Communications Technologies 258,
https://doi.org/10.1007/978-3-031-95068-1_3

the parliamentary elections in two countries, namely France and Slovakia. Politicians at the national level play a crucial role in involving European citizens in European and global issues. Our paper focuses on the legislative elections which took place in France (a founding and supporting country of the EU) in June 2022 and in Slovakia (a country of the former communist bloc) in September 2023. In the case of the latter country, we also present an analysis of the media discourse using the Sketch Engine software. Our analytical work is focused on the textual data collected in the Slovak national media, specifically in the electronic editions of the daily newspapers SME and Hlavný denník, during the last four weeks of the electoral campaign.

Keywords European project · War in Ukraine · Involvement of European citizens · National legislative elections · France · Slovakia · Populism · Extremist positions · Propaganda · Media · Electoral campaigns

1 The European Project Confronted with the Russian War in Ukraine

The Russian military aggression of Ukraine has deeply shaken the European project. It has become a turning point in the history of the European continent, to the extent that political discourse began to speak of a 'paradigm shift' or 'epoch shift' - from the German Zeitenwende (Lefebvre 2023), highlighting a significant turn in the politics of a previously prosperous and relatively peaceful Europe. It *"shakes the mental universe of the European Union (EU) and Europeans" because* it *"reflects old traumas and puts it back in history"* (Chopin, Naim, 2023). Russia's invasion of Ukraine is not only an illegal, unprovoked, and unjustified attack on a sovereign country defending its rights and democracy, but also a direct assault on fundamental European values: "*As a result of this war, the EU has been forced to rethink its sense of purpose and the values on which it has based its paradigm of political action*" (Chopin, Naim, 2023). Most importantly, the EU had to act fast and on a grand scale through humanitarian and financial aid. The EU also provided political support that just four months after the start of the invasion resulted in Ukraine[2] becoming an EU Member candidate - a "*unique and historic moment"* in its relations with the EU (Zelensky 2022). The adoption of several packages of sanctions against Russia, including targeted individual, economic, and visa measures, which complemented existing measures imposed on Russia since 2014 following the annexation of Crimea and the non-implementation of the Minsk agreements, should also be seen as unprecedented.

By imposing tough sanctions, the EU has demonstrated its ability to apply the "hard power" tools it needs to pursue its political interests on the world stage, as befits a true superpower (Borrell, 2023). Indeed, until then the EU had mainly utilized its "soft power" in various ways, above all in its ability to influence good governance in candidate countries (Gaillard, 2018), but also in neighboring states. Often criticized for its idealistic approach and geopolitical weakness, the EU had to "*come up with a new discourse and take decisions regarding hard power, thus realizing the transition from values to the defense of its interests*" (Chopin, Naima, 2023). As Maxime Lefebvre (2023) states, after 22 February 2022 "*the EU has indeed become a "liberal Great Power" that defends its values by resorting to hard power instruments (sanctions), even getting close to building a military alliance by supplying arms to Ukraine*". And although the EU has largely coordinated its sanctions with the US, whose military strategy relies primarily on NATO, Lefebvre (2023) finds remarkable that "*it is doing so with an unanimity that has surprised observers and perhaps the Russian president himself, and with strong public support*".

The war in Ukraine is a turning point for the EU in several areas. First, it is the biggest challenge yet to European security since the end of the Second World War. As Vice-President of the European Commission and EU High Representative for Foreign Affairs and Security Policy, Joseph Borrell (2023) states that the war directly "*threatens the principles on which international relations are based, in particular, the principles of the United Nations Charter and the Helsinki Final Act*". In the wake of the Russian invasion, several EU Member States have begun to fundamentally change their reluctance to participate in existing EU or NATO security and defense structures. Some have decided to reconsider their own neutrality or pacifism and others, such as Germany, have even undergone a change of identity (Yar 2023). Also notable is a package of almost €500 million for armaments and defense systems for Ukraine under the *European Peace Facility* (EPF), which was approved just three days after the start of the Russian invasion. According to Joseph Borrell (2023), that war of aggression allowed for the "*belated birth of a geopolitical Union*", even if one must admit that Europeans had been discussing raising the EU's awareness of its own security for some time. Although the ambitions of European strategic autonomy were already taking concrete contours just before the war (Yar 2023), more progress ensued in a short time after 24 February 2022 than in the entire previous decade (Borrell 2023).

Last but not least, the Russian war seriously threatened the EU's energy security, as up to one half of the Member States' natural gas imports came from Russian sources. Following the outbreak of the war, the world energy system was in shock, with major upheavals not only in the natural gas markets, but also in the coal and electricity markets, and with significant turbulence in the oil markets. Government leaders across the EU were immediately forced to look for other sources of energy supplies and to face the threat of exponentially increasing energy prices strongly feared by its population. However, according to Fatih Birol Executive Director of *the International Energy Agency* (IEA), that crisis has completely transformed the energy system, not just in the short term but "*for decades to come*": "*even within today's political parameters, the responses of governments around the world promise to make this a historic and definitive turning point towards a cleaner, more affordable and more secure energy system*" (World Energy Outlook 2022). Following a series of measures taken by the EU and its Member States, the war has allowed to substantially reduce the consumption of Russian fossil fuels and accelerate the EU's transition to renewable energy sources.

The war of aggression is first and foremost a huge humanitarian disaster, but it has also had a major impact on the world economy, fostering an increased level of insecurity. Indeed, before the attack on Ukraine, experts predicted a continued global recovery from the Covid 19 pandemic in 2022 and 2023 (Perspectives économiques de l'OCDE, 2022), but right after the outbreak of the war it became clear that those positive scenarios would not materialize due to reduced economic growth and increased inflation around the world, including in the EU (Rýsová, Kazanský, Smeriga 2023). According to several experts, those negative consequences have yet to manifest themselves. Russia and Ukraine have a significant impact on the global economy, primarily through the export of their natural resources, as they are key suppliers for a number of products. Jointly they account for around 30% of world exports of wheat, 20% of maize, mineral fertilizers, and natural gas, and 11% of oil (Perspectives économiques de l'OCDE, 2022). In addition, supply chains around the world depend on Russian and Ukrainian exports of metal ores which have risen sharply in price since the start of the war, without significantly disrupting their production or export volumes. Russia and Ukraine being the most important agricultural powers in the world, concerns have been raised about threats to global food security, particularly in countries dependent on Russian-Ukrainian imports, especially those in Africa which lack supply or production options, such as Somalia which imports 100% of its wheat from Russia and Ukraine, or Sudan 75% (Bernard, 2023). Thus, the war in Ukraine dealt another blow to the global economy at a time when many of the pandemic induced supply problems seemed to be fading allowing for renewed economic development.

It took some time before the EU was able to find concrete solutions to what was then the most significant crisis in its history, with Member States often forced to create their own solutions (Lefevre 2023). That challenging period was described by the outgoing President of the European Commission (2004–2014), José Manuel Barroso, as "*a 'decade of European resilience*', or '*ten difficult years, years of multiple crises, emergencies, and constant pressures, years when we had to face not only financial, economic and social crises but also the emergence of disagreements among Europeans, exacerbated by the dangerous simplification of some political discourses about Europe. During these years, Europe has never been so threatened by new fault lines. And yet every one of our Member States needs Europe more than ever*" (Barosso 2014).

Other crises followed, notably the migration crisis, Brexit, and the Covid 19 pandemic. The influx of refugees caused major contradictions and diplomatic tensions between European countries seeking a consensus position: while the European Commission has pushed for the introduction of quotas for each EU country as supported jointly by Angela Merkel and François Hollande, many countries mainly in Central and Eastern Europe have opposed migration flows on principle. At the same time, the EU had to handle the UK's decision to leave the European club following the referendum of July 2016. With Brexit, as an unprecedented process for which it is difficult to find relevant parallels in the past (Terem, Müller, 2018), the EU was not only losing a politically powerful state but also the second most important financial contributor and the main diplomatic partner of the US in Europe. According to Victor

Lefevre (2020), the EU has been torn apart many times before by its successes and failures, but it has indeed gone through an eventful evolution in the first twenty years of the twenty-first century. Faced with new external but also internal challenges, the EU *has "managed to adapt, improvise and resist*" (Lefevre 2023), thus confirming once again the prophecy of Jean Monnet, one of the pioneers of European integration, who stated in his memoirs that "*Europe will be forged in crises, and will be the sum of the solutions adopted for those crises*" (European commision, 2011) And history has since proved him right many times over: "*Since the 1950s, the explosions have multiplied, but Europe, more reed than oak, has adapted each time. The sanitary crisis and the war waged by the Russians in Ukraine have even encouraged it to go faster and further*" (ASFE 2022).

The EU has emerged from the Covid 19 crisis with the message that it will only be possible to address complex challenges thru solidarity and by working together. That experience quickly took a new shape on 24 February 2022, when the EU leadership decided within the first hours of the Ukraine crisis that they could only confront the aggressor together, "*otherwise any action by the twenty-seven will be toothless and within weeks a puppet regime of the Kremlin will be born beyond next to the Union's eastern borders*" (Yar 2023). The search for solutions contributed to the willingness of European actors to change some of the EU's rules of engagement. Synergies have also emerged in the case of the Franco-German engine of the EU, facilitated by the arrival of Olaf Scholz as Federal Chancellor, who unlike his predecessor has "*shown a much more pronounced desire*" to push the boundaries of the European project, and not just in the military sphere" (ASFE 2022). This change of approach was met with a similar ambition by French President Emmanuel Macron while assuming the presidency of the European Union Council. From his position, he expressed his ambition to breathe new life into Europe and to negotiate with his European partners to move towards "*a Europe that is influential in the world, fully sovereign, free in its decisions and master of its destiny*" (Déclaration 2021).

By defending its values against Russia, the EU is asserting itself as a 'liberal superpower', just as the US defends its own interests and values. The EU is bound to do so by the Maastricht Treaty, namely Articles 3–5 and 21–2 of the Treaty on European Union (TEU). According to Maxime Lefebvre (2023), that same approach led the EU to expand and include several countries of Central and Eastern Europe (2004, 2007, 2013), freed from the communist yoke, as well as to introduce the *European Neighborhood Policy* (2004), aimed at neighboring countries in Eastern Europe and the southern Mediterranean. That effort was later reinforced by the regional project *Eastern Partnership* (2009), designed to foster better conditions for EU's eastern partners to move closer to European standards. And for that same reason, the EU also decided to gradually distance itself from Russia's policy (Lefebvre 2023).

However, despite having formed a broad international coalition in support of Ukraine and isolated Russia while seeking to restore the international legal order after February 2022 (Borrell 2023), undoubtedly a key historical moment for the EU, according to Maxime Lefebvre (2023) "*the EU remains vulnerable in spite of that reactivity*". Indeed, agreeing on how to engage Russia was one of the EU foreign policy issues that most divided Europeans in the last few years before the outbreak of the war. Some countries favored maximum hardness in their stance, particularly Poland, Sweden, Estonia, Latvia, and Lithuania. Others favored a more realistic approach, such as France, Germany, Italy, Spain, and Finland. Still others were outspoken supporters of Vladimir Putin, such as Hungary under the baton of the populist politician Viktor Orbán (Lefebvre, 2023). The individual governments of the member countries have different attitudes not only to the sanctions against Russia, the effectiveness of which is questioned by several leaders, but also to the continuation of financial aid for Ukraine allocated from European sources, the sending of military equipment and, most recently, to the opening of accession negotiations with Ukraine, which received the status of a candidate country.

The enlargement of the European Union to include Ukraine and other countries represents a major operational challenge for the 27. The reform of European institutions has become a very important topic of discussion. Although Europe's multi-speed model is an old topic of debate, the prospect of further expansion is bringing it back into the spotlight. In September 2023, an independent Franco-German study group issued a report on reforms to be implemented in the coming years (Sailing on high seas, 2023). Experts suggest moving decision-making from unanimity to a qualified majority, especially in new areas such as foreign and fiscal policy. They also propose to increase the EU budget, which presupposes a debate on the Union's own resources - the expansion of the specific European tax base - and the expansion of national contributions, which is a sensitive topic.

From the point of view of the rule of law, the report recommends simplifying procedures in the event of a violation, as was the case in Poland and Hungary. The current Article 7, which normally allows a member country to be excluded from its voting rights in the Council in the event of a serious breach, has so far been ineffective due to the complexity of the procedure. During the meeting of the European Council held on December 14, 2023, in Brussels, Hungarian leader V. Orbán clearly conditioned aid to Ukraine on the release of all funds planned for Hungary, which were blocked because of attacks on the rule of law. This negotiation about democratic values is extremely sensitive and in the current political situation it can increase the level of populism in Europe, which already has a large negative impact on the future direction of the European project.

This introduction aims to contextualize the complex evolution of the EU as a geopolitical actor, highlighting its position regarding the conflict in Ukraine and its ability to react in times of crisis. It specifically focuses on the period from the outbreak of the armed conflict until the autumn of 2023. This timeframe has served as the basis for the analysis of the position adopted by the EU. It is crucial to note that, although the geopolitical situation has continued to evolve significantly beyond this period, the development of further events falls outside the scope of our current study.

2 Populism, Extremism, and Electoral Choice in Europe

Russia's invasion of Ukraine "*brought a new weapon to the arsenal of populists in Eastern Europe - the fear of war*". Almost two years after the outbreak of the conflict, with no end in sight, some politicians try to achieve their political goals by exploiting their citizens' fears that supporting Ukraine could help drag their country into the war (Mortkowitz 2023). A new wave of populism has hit the European political scene. Already the financial and economic crisis of 2008 and later the euro crisis had many economic, social, and political consequences that fostered a stream of radical left-wing populism into Europe, embodied at the time by the parties *Podemos* in Spain, *Syriza* in Greece, or *La France Insoumise* in France (Ivaldi, 2023). This type of populism, based on a radical critique of neoliberalism, rejected austerity measures imposed especially on southern European countries and supported egalitarian, humanistic and universalist values (Ivaldi et al. 2023). In several Western European countries, those crises even contributed to the rise of left-wing populist parties to power. After being founded in 2014, *Podemos* became the third strongest party in the Spanish parliament, winning deputies in all regional assemblies and occupying the city councils of the largest cities, such as Madrid and Barcelona. In turn, Jean-Luc Mélenchon's *La France Insoumise* managed to win almost 20% of the vote in the 2017 presidential election (Castaño, 2018).

However, most European parties which are considered populist by experts are extreme right-wing and nationalist groups. In most EU Member States, they mainly feed on the tension in defining a national identity while facing a growing wave of refugees from the Middle East since 2015 (Ivaldi, 2023). Although those far-right parties and movements on the European continent are different, we can identify common features in all of them, above all a very pronounced social conservatism, rejection of immigration, and nationalism (Statistics Research Department, 2023).

The success of nationalist, xenophobic and authoritarian populism also impacted the European Parliament elections in 2019. It gave rise to the parliamentary group *Identity and Democracy* created around the *Rassemblement National (RN)* of Marine Le Pen, who later managed to reach 41,45% of the vote in the second round of the French presidential elections in 2022 (Ivaldi, 2023). The extreme right has recently been successful in many other European countries. A major turning point was achieved when the post-fascist party *Fratelli d'Italia* won the Italian parliamentary elections in September 2022, after several other successes had already been recorded in Central European countries. In Hungary, Viktor Orbán's *Fidesz* party, in power since 2010, confirmed its victory again in the spring 2022 parliamentary elections by winning 59% of the parliamentary seats, allowing it to form a coalition with the Christian Democratic Party NKDP. In Poland, the ultra-conservative candidate with anti-LGBT rhetoric Andrzej Duda was re-elected in the 2020 presidential election, and the *Law and Justice* (*PiS*) party in power since 2015 won more than 43% of the vote in the 2019 parliamentary election (Statistics Research Department, 2023).

In February 2022, "*Russia's invasion of Ukraine embarrassed the main far-right movements in Europe, which oscillated between ideological loyalty to Vladimir Putin and solidarity with Kiev, pointing to Western responsibility in the conflict*" (AFP, 2023). According to Sébastien Maillard (2023), the Ukraine war actually divided the European extreme right into pro-Ukrainian and pro-Russian. While some parties, such as *Vox, Fratelli d'Italia, True Finns*, but also the Polish *PIS* sided with Ukraine, NATO, and no longer questioned the principle of European integration, others like the French *RN* and the German *AfD* remained faithful to their pro-Russian and anti-Brussels tradition. In Poland, where the nationalist right-wing government voiced its pro-Ukraine positions, the situation completely turned around just before the parliamentary elections held on October 15, 2023. National conservatives in the fight for voters surprised other European countries with an increasing number of negative statements towards Ukraine. Since every single vote counts in a tough election campaign, hardening rhetoric has become an important electoral tactic of *PiS* to stay in power (Bienvenu, 2023). Although that party managed to repeat its electoral victory, it did not maintain an absolute majority in the parliament. That created space for the liberal opposition of Donald Tusk to form a pro-European government in a coalition with the centrists and the left.

Most European populist movements are united by the rejection of traditional political patterns. Their radical nature places them on either the left or the right of the political spectrum, although there are also political parties with ambivalent positions that are not easily classifiable (Lastennet, 2021). According to J. Rupnik, this broad spectrum of populism haunts the whole of Europe. It is visible in the West–East dichotomy that prevails in the media space, and is justified according to him, at least in the search for identity and the reaction of the Visegrád Four (Poland, Hungary, the Czech Republic, and Slovakia) to the migration crisis. According to him, it would not be correct to define the nationalist and populist trend "*only as a specific feature of the post-communist space*" (Rupnik, 2018). Only a few European countries managed to avoid the rise of this phenomenon. For example, France can even be considered a forerunner in this respect: the *Rassemblement National*, formerly known as the *Front National*, has scored high in elections for more than 30 years, and the radical left party *La France Insoumise* has very quickly managed to win among the traditional governing parties (Lastennet, 2021). The fundamental strengthening of radical populism in Europe was also stressed by the victory of the *Leave* party in the British referendum on the country's withdrawal from the EU on June 23, 2016, or by the success of the right-wing Austrian radical and populist Norbert Hofer, who almost won the presidential election in of 2016 (Georgieva, 2017).

When looking for a common denominator of those populist parties across Europe, we find that the often-cited link between the power of populism and economic and social difficulties is not enough in itself. Some populist parties are also doing well in the most economically advanced and affluent EU countries, including Germany. Although the perceived or feared decline in the standard of living is among the basic factors that underlie the development of populist movements, the issues of sovereignty and identity are also gaining importance. Many populist leaders promote the theses of a "closed society" and a "homogeneous nation" (Lastennet, 2021).

Central European populist movements may have different forms, but they share many common features, such as the defense of values on the identity-sovereignty scale, which J. Rupnik explains as a link between the defense of the cultural identity of a nation while Europe is subjected to mass migration of non-Europeans, often Muslims. Many Member States want their sovereignty to decide who can enter and reside in their own territory: "*Central European countries with export-oriented economies are not opponents of economic globalization, but they are against migration globalization. In short: an open economy, but a closed society*" (Rupnik, 2018). All the countries of the Visegrád Four are almost completely united on this topic, as evidenced by their leaders' attitudes towards border protection and rejection of multiculturalism (Rupnik, 2018). In this context, the Czech political scientist Jiří Pehe states that Central Europe is still not completely united with what we call the West. According to him, various political events of recent years demonstrate that some countries of Central Europe such as Hungary or now Slovakia after the September 2023 elections, are in some way constantly leaning towards the East i.e., Russia: "*There is such a temptation, not only political, but also parts of the population of these Central European countries to be part of both civilizations*" (Na Západ jsme se úplně nevčlenili, 2023).

Many analysts believe that contemporary populism is a threat to democracy and the European Union (Georgieva, 2017). Populist movements often criticize and present the EU project as an anti-democratic and ultra-liberal project (Lastennet, 2021). From J. Rupnik's previous analysis, it also follows that although Central European states have successfully completed their transition to democracy by joining the EU, populist politicians managed to come to power in several of these countries after that transition was complete: "*The Union has long embodied the anchoring and irreversibility of democracy, while […] in Budapest and Warsaw they position themselves as the vanquishers of the Brussels Commission*" (Rupnik, 2018). Populist governments are fundamentally hostile, if not to European integration in general, then at least to the federal nature of the EU and some of its basic values, such as the protection of human rights and minorities, or environmental issues. Such increasing trends threaten the cohesion of the EU and could lead to "European disunion" if nothing is done (Munier, 2023).

The programs of those governments have thus become centrifugal forces for the EU, the Ukraine crisis serving as one very good example. As the center of Europe has moved eastward, the EU and many member countries are now paying more attention to national elections organized in 2023 in Central and Eastern European countries. They include the presidential elections in the Czech Republic (January) and the parliamentary elections in Bulgaria (April), Slovakia (September), and Poland (October). The Slovak parliamentary elections of September 2023 which elected Róbert Fico as Prime Minister and open sympathizer of Russia, were supposed to "*strengthen the axis emerging in the east of Europe, undermining the effort to build a pro-Ukrainian European front. Hungary was the first to weaken its support for Ukraine. We should now be worried about new contradictions that may appear within the Union*" (Gradvohl, 2023).

3 What Mobilization of Voters for the European Project During National Elections in Slovakia and in France

Given that the war in Ukraine has fundamentally reshuffled the political cards across Central European countries, we need to pay attention to the debates and fault lines that cut across the political formations in those countries and influence their social development, especially those that directly or indirectly concern the geopolitical direction and strategic stance towards Ukraine, Russia, the EU, and NATO. We wish to compare them with the debates that resonate in other EU Member States, especially in France. To this end, we decided to observe, compare, and analyze the political discourse that is mediated, reflected, and filtered by the media during the national parliamentary election campaigns, bearing in mind that national-level politicians play a crucial role in engaging citizens in European and global political issues, and that it is the media that increase citizens' interest in politics and their participation in the political process. Our analysis focuses on France (as a founding and sustaining country of the EU), where parliamentary elections were held in June 2022, and on Slovakia (as one of the Central European countries) and its parliamentary elections held in September 2023. In the case of the second-named country, we also present an analysis of the media discourse through the sketch engine software, the results of which are presented in the last part of the paper. Our analytical work is focused on the sets of texts collected in the Slovak national media, more precisely in the electronic editions of two daily newspapers for the last four weeks of electoral campaign.

Legislative elections are a key moment in the political life of each of the EU's 27 Member States, as they allow citizens to directly choose their representatives in the lower (or only) house of parliament and thus determine the composition of their government. While voters decide the future direction of the EU mainly through European elections, national elections are also of great importance to resolve many European issues, since governments often determine their foreign policy direction on the basis of a government program statement that is shaped immediately after the parliamentary elections. Interestingly, the last European elections in France in 2019 demonstrated a largely 'nationalized' vote: "*Both in terms of campaign calendar and content, the European Parliament elections were dominated by domestic issues, thus largely overshadowing the European dimension of the debate*" (Élections européennes, 2019). However, the reverse can also be true: elections at the national level (both parliamentary and presidential) can generate important debates on several European issues that are directly related to citizens' lives.

For France, the comparison is even truer, as it is a parliamentary democracy of the semi-presidential type (Augustín, 2021). Christian Lequesne proves that point (2017) by asking the question: "*How could we actually think today about fundamental issues where macroeconomic policy, defense, environmental protection or labor mobility are included without taking into account the European dimension?*". In doing so, he points to another important argument, namely Euroscepticism, which is widespread in all EU countries. Indeed, the rejection of the European project often appears as a very explicit element in the rhetoric of candidates from both ends of the political spectrum who to distinguish themselves, bring European themes into the campaigns for both presidential and national parliamentary elections (Lequesne, 2017). Given its important role in European integration, France is watched with great interest by all European partners at times of national elections, and even more so in times of crises. This was already the case in 2017 in the post-Brexit context, when the future French president was largely responsible for relaunching the European project. It was also the case in 2022 when Europe, reeling from a sanitary crisis, was caught off guard by Russia's military intervention in Ukraine. Moreover, the 2022 election campaigns in France (both presidential and parliamentary) were conducted during the ongoing French presidency of the Council of the EU, which Emmanuel Macron assumed for a six-month period starting at the end of December 2021. It could therefore be assumed that that French media discourse would be suffused with European themes during the first half of 2022.

In the run-up to the French presidential elections on 10 and 24 April 2022, several pressing issues such as economic recovery, maintaining purchasing power, a new industrial policy, and ecological transformation emerged as important issues in the European debate as the health crisis was waning. Later they became important themes in the French electoral campaign. Emmanuel Macron, one of the 12 candidates in the presidential election, has clearly emerged as the candidate who most supports the European Project, describing the second round of the French presidential elections as a "referendum for or against the EU". On the other hand, he was opposed by candidates who in the past had openly questioned the very principle of European integration, namely Marine Le Pen traditionally the embodiment of the French far right, and Jean-Luc Mélenchon as the main representative of the new far left.

The election campaign started in March 2022, that is, only a very short time after the beginning of the Russian invasion of Ukraine on February 24, 2022. "*Already swallowed up by the Covid crisis*" (Georis, 2022), this campaign was, according to Jean-Pierre Camby (2022), destabilized by the Ukrainian drama for two reasons: firstly, exchanges between candidates were shorter and not as contradictory as usual, and secondly, a larger place was taken by international relations in the national debate. According to Holeign (2022), international issues are perceived as less divisive on the ideological level than other economic or social questions: "*Under the Fifth Republic, we can only be struck by the continuity of French foreign policy, which seems little affected by the alternation and the right-left divide. If changes have been observed depending on the presidents who have held office, the fundamentals initiated by General de Gaulle have varied little, unlike economic and industrial policy for example.*" However, this subject has a strong symbolic dimension which, although difficult to measure, reinforces the trends at work in the campaign. In a context where foreign policy presents itself as a "reserved domain" of the presidential function (the head of state is also head of the armed forces, in charge of nuclear deterrence), the ability of a candidate to impose his views on these themes can help establish his credibility and stature (Holeign, 2022).[1]

Voting for a candidate who will have to defend the interests and integrity of France in the event of an international crisis or war is then an advantage which clearly also resonated during the 2022 campaign which happened to start when the war in Ukraine did (Camby, 2022). The institutional position of the outgoing President gave him a competitive advantage over other candidates, which revealed itself as a "Rally-Round-the-Flag effect", a phenomenon whereby part of the population displays its solidarity with the government during a crisis. The situation was even more favorable for him due to his dual role as President of the Republic and head of the French presidency of the European Union when he perfectly assumed this role by pleading once again for the construction of a European defense. According to Bruno Cautres "*these are his themes: energizing Europe to guarantee peace and security of the French. Since his election, he has tried to set milestones on the European defense theme. Everyone recognizes his strong commitment*" (Cauvrai in Quentin, Boiteau, 2022), even if he is not the only one in France to advocate for the creation of a European army.[2]

The war in Ukraine has reshuffled the cards in the presidential election, not only by temporarily reinforcing Macron's position and his party *La République en Marche (LREM)*, but also by relaunching the candidates of extremist parties, namely Marine Le Pen of the *Rassemblement National (RN)* and Jean-Luc Mélenchon of *La France Insoumise (LFI)*, as demonstrated by several opinion polls carried out at that time. Between the start of the Russian aggression and March 15, voting intentions in favor of the outgoing President Emmanuel Macron jumped from 25 to 30%, even if a little later they lost 3 points down to 27% (Georis, 2022).[3] At the same time, voting intentions for his main rival Marine Le Pen increased from 18 to 21%, while their expected gap in the second-round duel narrowed significantly going from 58–42% to 51.5–48.5% (Georis, 2022). At the same time Eric Zemmour of *Reconquete* (far right) experienced the opposite evolution. In one month, voting intentions in his favor fell from 14 to 10%, because the French had not forgiven him for his pro-Putin and anti-NATO outings, as well as his refusal to accept immigrants from Ukraine (Georis, 2022). In fact, he was the only one opposed to welcoming Ukrainian refugees in France, although he later softened his position.

---

[1] In fact, according to this author, electoral sociology in France shows that international questions occupy a paradoxical place: on the one hand, they are often the poor relation of the analysis, traditionally centered on "internal" issues; on the other, it would be wrong to deduce that international issues play an incidental role in the presidential election (Holeign, 2022).

[2] This is also a desire of traditional parties like the Socialist Party or The Republicans. But the outgoing president is undoubtedly the one who insists the most on this measure which would gain the support of 62% of French people according to a 2019 Odoxa survey (2019) organized between March 13 and 14, 2019, two months before the European elections. During the last European elections, it was one of the flagship measures of the "*Renaissance*" list led by *LREM* (Quentin, Boiteau, 2022).

[3] According to a Harris poll for *Challenges* available at Georis (2022).

The war in Ukraine led candidates to revisit their previous positions on relations with Russia, national defense, or the acceptance of refugees, even if firmly anchored before. Although immigration traditionally holds a dominant place in speeches of far-right candidates in France, in 2022 the *RN* candidate turned to purchasing power as the main theme to attract voters. And it is also purchasing power which became the main issue for French voters, according to the Elabe survey (2022) organized between March 31 and April 1, 2022.[4] That theme was thus cited by 55% of those questioned (as 1st, 2nd, and 3rd possible answers among 17 items), followed by healthcare (27%), security (26%), pensions (24%), immigration (22%) and environment/ecology (20%). But remarkably almost one in 10 cited the war in Ukraine (12%) or the importance of France's place in the world (9%). Purchasing power is an important issue for potential voters of the main candidates. But there are notable differences regarding the other themes. Here are the 3 themes most cited by potential voters: Jean-Luc Mélenchon: purchasing power (55%), inequalities and social injustice (39%), environment and ecology (32%); Emmanuel Macron: purchasing power (51%), health (32%), pensions (25%), tied with the war in Ukraine at 25%. Note that France's place in the world is progressing in this electorate (17%); Marine Le Pen: purchasing power (63%), immigration (45%), security (37%); Eric Zemmour: immigration (67%), security (43%), purchasing power (43%). France's place in the world is progressing in this electorate as well (14%) (Elabe, 2022).

Purchasing power emerged as one of the priorities also during the battle between the two rounds, where Marine Le Pen linked it to France's foreign policy by emphasizing the need for "*strategic rapprochement*" between Russia and NATO after the war in Ukraine and by the desire of the *RN* neither to focus on "*submission to Moscow*", nor to "*following the Biden Administration*" while showing reluctance to the delivery of arms to Ukraine, as well as economic sanctions against Moscow in the name of protecting the purchasing power of the French (Le Pen in France 24, 2022).

The war in Ukraine marked the 2022 presidential election also on social networks where candidates and activists clashed more than ever, leaving behind an unprecedented 121 million messages that propelled security to first place among the most commented campaign themes (with more than 4 million posts), far ahead of purchasing power (1.4 million posts) and immigration (851,584 posts). Of course, the war in Ukraine was also one of the most commented online topics during that period (Visibrain, 2022).[5]

The presidential election represents an important political moment for the French, as reflected by the traditionally high voter turnout of between 70 and 85%. In 2022 however, both voting rounds recorded lower participation, 73.69% in the second round (down by 4.1% compared to 2017) and 71.99% in the first round (down by 2.6%). In that election, the French witnessed a historic turning point towards the far

[4] According to the Elabe Survey for BFMTV, L'Express, and SFR published on April 2 on their site.

[5] Analysis was carried out between January 1, 2022, and April 25, 2022, by Visibrain, the social media monitoring tool (Visibrain, 2022).

right as it reached the second round for the third time out of five elections in the last 20 years. More importantly, the far right managed to more than double its score during that period (Clavier, 2022).

The French parliamentary elections (12 and 19 June 2022) were held just weeks after the presidential elections, from which the then President Emmanuel Macron of *La République en Marche (LREM)* emerged victorious, elected for a second term after a duel with Marine Le Pen of the *Rassemblement national (RN)*. Jean-Luc Mélenchon of *La France insoumise (LFI)* finished third in the first round and immediately called on the French people to elect him prime minister before the second round had taken place. He offered French voters a "cohabitation", thus linking the presidential and legislative elections. To succeed, he created the *New Ecological and Social Popular Union (NUPES)*, which brought together his *LFI* with other left-wing parties. The parties of the then government also grouped themselves into a formation called *Ensemble*. Together with these two main players, *Les Républicains* and the *Rassemblement National* also sought to achieve the best possible result. Although *Ensemble* remained the main force in the National Assembly with 245 deputies, it failed to win an absolute majority. *NUPES* doubled the representation of left-wing parties compared with the previous election period. The stunning surprise came from Marine Le Pen's *RN* party which grabbed 89 seats while it had only 7 in 2017, which indicates a significant increase in political support for the RN in the legislative elections despite the "discriminatory" nature of the two-round electoral system (Kucharčík, 2022).

For the first time since the 1988 parliamentary elections, a president-elect could only rely on a relative majority after a presidential election. The 2022 legislative elections marked a departure from the scenario that had appeared after Chirac's electoral calendar reform. Unlike previous elections, the voters did not fully support the victorious presidential party, contrasting with the coherent confirmatory support seen in previous elections, which gave the presidential majority a comfortable, absolute majority in the lower house (Augustín, 2023). The current legislature is thus one of the most fragmented in the history of the Fifth French Republic, and the election campaign mirrored that. According to Frédéric Faravel (2023), it opened "*a confused and uncertain political period which, more than ever before, reflects the state of democratic crisis of our institutions and, more broadly, of the whole country*". In 2022, the parliamentary elections in France attracted only 47.5% of voters in the first round and 46.2% in the second round, the second lowest score since 2002. That marked a significant turning point in the French political system with the adoption of a law shortening the presidential term from seven to five years thereby impacting the electoral calendar, i.e., parliamentary elections will now be held immediately after the presidential elections. In practice that change means that French voters would probably feel "demobilized" and lose interest in the parliamentary vote since according to their understanding, the redistribution of power is already decided by the presidential election. The year 2022 was in that sense different from previous ones, as for the first time since the constitutional change introduced in 2002, voter turnout in the legislative elections increased.

In the Slovak parliamentary elections organized on 30 September 2023, two main camps clashed during the election campaign. One focused on the EU, embodied by Michal Šimečka who was elected as an MEP in 2019 and subsequently as a vice-president of the EP in 2022. He is currently leading the *Progressive Slovakia* (*PS*) party, a liberal and pro-European party. The other is flirting with nationalism and Euroscepticism. It is led by Róbert Fico, the chairman of *Smer-SSD*, a populist party with pro-Russian attitudes, close to Viktor Orbán. Thanks to his victory in the Slovak parliamentary elections, R. Fico managed to form a new governing coalition with an anti-Ukrainian and Eurosceptic flavor, together with the left-wing party *HLAS-SD* (Social Democracy) and the far-right party *Slovak National Party* (*SNS*). That alliance with the *SNS* meant that his *Smer-SD* party (but also the *HLAS* party), which officially espouse left-wing values, was once again excluded from the *European Socialist Party*, just as fifteen years earlier when he had formed a similar coalition with the *SNS*. His party's rise to power is particularly worrisome in view of previous statements made by its leader Andrej Danko, who believes that the territories occupied by Russia were not 'historically Ukrainian', and whose political positions are openly pro-Russian. Indeed, in the past he has referred to Russian Foreign Minister Sergei Lavrov as 'a dear friend' and boasted of a selfie taken with the Speaker of the Russian Duma, Vyacheslav Volodin.

This is Róbert Fico's fourth government mandate in Slovakia, and while in 2020 his political future looked bleak, as confirmed by pre-election polls in 2021 showing only about 10.9% for Smer-SD, today that declining party is once again enjoying optimistic Figs. (22.94%). What is alarming is that its leader R. Fico has fundamentally changed. He has become radicalized. He has adopted the same illiberal elements as his Hungarian counterpart, such as anti-elite, nationalist, and Eurosceptic discourse, anti-LGBT rhetoric. He has referred to the President of the Slovak Republic, Zuzana Čaputová, as "*Soros's child*" (TASR, 2022). That is a departure from his first mandates, where, while opposing European quotas for migrants, he described himself as a pro-European politician "*wishing to include Slovakia in the hard core of the EU if it were structured according to a more nuanced logic. […] Compared to yesterday's Robert Fico, today's Robert Fico is clearly different politically, ideologically, and humanly*" (Macek in Palacin, 2023).

The Slovak political scene has thus fundamentally changed after the 2023 parliamentary elections. The voter turnout rate of 68.06% was the highest since 2002 when 70.06% of voters went to the polls, and thus higher than in 2020 (65.8%), when the election campaign of Igor Matovič, leader of the *OĽANO* (*Ordinary People and Independent Personalities*) party, mobilized Slovaks with the promise of fighting corruption of the then political establishment. Although this political formation was able to benefit from the broad mobilization of citizens after the murder of journalist Ján Kuciak in 2018 highlighted the *Smer-SD* party's links to the criminal underworld, the new governing coalition consisting of four different political formations failed to cope with the Covid 19 pandemic, thus triggering a serious political crisis and a wave of government resignations. The then Prime Minister Igor Matovič was forced to resign by his coalition partners and was replaced by the Deputy Prime Minister Eduard Heger. However, in September 2022, because of divisions in economic policy, the *SaS* party (*Freedom and Solidarity*) decided to leave the ruling coalition, creating a minority government, which gradually led to the calling of early elections at the end of September 2023. Three years of political instability thus led to the re-mobilization of *Smer*, *Hlas*, and *SNS* voters, who felt dissatisfied with previous governments and feared a deteriorating social situation.

The process of forming the Slovak government after the snap elections was completed by October 25th. As the European Council meeting began in Brussels, R. Fico traveled as a fresh prime minister ready to prove that he was serious about his pre-election promises of a "sovereign" approach to Ukraine. At the two-day meeting of heads of government, where the newly elected prime minister returned after more than five years, and also during his meeting with the president of the European Commission Ursula von der Leyen, he stressed that he would promote Slovak policy and defend Slovak interests at EU summits.[6] Several of his statements were important in terms of Slovakia's approach to the Russian aggression in Ukraine, including the fact that Slovakia will not militarily assist the already invaded Ukraine. Although he and V. Orbán have not yet blocked the adoption of a financial aid for Ukraine in the amount of 50 billion euros for the years 2024–2027 (the details of the scheme are still to be negotiated), R. Fico has already expressed his views on the conditions under which he is willing to support this package.

The Slovak campaign was marked by a high level of disinformation in the online space. Although political forces have already regrouped, pro-Russian propaganda continues to spread through various channels in Slovakia even after the elections. The success rate of disinformation in Slovakia is relatively high and, according to sociologist Michal Vašečka (Poloha, 2023), even higher than in neighboring countries. To some extent, that can be linked to Pan-Slavism which has strong roots in Slovakia. But the strong pro-Russian sentiments should not be explained solely by the Pan-Slavic sentiments of Slovaks. According to that sociologist, it is also necessary to perceive "*a large order for authoritarian solutions. Russia has always been a country with a strong authoritarian current. It has never been a full-fledged democracy*" (Vašečka in Poloha, 2023). That thesis is supported by expert analyses and surveys, according to which the spread of propaganda is pervasive in Slovakia while its support for Ukrainians is one of the lowest in Europe. One of the key factors is that Ukrainian refugees are perceived negatively. According to Globsec experts, Slovaks fear that by helping and funding refugees they are burdening their own economic situation. One of the factors influencing people's perceptions of Ukrainians is party politics: "*Voters of parties that promote hostile rhetoric towards refugees are much more likely to perceive the presence of Ukrainians unfavorably.*" (Kazaz in Murajdová, 2022). Therefore, it is desirable to continually increase digital literacy levels (Javorčíková et al. 2021) to prevent fake news and disinformation in order to critically approach social media and their content (Hurajová, Hladíková, 2022).

[6] His expressed position is to demand a thorough analysis of any sanctions to be prepared against Russia so that they do not harm Slovakia. In this context, he mentioned that if there is talk of possible sanctions on the Russian nuclear sector, Slovakia will never agree to it (TASR, 2023).

4 Political Discourse Analysis, the Media and Campaign Communications in Slovakia

Our analytical work focused on the sets of texts collected in the Slovak national media, more precisely in the electronic editions of the daily newspapers we monitored: *SME* and *Hlavný denník*, for the last four weeks of electoral campaign. We manually sorted the articles published in the media based on thematic criteria. Our search was organized using pre-established keywords related to the war in Ukraine. We then analyzed the resulting set of texts using the Sketch Engine software.[7] A qualitative and, incidentally, quantitative analysis of the thematic vocabulary was carried out, focusing on the frequencies of lexical elements and their collocability. The lexicometric results of our analysis were then carefully read, within the framework of the theory of enunciation. Our focus was on subjectivity in political language, particularly on the identification and interpretation of implicit content in the service of persuasion.

The first periodical we monitored was *SME*, a daily newspaper representing right-wing views and attitudes and categorized as a mainstream media. The *SME* daily is the second or third most read, mainstream electronic daily in Slovakia.[8] Our second choice was *Hlavný denník*, representing the opposite spectrum of opinion (extreme left), categorized as an alternative media by its founders and as a conspiracy media by external evaluators. Our choices of media were conditioned by their fundamentally different political focus, and our assumption that each would write about the conflict in Ukraine during the pre-election campaigns in a different but significant way. According to available sources, *Hlavný deník* is not included in the readership statistics.

In the *SME* and *Hlavný denník* newspapers, we focused on articles about the pre-election campaign of political parties, especially how they represented the military conflict in Ukraine. We searched through the electronic archives of both media and stored potentially relevant communications in text form in our database. We then filtered the available texts using the following criteria: (1) date of publication and (2) tag (keyword) assigned to the article and, additionally, the thematic section in which the text was had been published. In the *SME* newspaper, we followed articles tagged with the terms "early parliamentary elections" and "Ukraine". In *Hlavný denník*, we selected texts tagged with the term "2023 elections". In accordance with the common practice of the electronic press, each published text carried one or more tags which made them searchable in our database.

Each of our chosen media has its own way of letting readers access their reporting, due mainly to different marketing strategies. While it is not possible to read the current issue or access the archive of *SME* without a paid subscription, access to the current issue and entire archive of *Hlavný denník* is free. Working with the Hlavný denník archive is easier than with that of *SME*. The *Hlavný denník* archive is searchable

---

[7] Available at: https://www.sketchengine.eu/what-can-sketch-engine-do/

[8] Statistics of the readability of Slovak online media are available at: https://e-public.gemius.com/sk/rankings/14345).

Table 1 Noun frequencies in SME

slovensko (Slovakia)	333	*vláda* (government)	140	*percento* (percentage)	107
rok (year)	303	*článok* (article)	127	*krajina* (country)	105
strana (party)	241	*deň* (day)	126	*smer* (smer)	98
ukrajina (Ukraine)	212	*odber* (subscription)	125	*brífing* (briefing)	96
voľba (election)	176	*štát* (state)	122	*začiatok* (beginning)	91
človek (person)	172	*fico* (Fico)	115	*volič* (voter)	89
správa (report)	160	*klik* (click)	112	*cena* (price)	88
formulár (form)	147	*časť* (part)	108	*prezident* (president)	88

using a date range and keywords. In contrast, the *SME* archive is only searchable using keywords and the results are then presented sorted by relevance or by date. Instead of a subscription, *Hlavný denník* appeals to and invites readers to provide voluntary financial support with a suggestion at the end of each article.

Applying the above-mentioned procedure and selection criteria, we gathered 55 texts from *SME* and 136 texts from *Hlavný denník*. Although the number of published articles seems to imply a higher frequency of publication in Hlavný denník, however, other parameters do not necessarily confirm a higher level of activity. In our database *SME* data contains 540,825 characters (73,751 words) whereas *Hlavný denník* data contains 529,285 characters (80,752 words). Both can therefore be considered comparable quantitatively. The compiled text files from the media environment do not have the character of parallel corpora, but come from the same period, have the same thematic focus and the social context in which they were created. They are the basis for the identification of language elements, subsequently for their extraction, further processing, and comparative evaluation.

Texts from both newspapers were selected and archived manually, preserving metadata and image attachments. Subsequently, both files were processed using the semi-automated, online tool called Sketch Engine which can analyze units at the word, word grouping, sentence, and full text level.

We first analyzed the working corpora of texts from each journal using the Wordlist function. We thus obtained data on the most frequent nouns, adjectives, and verbs. Subsequently, we focused our attention on nouns (Table 1).

In *SME*, Sketch Engine identified a total of 4,778 entries with a total frequency of 25,468 occurrences.[9] The lexeme *war* is in 28th place and the lexeme *Russia* ranked 34th (Table 2).

The lexemes *war* and *Russia* are both absent from *Hlavný denník*. We see more references to domestic politicians active in the campaign, systematically referring to the presumed strongest party, which in its discourse clearly defines itself against cooperation with the *Smer* and *Hlas* parties, i.e., *Progressive Slovakia* (PS). Sketch

[9] Frequency count is established on the basis of total numbers of occurrences of nouns in the analyzed corpus.

Table 2 Noun frequencies in hlavny dennik

slovensko (Slovakia)	510	*hlas* (hlas)	210	*štát* (state)	121
voľba (year)	485	*smer* (smer)	210	*matovič* (Matovič)	121
strana (party)	433	*robert* (Robert)	167	*republika* (Republika)	120
rok (year)	283	*Ukrajina* (Ukraine)	147	*kampaň* (campaign)	116
človek (person)	278	*kraijna* (country)	145	*parlament* (parliament)	109
fico (Fico)	277	*ps* (Progressive Slovakia)	137	*hranica* (border)	102
vláda (government)	255	*prieskum* (survey)	128	*médium* (meduim)	101
percento (percentage)	236	*volič* (voter)	126	*predseda* (prime minister)	100

Engine identified a total of 4,603 substantive items in this subcorpus with a total frequency of 26,062 occurrences.

The most frequent noun in both Slovak dailies is *Slovakia.* In *SME* it is followed by *year, party, Ukraine, choice, person.* Among the references to political parties and politicians, the lexemes *Fico, Smer, president, Kollár, republic* dominate. In *Hlavný denník* the most frequent nouns are *election, party, year, person, Fico.* The lexeme *Ukraine* appears in the 12th position. Among the references to political parties and politicians, *Fico, government, Hlas, Smer, Robert, PS* (political party *Progressive Slovakia*) appear.

We note that in the analyzed Slovak media the topic of the war conflict in Ukraine appears to be very important at the time of the peak of the pre-election struggle. At the same time, the way this topic is presented differs significantly depending on which political party the media or the speaker stands for. While a pro-European stance is associated with the discursive image of Ukraine as the (future or inevitable) winner of the conflict, Ukraine's victory is not mentioned at all in media communications that represent anti-European positions.

In the Slovak election campaign, the topic of Ukraine is very closely linked not only with Euroscepticism, but also with the topic of the purchasing power of the Slovak population. This is evidenced by the strong representation of the sub-theme of Ukrainian grain exports in connection with the crisis, which was caused not only in Slovakia but also in the whole of Central and Eastern Europe by problems with the measures introduced by the European Union to free Ukraine from its stocks. As a result, tons of wheat (but also corn, rapeseed and sunflower) reached the countries neighboring Ukraine in a few months, which were to be sent elsewhere, especially to North Africa and the Middle East, to fight against food insecurity. In fact, most accumulated in the silos of transit countries and destabilized local markets, causing a wave of resistance and condemnation of unfair competition not only by Slovak farmers, but especially by Poles, Romanians, and Bulgarians, which culminated in the spring of 2023:

(1) Zhoršujúce sa vzťahy Poľska a **Ukrajiny**: Pre poľskú vládnucu stranu Právo a spravodlivosť je pred voľbami prioritou zabezpečiť, aby boli farmári spokojní,

a to aj za cenu hroziaceho trvalého poškodenia spojenectva s Ukrajinou. Spor o obilie by mohol byť signálom ďalších. (SME, 26 September 2023)

[Deteriorating Poland-**Ukraine** relations: for Poland's ruling Law and Justice party, the priority ahead of the elections is to keep farmers happy, even at the cost of threatening to permanently damage the alliance with Ukraine. The grain dispute could signal more to come.]

(2) Poľsko, jeden z najvernejších spojencov Ukrajiny, oznámilo ukončenie svojich transferov zbraní do krajiny deň po tom, čo prezident obvinil Varšavu, že hrá do karát Rusku tým, že zakazuje dovoz ukrajinského **obilia** s cieľom chrániť záujmy poľských farmárov (Hlavný denník, 21 September 2023)

[Poland, one of Ukraine's staunchest allies, announced an end to its arms transfers to the country a day after the president accused Warsaw of playing into Russia's hands by banning Ukrainian **grain** imports to protect the interests of Polish farmers]

It is precisely in this combination of themes that the importance of economic subject matter can be seen. However, this topic is also treated differently. While in the alternative media it is presented using highly colloquial language with a strong subjective charge, in the mainstream media we find more neutral statements formulated with a higher degree of objectivity.

Overall, we conclude that the election campaign in Slovakia, as in France, is largely based on the theme of external threats, or the need to increase the security of the country and the protection of its citizens.

(3) Treba ďalej podstatne výraznejšie nasadiť slovenskú armádu na **ochranu** štátnych hraníc. A samozrejme, odvolať neschopného policajného prezidenta Hamrana, ktorý pri ochrane bezpečia slovenských občanov neustále zlyháva. V HLASe sme presvedčení, že len slovenskí občania môžu rozhodovať o tom, kto má na Slovensko právo prísť a kto tu má právo ostať. (Hlavný denník, 16 September 2023)

[Furthermore, the Slovak army needs to be deployed significantly more to **protec**t the national borders. And, of course, remove the incompetent police president Hamran, who has consistently failed to protect the safety of Slovak citizens. We at HLAS believe that only Slovak citizens can decide who has the right to come to Slovakia and who has the right to stay.]

(4) Pomoc Ukrajine vnímame ako priamu podporu **bezpečnosti** pre našich občanov a budeme v nej pokračovať. Ak by Ukrajina podľahla agresii, budeme mať agresora na našich hraniciach a zaistenie našej bezpečnosti nás bude stáť oveľa viac. (SME, 18 September 2023)

[We see our assistance to Ukraine as a direct support to the **security** of our citizens and we will continue to do so. Should Ukraine succumb to aggression, we will have an aggressor on our borders, and it will cost us much more to ensure our security.]

This thematic line often appears explicitly, but much more often it is conveyed to the recipients of media texts implicitly, as a subliminal one arising from the context. In this case, too, it is possible to clearly identify opposing views. In the mainstream media, the issue of security is more often associated with the strengthening of Slovakia's integration in the European Union and NATO. On the contrary, in the media that describe themselves as alternative, representing mainly the views of supporters of the political parties *Smer-SD*, *Hlas*, *SNS*, *Republika* (far-right party, led by Milan Uhrík, a member of the European Parliament elected in 2019), and others, the prevailing opinion is that if we want to strengthen security, it is not advisable to accept all the common decisions of the EU, nor to strengthen our involvement in the common defense. According to these media, the solution is to lean towards pro-Russian attitudes, which was in clear contrast to the long-term line of Slovakia's official foreign policy.

5 Conclusion

More than a year and a half after the invasion of Ukraine by Russia, we are facing a fundamental question: we do not know how this military conflict will develop further and what impact it may have on the future development of the EU and the entire international order for which this war represents a demanding test. Faced with alarming prospects, pessimistic opinions and the dismissive attitudes of many politicians and citizens in the EU member countries, which have not yet had time to recover from the effects of previous crises, are beginning more and more voiced. However, it turns out that this theme is important for EU citizens, as confirmed by many national elections held within the years 2022 and 2023. In this paper, we focused on the parliamentary elections in France (2022) and Slovakia (2023), where we observed how the theme of the war in Ukraine became part of pre-election political discussions.

In the French elections, both presidential and parliamentary, which are largely interconnected due to the French political system and the electoral calendar, the theme of the war was understandably not the most important. The campaigns were primarily dominated by the theme of strengthening the declining purchasing power, which was significantly represented in the pre-election political debates across the entire political spectrum, but mainly of both ends of this spectrum, i.e., at Marine Le Pen from *RN* and Jean-Luc Mélanchon from *La France Insoumise*. At the same time, this theme was one of those that, according to public opinion polls, attracted the most attention of the electorate, which also applies especially in the case of these two political parties. Other politicians, in addition to the standard themes related to the daily lives of French citizens, attracted their electorate with other themes dominated by security (especially Eric Zemmour from the extreme right party *Reconquete*), the war in Ukraine or France's position on the international political scene. Commenting on the political opinions of individual candidates on social networks also corresponds to this fact, with the war in Ukraine and security being the themes with the highest number of comments and posts, only in a different order and to a different degree of importance compared to the other discussed themes.

When comparing the situation with Slovakia, it is necessary to underline that the parliamentary elections in this country were held only 15 months after the French parliamentary elections and 19 months after the start of the Russian war of aggression. It was the period when the unprecedented level of support expressed to Ukraine began to weaken (especially as regards military aid). This tendency was evident not only among politicians mainly from Central Europe (apart from Slovakia, also in Hungary and Poland) but also among the citizens of these states. The results of the quantitative–qualitative research of media texts from the Slovak dailies clearly show two different vocabularies and two different perspectives reflecting the political orientation of the producers of political discourse as shown in the media. In this sense, we can clearly see the division of opinions and the deep internal schism in Slovak society. The difference is shown through the same themes (the conflict in Ukraine, economic prosperity, and internal security) that resonated the most in the French election campaign.

The rise of populism and extremism in Europe is a phenomenon that is often noticed by both experts and politicians. However, the question remains to what extent and in what way the development of the situation in Ukraine is also a sign of this phenomenon. The explanation of this issues is extremely important, mainly within the context of the upcoming European elections, which will be held in the spring of 2024.

The present contribution is financially supported by the VEGA project No. 1/0661/23: "*The impact of the accession of the Slovak Republic to the European Union on the implementation of Slovak national interests and their expression in foreign policy (2004–2024)*".

Bibliographic References

AFP (2023) Guerre en Ukraine: après un an, l´extrême droite française reconnaîitsa "naiveté" face à Poutine. https://www.lalibre.be/international/europe/2023/02/24/guerre-en-ukraine-apres-un-an-lextreme-droite-francaise-reconnait-sa-naivete-face-a-poutine-DYVSWZSHHNHMJHEIDLAQSTPIFY/

ASFE (2022) L'Europe se fera dans les crises et elle sera la somme des solutions apportées à ces crises. https://alliancesolidaire.org/2022/05/12/leurope-se-fera-dans-les-crises-et-elle-sera-la-somme-des-solutions-apportees-a-ces-crises/

Augustín M (2021) Fenomén Macron: Smerovanie k prezidentskej republike? *Medzinárodné vzťahy: vedecký časopis pre medzinárodné politické, ekonomické, kultúrne a právne vzťahy*, Bratislava: Fakulta medzinárodných vzťahov EU 19(1):89–110 ISSN 1336–1562

Augustín M (2023) Failing to Change the Political Landscape: La République en Marche in the Process of Its Institutional Rooting. Ljubljana: CAAPPI – Centre for the Analysis of Administrative-Political Processes and Institutions. J Comp Polit 16(2):5–19. ISSN 1338–1385

Barroso M (2014) Une décennie de résilience européenne, Les années Barroso. 2004–2014, sous la direction de BUSSIèRE Éric, MIGANI Guia. Paris: Tallandier, Hors collection, pp 7–32. https://www.cairn.info/les-annees-barroso--9791021008717-page-7.htm

Bernard H (2023) Institut Montaigne, Les conséquences de la guerre en Ukraine pour la sécurité alimentaire de la France et de l'Afrique. https://www.institutmontaigne.org/expressions/les-consequences-de-la-guerre-en-ukraine-pour-la-securite-alimentaire-de-la-france-et-de-lafrique

Bienvenu H (2023) En Pologne, le soutien à l'Ukraine, victime collatérale de la campagne pour les législatives, Le Monde, 5. https://www.lemonde.fr/international/article/2023/10/05/en-pologne-le-soutien-a-l-ukraine-victime-collaterale-de-la-campagne-pour-les-legislatives_6192579_3210.html

Borrell J (2023) L'Europe dans l'interrègne : notre réveil géopolitique après l'Ukraine, RED 1(5):116–120. https://www.cairn.info/revue-red-2023-1-page-116.htm

Camby JP (2022) Une élection présidentielle française fortement perturbée du fait de la guerre en Ukraine, mais non reportable, La lettre juridique, mars 2022, [Focus]. https://www.lexbase.fr/article-juridique/82239746-focus-une-election-presidentielle-francaise-fortement-perturbee-du-fait-de-la-guerre-en-ukraine-mais

Castaño P (2018) Populismes de gauche en Europe : une comparaison entre Podemos et la France insoumise. Mouvements 4(96):169–180. https://www.cairn.info/revue-mouvements-2018-4-page-169.htm

Chopin T, Naim A (2023) Les « valeurs européennes » à l'épreuve de la guerre en Ukraine, Policy paper, Paris : Institut Jacques Delors, 9 mai

Clavier H (2022) un printemps électoral historique en France, Le Taurillon. https://www.taurillon.org/2022-un-printemps-electorale-historique-en-france?lang=fr

Déclaration de M. Emmanuel Macron, président de la République, sur les objectifs en matière de frontières, de climat, de numérique, de social et de valeurs de la présidence française du Conseil de l'Union européenne, à Paris le 9 décembre 2021. https://www.vie-publique.fr/discours/282826-emmanuel-macron-09122021-union-europeenne

Élections européennes : une victoire des français pro-européens, *Le Mouvement Européen*. https://mouvement-europeen.eu/elections-europeennes-une-victoire-des-francais-pro-europeens/

European commision (2011) José Manuel Durão Barroso President of the European Commission Meeting with EU Heads of Delegation Brussels, 30 November 2011. https://ec.europa.eu/commission/presscorner/detail/en/SPEECH_11_838

Faravel F (2023) Élections législatives de 2022 : entre approfondissement et contradictions, une nouvelle étape de la crise démocratique. Institut Rousseau. Le laboratoire d'idées de la reconstruction écologique et républicaine. fhttps://institut-rousseau.fr/elections-legislatives-de-2022-entre-approfondissement-et-contradictions-une-nouvelle-etape-de-la-crise-democratique/

France (2022) Marine Le Pen prône un rapprochement stratégique entre l'Otan et la Russie après la guerre, France 24 Accessed on Apr 13 2022. https://www.france24.com/fr/france/20220413-marine-le-pen-pr%C3%B4ne-un-rapprochement-strat%C3%A9gique-entre-l-otan-et-la-russie-apr%C3%A8s-la-guerre

Gaillard M (2018) Le "soft power" de l'Union européenne, Vie publique. https://www.vie-publique.fr/parole-dexpert/38553-le-soft-power-de-lunion-europeenne

Georis V (2022) Comment la guerre en Ukraine a influencé la présidentielle francaise, L´Echo, 6 avril. https://www.lecho.be/dossiers/conflit-ukraine-russie/comment-la-guerre-en-ukraine-a-influence-la-presidentielle-francaise/10378833.html

Gradvohl P (2023) Les élections législatives en Slovaquie dimanche dernier ont porté au rang de premier ministre Robert Fico, connu pour être un franc sympathisant de la Russie, Radio France. https://www.radiofrance.fr/franceculture/podcasts/les-enjeux-internationaux/un-axe-pro-russe-est-il-en-train-de-diviser-l-europe-7487279

Gueorguieva P (2017) Populismes et populistes en Europe centrale et orientale, Hermès, La Revue 1(77):117–125. https://www.cairn.info/revue-hermes-la-revue-2017-1-page-117.htm

Holeindre JV (2023) Chapitre 10. La guerre fait-elle l'élection ? Le rôle du conflit en Ukraine dans l'élection présidentielle 2022, Le vote clivé. Les élections présidentielle et législatives d'avril et juin 2022, sous la direction de Perrineau Pascal, Fontaine, Presses universitaires de Grenoble, Libres cours Politique, pp 113–120. https://www.cairn.info/le-vote-clive--9782706152979-page-113.htm

Hurajová A, Hladíková V (2022) Educating for good character: from critical thinking to intellectual character virtues, media literacy and academic research 5(1):178–191. https://www.mlar.sk/wp-content/uploads/2022/05/10_Anna-Hurajova-Vladimira-Hladikova.pdf

In: DIEM: Dubrovnik International Economic Meeting–Dubrovník: University of Dubrovnik, 2024 9(1):143–165. ISSN 1849–3645
Ivaldi G (2023) Populismes européens : poussée de la droite radicale et fin du moment "hybride". Futuribles mai-juin (454) :73–86. https://shs.hal.science/halshs-04087391/document
Javorčíková J, Badinská M, Ližbetinová L, Brett D (2021) The need for integration of reading, critical thinking and academic reading skills : a quantitative analysis of Slovak undergraduates' reading performance. J Lang Cult Educ 9(1):12–29 ISSN 1339–4045
Kucharčík R (2022), Národný front / Národné združenie na politickej mape Francúzska po prezidentských voľbách 2002: National Front / National Rally on the Political Map of France After Presidential Elections 2002. *Ekonomické, politické a právne otázky medzinárodných vzťahov 2022: zborník z medzinárodnej vedeckej konferencie konanej 3. a 4. júna 2022.* Bratislava: EKONÓM, 2022, pp 208–215. ISBN 978–80–225–4958–5. ISSN 2585–9404
Lastennet J (2021) Le populisme en Europe, Toute l´Europe. https://www.touteleurope.eu/vie-politique-des-etats-membres/le-populisme-en-europe/
Lefebvre M (2023), L'Union européenne face à la guerre en Ukraine : la puissance libérale et ses limites, Schuman Papers n°651. https://www.robert-schuman.eu/questions-d-europe/0651-l-union-europeenne-face-a-la-guerre-en-ukraine-la-puissance-liberale-et-ses-limites
Lefevre V (2020) L'histoire des crises européennes qui ont soudé l'Europe*, Taurillon en Flam's* n°18 / Septembre - Novembre 2020. https://www.taurillon.org/l-histoire-des-crises-europeennes-qui-ont-soude-l-europe?lang=fr
Lequesne C (2017) L´enjeu européen dans la campagne présidentielle. https://www.sciencespo.fr/ceri/fr/content/l-enjeu-europeen-dans-la-campagne-presidentielle
Maillard S (2023) La guerre en Ukraine a facilité un rapprochement des droites, Sud-Ouest. https://www.sudouest.fr/international/europe/sebastien-maillard-la-guerre-en-ukraine-a-facilite-un-rapprochement-des-droites-15987678.php
Mortkowitz L (2023) Les populistes en Europe de l'Est usent de la politique de la peur avec la guerre en Ukraine, Le monde, AFP. https://www.ledevoir.com/monde/europe/785204/guerre-en-ukraine-les-populistes-en-europe-de-l-est-usent-de-la-politique-de-la-peur-avec-la-guerre-en-ukraine
Müller V (2023) Národný záujem v Európskom parlamente: homogénnosť hlasovania slovenských poslancov Európskeho parlamentu. In Politické Vedy. 26(1):8–25. ISSN 1335 – 2741. En ligne sur https://doi.org/10.24040/politickevedy.2023.26.1.8-25
Munier F (2023) Face aux progressions populistes, le risque de la désunion européenne. https://www.pourleco.com/monde/face-aux-progressions-populistes-le-risque-de-la-desunion-europeenne
Murajdová A (2022) Prieskum: Slováci vnímajú Ukrajincov najhoršie v regióne. Zdá sa nám, že im pomáhame až príliš, TV Noviny. https://tvnoviny.sk/domace/clanok/383400-prieskum-slovaci-vnimaju-ukrajincov-najhorsie-v-regione-zda-sa-nam-ze-im-pomahame-az-prilis?campaignsrc=tn_clipboard
Na Západ jsme se úplně nevčlenili. Střední Evropa se v krizích rozhoduje emocionálně, míní politolog Pehe (2023), iRozhlas CZ. https://www.irozhlas.cz/zivotni-styl/spolecnost/zapad-evropa-krize-cesko-politolog-jiri-pehe-populismus-visegrad_2309291045_aur
Odoxa (2019) Les Français disent « OUI » à une armée commune européenne, et, en attendant, souhaitent une augmentation des dépenses militaires, publié le 30 mars. https://www.odoxa.fr/sondage/francais-disent-oui-a-armee-commune-europeenne-attendant-souhaitent-augmentation-depenses-militaires/
Palacin H (2023) Législatives en Slovaquie : des élections à fort enjeu pour l'Union européenne, Toute l´Europe. https://www.touteleurope.eu/vie-politique-des-etats-membres/legislatives-en-slovaquie-des-elections-a-fort-enjeu-pour-l-union-europeenne/
Poloha L (2023) Zahraničné médiá o Slovensku: Ruská propaganda má Slovákov úplne v hrsti, *Startitup.* https://www.startitup.sk/o-slovensku-pisu-v-zahranici-sme-najslabsia-cast-europy-ruskej-propagande-sa-tu-mimoriadne-dari/

Présidentielle 2022 les enjeux européens du débat Emmanuel Macron / Marine Le Pen, Toute l´Europe. https://www.touteleurope.eu/vie-politique-des-etats-membres/live-presidentielle-2022-les-enjeux-europeens-du-debat-emmanuel-macron-marine-le-pen/
Quentin A, Boiteau V (2022) "Effet drapeau. Guerre en Ukraine: quel impact électoral pour les candidats à la présidentielle en France? *Libération*, 3 mars. https://www.liberation.fr/politique/elections/guerre-en-ukraine-quel-impact-electoral-pour-les-candidats-a-la-presidentielle-en-france-20220303_OV43V3UQYZGENLHXNVLRXI3VMQ/
Rošteková MH, Molnárová E (2022) Désinformation et politique éducative, Hermès, La Revue 2(90):157–161. https://www.cairn.info/revue-hermes-la-revue-2022-2-page-157.htm
Rapport Schuman sur l'Europe, l'état de l'Union 2023, éditions Marie B, Paris, avril 2023
Rupnik J (2018) Spécificités et diversité des populismes en Europe centrale et orientale. https://www.sciencespo.fr/ceri/fr/content/dossiersduceri/specificites-et-diversite-des-populismes-en-europe-centrale-et-orientale
Rupnik J (2020) Populismes et révolution conservatrice en Europe de l'Est, Vie publique. https://www.ledevoir.com/monde/europe/785204/guerre-en-ukraine-les-populistes-en-europe-de-l-est-usent-de-la-politique-de-la-peur-avec-la-guerre-en-ukraine
Rýsová L, Kazanský R, Smeriga P (2023) Education, science and research - key elements supporting dynamics of economic growth in the EU in the period after overcoming the effects of the Covid-19 pandemic
Sailing on high seas: reforming and enlarging the eu for the 21st century. Paris-Berlin. Report of the Franco-German working group on the EU institutional reform. 18 September 2023. https://www.diplomatie.gouv.fr/IMG/pdf/20230919_group_of_twelve_report_updated14.12.2023_cle88fb88.pdf
Sondage Elabe pour BFMTV, L'EXPRESS et SFR (2022) published on April 2, Français et l'élection présidentielle 2022. https://elabe.fr/wp-content/uploads/2022/04/les-francais-et-lelection-presidentielle-2022.pdfLes
Stanovisko predsedu vlády SR Roberta Fica z 27.10.2023. https://www.vlada.gov.sk/30034/stanovisko-predsedu-vlady-sr-roberta-fica/
TASR (2022) Prezidentka Čaputová je Sorošovo dieťa a zaslúži si lízatko od Bidena. https://sita.sk/prezidentka-je-sorosove-dieta-zasluzi-si-lizatko-od-bidena-tvrdi-fico-video/
TASR (2023) Fico na summite zdôraznil, že v EÚ bude presadzovať slovenskú politiku. https://www.teraz.sk/slovensko/fico-na-summite-zdoraznil-ze-v-eu-bu/750665-clanok.html
Terem P, Müller V (2018) Transparency in EU trade negotiantions: parallels and differences between TTIP and Brexit. In: Proceedings of the 4th International conference on European integration 2018 (ICEI 2018) May 17–18, Ostrava, Czech Republic, Ostrava: VŠB - Technical University of Ostrava, 2018. ISBN 978–80–248–4169–4. ISSN 2571–029X, pp 1514–1521
Visibrain (2022) Présidentielle 2022:bilan de la campagne sur les réseaux sociaux, publié le 25 Avril 2022. https://www.visibrain.com/fr/blog/bilan-reseaux-sociaux-de-la-presidentielle-2022/
World Energy Outlook 2022 shows the global energy crisis can be a historic turning point towards a cleaner and more secure future (2022), IEA, 27 October 2022. https://www.iea.org/news/world-energy-outlook-2022-shows-the-global-energy-crisis-can-be-a-historic-turning-point-towards-a-cleaner-and-more-secure-future
Yar L (2023) Ako vojna na Ukrajine ovplyvnila európsku bezpečnosť? 4 najzásadnejšie zmeny, Euractiv. https://euractiv.sk/section/obrana-a-zahranicie/news/ako-vojna-na-ukrajine-ovplyvnila-europsku-bezpecnost-4-najzasadnejsie-zmeny/
Zelensky Volodymyr on Twitter (2022), It´s a unique and historical moment in Ukraine-EU relations. https://twitter.com/ZelenskyyUa/status/1540038995178037249

Digital Transformation and Innovation in the Development of Rural Areas in Romania

Simona Biriescu and Iulia Para

Abstract Like many European Union countries, Romania aims to develop its villages. This initiative is considered innovative for developing rural areas in terms of agriculture and beyond. Although "smart villages" are a political instrument, in practice, this can include a wide range of measures, such as using digital tools to improve general municipal services in rural areas or employing online platforms to promote democratic participation in local and regional policies. This paper focuses on two objectives, namely: the first objective, digital transformation, which involves promoting lifelong learning and preventing the expansion of geographic inequalities. Lifelong learning is essential for all workers and citizens to adapt to changes in the workplace and society, especially in rural areas. For lifelong learning systems to be effective, they must be based on measures that guarantee access to quality education and training, anticipate changes in skill demand and ensure coherence between education and training systems for labour market needs. It is also essential to adopt policies that create synergy between the positive local effects of skills and technologies so that regional inequalities are not amplified. The second objective, digital innovation, concerns public services, agricultural development, mobility, digital technologies and uses, or resource management in rural areas. These adopt an entire "culture of innovation," with prospects for transforming territorial innovation policies and recognizing and promoting new ideas that emerge in rural areas. Both in innovation and transformation, new technologies play an essential role. Technologies are profoundly transforming the world of work and societies in rural areas. To adapt to these changes, rural populations need to be able to learn and develop their range of skills throughout their lives. Measures to assist and encourage the rural population to achieve this are still under discussion by the government. It believes that digital transformation is related to future skills needs. In Romania, digital transformation in rural areas began a few years ago in some communes and villages, gradually starting with

S. Biriescu (✉) · I. Para
Faculty of Economics and Business Administration, West University of Timişoara, Timişoara, Romania
e-mail: simona.biriescu@e-uvt.ro

I. Para
e-mail: iulia.para@e-uvt.ro

G. Pellat et al. (eds.), *Data-Centric Business and Applications*, Lecture Notes on Data Engineering and Communications Technologies 258,
https://doi.org/10.1007/978-3-031-95068-1_4

rural schools by equipping them with computers, tablets, and smart boards. Based on the title of this paper, digital platforms are the solution in rural areas, as well as well-organized databases with the necessary information. However, traditional territorial development tools are also needed in rural contexts. The vision of economic development through the lens of sectors of activity, economic sectors, equipment, or infrastructure seems, in fact, ill-suited to rural territories. Some voices argue that innovation is more suited to urban areas. In this context, lifelong learning is essential to policy responses to the uncertainty surrounding future skills needs. As workers and citizens of rural areas, individuals must be able to constantly evolve their skills portfolio in response to the emergence of new needs. This means transformation, innovation, development, digitization, and, last but not least, adaptation.

1 Introduction

Strategically reinvigorating the rural milieu is pivotal for astute territorial advancement, especially when approximately 50% of Romania's populace resides in these areas. A myriad of rural regions continues to exhibit economic, demographic, and social challenges due to the compounded influence of diverse factors. Thus, discerning avenues for their digital evolution and innovation is imperative.

The objective of this study is to undertake comprehensive analyses of rural territories, evaluating the inherent local resources against the backdrop of developmental requisites, external opportunities, and potential threats. Subsequently, the aim is to conceptualize strategies and developmental initiatives that catalyze local digital transformation and innovation. It is necessary to have:

1. Fundamental concepts regarding the rural area and its digital transformation.
2. The use of methods, techniques, and tools to conduct studies on various rural areas, evaluating local resources and potential.
3. Analysis, evaluation, and formulation of development strategies and projects for the digital transformation of rural areas and innovations in these zones.

In certain rural regions that are more challenged and underdeveloped, where initial digitization efforts have been implemented, there is an imperative need for further digital transformation. Digital transformation might also pertain to rural territories that hold developmental potential but necessitate guidance in leveraging their available resources effectively. It becomes paramount to incorporate information regarding the rural terrain, intrinsic characteristics of the rural environment, as well as European and national policies/programs dedicated to digital progression. Employing appropriate methodologies and tools for the assessment of local assets and pinpointing developmental trajectories is also crucial. Simultaneously, one of the primary objectives of this research is to enhance the cognizance of local inhabitants concerning the significance of digitization in rural contexts, advocating for a harmonized and cohesive territorial evolution.

At the same time, the role of human capital and local promoters/agents will be emphasized, who become active factors that can induce rural development and digitization with the participation of local authorities. The purpose of this approach is for residents to accept digitization and for rural area authorities to be able to create specialized development directions based on existing potential. Among these, we mention:

- The concept of rural space. Diversity and multifunctionality of the rural space through digitization.
- Development of rural space through digitization and innovation.
- Subsistence and semi-subsistence farming.
- Rural digitization.
- Robotics.
- Diversifying economic activities and integrated resource utilization.
- Local agents/promoters and community development.
- Forms of cooperation. Digitized entrepreneurship.
- Local governance and its role in revitalizing rural areas.
- Transformation and digital innovation. Formulating objectives and revitalization strategy through artificial intelligence.
- Rural development projects. Identifying and accessing funds through developed IT programs.
- Rural development experiences from the European Union.

The Romanian rural landscape is distinguished by its multifaceted and heterogeneous local conditions, shaped by an intricate array of factors over recent decades. Certain rural regions exhibit a myriad of challenges and irregularities, necessitating the identification of revitalization strategies anchored in indigenous resources.

Nationally, a succession of territorial discrepancies has been recognized when delineating rural development priorities for the timeframe 2018–2024 (Bold et al. 2003).

2 Theoretical and Conceptual Context

In Romania, a significant portion of the rural area is in decline and faces complex problems, the effects of which will intensify in the medium and long term in the absence of concrete measures and programs. The development potential of an area is greater as its infrastructure is more developed. Moreover, economic growth puts pressure on the existing infrastructure, leading to a more pronounced need for its development. Thus, building and maintaining infrastructure affect their rural poverty demographics, creating numerous jobs and boosting economic development (Dumitrescu and Fuciu 2010)." The first vicious cycle concerning demography (Fig. 1).

Education, the second vicious circle (Fig. 2). The labour market, the third vicious circle (Fig. 3), begins with weak job opportunities in many rural areas (Eupen et al. 2012).

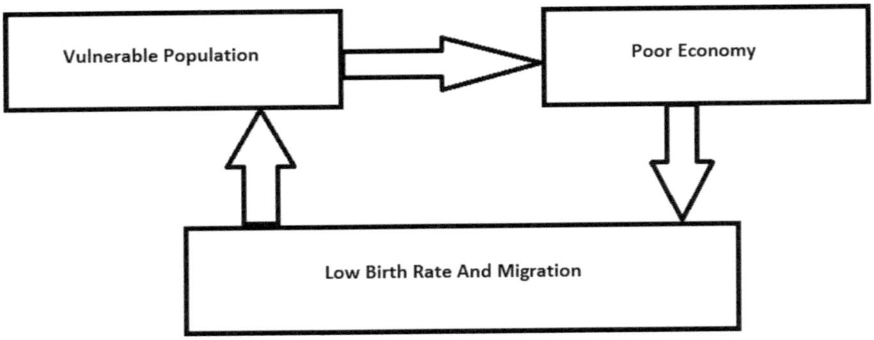

Fig. 1 The vicious circle of poverty in rural areas (Own realization)

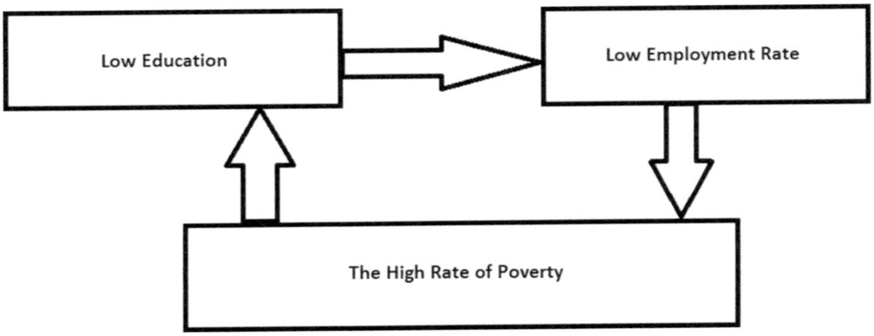

Fig. 2 The vicious circle of a poor education in rural areas (Own realization)

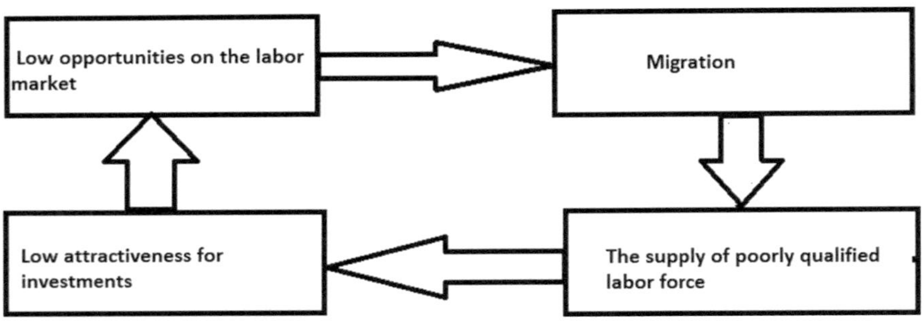

Fig. 3 The vicious circle of poverty on the labor market in rural areas (Own realization)

A more adequately qualified workforce is a deterrent for investments, resulting in a continuous deterioration of the labour market situation (Michael 2015). CAP reform objectives: In this context, for economic recovery and addressing some of the existing issues, the funds dedicated to rural development can be utilized, capitalizing on available resources and customizing measures and action directions based on local specifics (Kerekes et al. 2010).

National Rural Development Program (PNDR) 2014–2020, rural development policy objectives: Regarding the rural development policy (pillar II), in line with European provisions, in Romania, implementation will be carried out through the National Rural (MacKinnon 2002).

1. Rural Development Strategy, 2018
2. MADR, 2018
3. Bertolini et al., 2008 citat în Kerekes et al. 2010, pg. 64- 65
4. Bertolini et al., 2008 citat în Kerekes et al. 2010, pg. 64- 65
5. European Comission, 2015, pg. 3
6. PNDR, 2015, pg. 112

Innovation and the knowledge transfer associated with it are vital development tools for rural Europe. Innovation is a driver of social, economic, and ecological progress. Therefore, innovation in rural development should not be considered in any way as an elitist concept. New approaches to business development, environmental resource management, and/or local community activities are relevant and have potential for everyone interested in Europe's rural areas and Romania's. This fact is reflected in the European programs and policies for 2023 - 2027. Furthermore, innovation is introduced as a cross-cutting theme in the National Rural Development Program *(PNDR 2023–2027).*

Innovation and knowledge transfer support will increase and strengthen in 2023–2027. Of particular importance is the growing interest in innovation. Innovation is a broad topic but, generally speaking. Arguing that it is in society's interest to support SMEs, an essential action is to offer support services to small farms, family associations, and SMEs, facilitating their access to innovative businesses. For Romania, European funds remain one of the most essential tools for reducing regional disparities, promoting economic progress at the regional and local levels, strengthening competitiveness, and increasing employment, objectives aligned with the European Union's benchmarks. In rural areas, innovation and knowledge base will be strengthened through cooperation between the agricultural, food, and forestry sectors, in particular, as well as through the creation of clusters and networks and the use of consulting services. Improving the agricultural sector's performance through innovative businesses will positively affect micro and macroeconomic stability. It will contribute to balancing food consumption and agri-food security, increasing the proportion of commercial farms and job creation, including absorbing the labour surplus from agriculture. Many of the identified needs are directly related to increasing competitiveness. Also, diversifying agricultural activities offers income growth opportunities. It is essential to support the renewal of farm managers' generations to stimulate investment levels and innovation, directly affecting rural inhabitants' incomes. Knowledge about farm modernization and agricultural business viability, forest management, adaptation to climate change, and adaptation of innovative technology from European and international networks to Romanian conditions is required.

Innovation through social enterprises. Social enterprises employ vulnerable individuals for social reintegration, reinvest all profits in social activities, carry out social activities identified in a specific community as needs that otherwise could not be met, or any combination of these forms. Activities of social enterprises in rural areas address the most deeply rooted problems these communities face, trying to bring as many positive and innovative solutions to the services they provide. These enterprises' activities positively affect individuals and communities, influencing their behaviour simultaneously. The development of rural areas depends on innovation and innovation systems. Innovation is widely recognized as a significant source of productivity, competitiveness, and economic growth improvement in all advanced and emerging economies. The most critical role of innovation is job creation, income generation, directing social development, and poverty reduction. When we talk about innovation in rural areas through social enterprises, we are not just referring to new inventions or modern technologies; it can manifest in the following ways:

- *Adoption of new working methods within the enterprise.* Different sectors and stakeholders can be brought together with new relationship methods, supporting disadvantaged groups or finding solutions to social, economic, and environmental challenges. These new working methods are specific to social enterprises because the disadvantaged individuals or vulnerable groups they work with or address require a different approach.
- *Development of new products and services.* This often results from testing innovative working methods and can be made possible by applying techniques, partnerships, technology, processes, research, and new thinking.
- *Adapting old approaches to new circumstances and digital transformations.* These innovative actions are often facilitated by knowledge transfer and are recognized as effective in the rural evolution process. The impact of social enterprises goes beyond generating jobs for vulnerable individuals, leaving a mark on the communities in which they operate, changing mindsets, creating cohesion, and the very presence of social enterprises in rural areas represents an innovation.

New Rural Economy: Innovation in the Rural Environment Technology, efficient investments, and high-performance management are the three main vectors for the development of profitable agriculture in the future and a sustainable rural economy. Given the competition in agriculture, both at the level of Romania and at regional and European levels, as well as significant price fluctuations from inputs to crop sales, there is a clear need to focus on precision, innovative agriculture to ensure maximum productivity and efficiency in any market or weather conditions. It is vital to generate new ideas for development and improve existing technologies.

To develop an efficient, innovative system, the following premises are taken into account:

1. *Creating a stimulating environment for private sector initiatives* supporting market mechanisms. For this purpose, through the "Research and Innovation Strategy 2023–2027", the main lines of action will be offering loans, establishing venture capital funds and guarantee funds, and managing intellectual property.

Additionally, creating a system of micro-loans and subsidized interest loans for working capital and development investments, favouring innovative SMEs in the rural environment.

2. *Supporting smart specialization, using tools* creative.

"Potential priorities for innovative projects in rural areas could be:

1. Creating optimized food products.
2. Developing new products
3. Sustainable development of field crops tailored to the impact of global climate change, aiming to diversify the national offer of varieties specific to the excessively temperate climate, increase international competitiveness, and adapt to EU norms and markets.
4. Sustainable development of the forestry sector and increasing its competitiveness. This can contribute to ensuring the stability, biodiversity, and increased functional efficiency of forests, generating resources and services and enhancing the role of the European Technological Platform for the Forestry Sector.
5. Utilizing agri-food biotechnologies."

Local Economy Based on IT Knowledge–New products, new markets, and internet access a necessity for rural development. The broadband internet penetration rate at fixed points was only 23.47% in rural areas. This is due to the need for a modern infrastructure for information technology and communications (ICT) and the low income of the population. The programming document analysis shows that the need for ICT infrastructure development in rural areas has been considered, with funds to be allocated in this area. Therefore, the need for a communications network in rural areas and broadband internet services in central points of interest, such as communal centres, libraries, and local administrations, is extensively addressed in the document. Implementing these projects will improve information dissemination, better knowledge exchange, and enhance innovation in farms and forestry operations.

In Romania, there is a need to develop the organic farming system and its adaptation to market needs, raising awareness of the importance of practising organic farming in rural areas as a solution for local revitalization. There is also the promotion of high-quality Romanian products, certified organic products, authentic traditional products, and valuable local and regional brands. Romania mainly exports cereals, honey, forest fruits, and fewer processed goods like cheeses, wines, or bakery products. European countries are Romania's primary export markets for organic products. Germany, Austria, and Belgium are the leading countries importing raw materials produced in the local organic system. They are also the primary suppliers of processed organic products for the Romanian market. Thus, creating synergies of rural tourism-ecotourism-organic farming-vine/wine-craftsmanship-folklore aims to export solid regional brands, focusing on excellence and traditions, promoting the identity of places and being a significant source for sustainable regional development.

Regarding the development of sustainable rural tourism services, another perspective segment worth watching is that of folk craftsmen, local artisans, and local, regional, or even international fairs where they showcase their products. Seminars, workshops, and presentations aim to gain new markets for these products, increase the number of entrepreneurs, improve entrepreneurial skills, and increase the market visibility of craftsmen and their products. These products, made using simple, traditional technologies with a significant manual processing component, preserve the authentic Romanian essence and national identity.

Creative Economy and Rural Development mean knowledge exchange and information transfer at the European level. Creativity and innovation are two essential components for successful rural development. They address the critical challenges of villages and can support the development of the European and Romanian rural environment. Creativity is a supreme source of innovation: it turns ideas into products and services. It is always present when innovation leads to economic outcomes. Transitioning to a creative rural economy, innovative practices, and creative solutions are a springboard to economic growth and the well-being of European rural regions. For example, innovation is relevant for rural development activities that aim to identify new markets, develop new partnerships, adapt to various circumstances, create new products, introduce new services, or test and implement new cooperation methods.

Rural creativity in economic sectors spans various activities, relying on individual creativity to produce original goods and services. This includes crafts, art, antique trading, architecture, fashion, publishing, advertising, music, performing arts, film and photography, radio and television, the internet, software design, and digital media. They contribute to rural development sustainably and have become a key measure of local competitiveness. In the new programming period, innovative projects from non-agricultural sectors, those concerning investments in physical assets, and those that promote rural creativity and innovation will be supported and funded under the National Rural Development Program *(NRDP 2023 – 2027).*

Rural Innovation. Rural development policies are built upon a programming history that acknowledges the critical role and benefits of innovation and creativity for villagers. Innovative approaches are promoted through national and European policies and programs. According to them, embedding innovation measures in rural development is a priority. Nationally, the NRDP 2023 – 2027 includes actions aiming to promote innovation and identify new opportunities to enhance economic development and the social prosperity of communities involved in rural development. Innovation is deemed crucial in facilitating a framework for sustainable development, which strikes a balance between economic growth, production, and protecting "public interest goods" like biodiversity and other environmental resources.

Also, creative thinking is vital for rural development participants and decision-makers in addressing core challenges: competitiveness, quality of life, diversification, and territorial cohesion. Whether cross-border or transnational, territorial cooperation can foster creativity and innovation through numerous contacts, exchanges, and jointly developed projects. Rural development policies promote cooperation and support the creation of cooperation networks and operational groups to exchange experiences and best practices and to develop new opportunities.

Business networks in the rural environment mean exchanging knowledge and expertise in rural development.

Knowledge transfer and information play a critical role because they bolster innovation and facilitate the development.

Businesses of the future, the engine of sustainable rural development. "The sustainability of agricultural production faces diverse pressures from civil society and markets. Sustainability means environmental concern, traceability, water and energy use. Entrepreneurs are often engrossed in the present, are future-oriented, but also require analyses regarding the strategy to adopt in the medium and long term," *asserts Prof. Dr. John Rich, Regional Consulting Senior.*

Organic farming is a future business; a current consumer trend is leaning towards natural and fresh products as possible and local producers over imported products.

Platform Smart City 2023

"Smart City e-governance platforms are innovative solutions developed to enhance the quality of urban life by leveraging technology and data for the efficient management of public services and the interaction between citizens and local administration" (*NRDP 2023 – 2027*). We need to make it feasible through technology for these platforms to be developed for municipalities and villages. Such a platform integrates advanced technologies like the.

These platforms have certain features, namely:

1. Monitoring and managing infrastructure: *Smart City platforms allow real-time monitoring of critical urban infrastructure such as water, gas, electricity networks, and traffic management. This way, potential defects or critical situations can be quickly detected, and repairs can be efficiently planned.*
2. Intelligent public services*: Residents can access public services such as paying taxes and duties, requesting permits or certificates, scheduling medical institutions,* etc.*, through an online platform or mobile application. This reduces queues and waiting times, increasing service efficiency.*
3. Traffic and parking management: *The platform can help reduce traffic congestion through smart signalling systems, real-time traffic monitoring, and providing information about available parking.*
4. Civic participation: *Citizens can provide feedback and suggestions to the local administration through a dedicated digital channel. This encourages transparency and citizen involvement in decisions that affect the community.*
5. Data analysis for informed decisions: *The platform collects and analyzes data from various sources, providing local administration with valuable information for strategic decisions, such as urban development planning or resource allocation.*
6. Security and monitoring systems: *Through IoT and video surveillance systems, the platform can enhance city security by quickly detecting incidents or suspicious behaviours.*
7. Energy efficiency and environment: *Technology can help reduce energy consumption and monitor pollution, thus contributing to a cleaner and more sustainable environment.*

The Smart City e-governance platform aims to create a more connected, efficient, and friendly urban environment by using technology to meet the needs and requirements of citizens and local administration. This can bring significant benefits to city life quality, promoting sustainable urban development and more active citizen participation in decision-making processes. The development vision for the next few years includes:

1. Increasing the connectivity to transport infrastructure and energy and ICT networks contributes to locating new companies and visitors.
2. Enhancing accessibility, use, and quality of ICT in rural areas.
3. Implementing ICT solutions for community involvement, communication, and transparency in delivering public services adapted to social changes and local community needs.

It is vital for administrations in villages and communes to understand what smart technologies, the internet, green energy, and digitalization are. Some have understood, but they need to be more in number. The Smart City concept has remained a mere formality for many Romanian cities. However, for instance, Ciugud has long been a genuine Smart Village - specifically, the first rural community that significantly digitized the relationship between citizens and local administration. Ciugud was the first commune to announce that it would collect local taxes and duties online. It has smart schools and kindergartens, attracting families from other parts of the country, and even launched its digital currency, *ciugubanul*.

It seems improbable that all this happened in a Romanian rural community, so the commune chose a motto that perfectly illustrates a specific state: "Ciugud to Be True". The school in this locality has transformed into a digital school, and this happened before the pandemic. Digitalization has even penetrated the environment. A mini wind power plant with Norwegian funding was built in the commune almost ten years ago when energy crises were just theoretical.

This year, we believe that the Romanian village is on the path to increased digitalization in the coming period, considering that even in Europe, the term "Smart Village" is relatively new, introduced only five years ago.

Smart Village means - intelligent local administration, through which digital services are offered to interact with citizens and local authorities (Fig. 4). Thus, tax payments can be made electronically, and the issuance of documents by the local authority can be done remotely. Educating residents to use smart solutions is vital in villages. A smart school in the Romanian rural environment, where much of the teacher-student interaction is done using technology x.

To succeed with the Smart Village project, it is necessary to attract European funds. According to the Smart Village project, "the 224 projects identified at the national level are included in just 20 solutions. 124 projects are implemented (i.e., 55% of the total), 44 are in the implementation process, and 56 are in the project phase. It was also identified that, in terms of financing such projects, there is a high dependence on non-reimbursable European or national funds" *(RDP 2014 – 2020)*.

We dream of online school and digitalization, but there are villages and hamlets where we need utilities, sewerage, and roads. Everyone talks about modernization and digitalization, staying caught up to other countries and staying with our residents.

Some changes are necessary. Firstly, the state needs to generate.

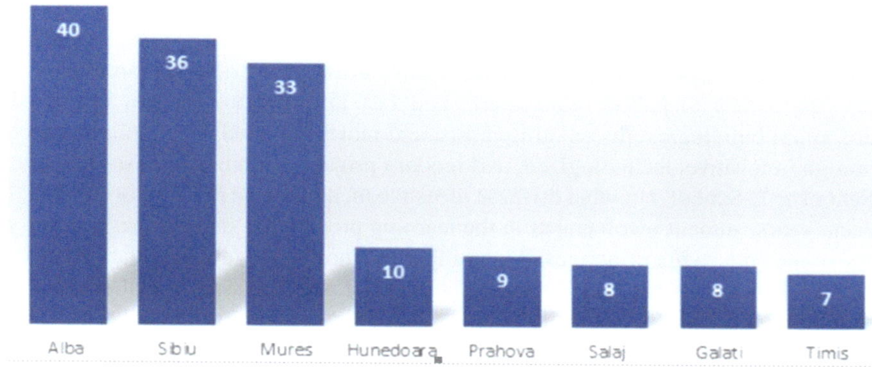

Fig. 4 Ranking of smart village projects by county – PNDR, 2020

- trust in digital services,
- stimulate innovation,
- support infrastructure construction,
- develop policies in this regard,
- support a faster transition to a digital society,
- adopt electronic services that the state offers citizens (reducing bureaucracy by increasing the number of forms that can be completed online, connecting various public institutions online, and digitizing the interaction between citizens and public administration).

Secondly, the business environment must:

- support service innovation,
- increase the number of users and adopt and integrate smart city concepts into society.

Technology should be used to make life easier and positively change the society in which we live. One idea would be to eliminate the eternal business stamp. Another would be issuing electronic invoices or using videoconferences.

If 89% of Romania's population currently has access to the Internet, the credit goes to private operators. Recently, the reduction of costs and the accessibility of the Internet have led to increased education and consumption of digital products. The Internet and the digital age have created a bridge between people and have brought opportunities that need to be maximized. It is our responsibility as citizens, innovators, promoters, and supporters of digitization to contribute to the technologization of society through various actions, from pressuring the state to educating ourselves and the generation that did not grow up in the digital world on innovation and finding smart solutions to ease everyone's work and increase the standard of living and well-being of everyone, both in urban and rural areas.

If we want quality education, then we must understand that each of us desires access to quality education, quality medical services, optimal public transport infrastructure and suitable roads, public safety, jobs, affordable prices for products and services, access to public, transparent, and fast information, modern infrastructure, smart buildings, reduced utility costs, and more. We need educational projects through innovative, technologized, and modern projects. Some call these initiatives ISchool or E-School, but what do these ideas mean, and where did they originate? In recent years, student involvement in the learning process has declined significantly. One of the underlying principles of learning is the motivation law. Motivation causes learning to occur and sustain itself. Studies have found that aspects that could help students be motivated are engaging classes, a practical applicative approach, and the atmosphere during lessons.

Both the school, the parents, and the students want performance. Furthermore, performance can be achieved through technology, which is a significant factor. The facilities of such a project are access to an online catalogue, an online communication platform, access to check students' grades, access to the platform from a PC, tablet, or phone, and real-time notifications via email or SMS. Modernizing these procedures and methods through technologization aims to improve the teaching–learning process. *(A promoter of this concept is TNT Computers, which has developed a complete educational management solution – ISchool:* http://iscoala.ro/*).*

Smart Villages. EU Actions for Smart Villages - "It is necessary to invest in people, ideas and businesses, local communities, and neighbourhoods (Fig. 5). We need to support digital infrastructure and offer rural citizens the opportunity to develop online and offline solutions that strengthen rural vitality and resilience through social innovation and specialization for precision technology."

Smart villages are inhabited by hardworking, creative, intuitive people, referring to the inhabitants of rural areas, who take the initiative in finding practical solutions—both to the severe challenges they face and especially to the new opportunities for transforming rural areas and using digital technologies when appropriate, not just because they are trendy. Smart villages often use digital technologies.

Some initiatives occur at the village level, but most involve neighbouring areas, groups of villages, small towns, and city connections. Being smart means creating new forms of cooperation and alliances between farmers and others involved in agriculture and the rural environment, between municipalities, the private sector, and civil society, from bottom-up to top-down (Fig. 6). There is no standard model or solution for smart villages—it's about locals using local goods and resources, leveraging available knowledge, and taking the initiative. In the USA, the term "smart villages" is replaced by "intelligent communities", currently used in the USA.

Year 2023, efforts are being made to fully utilize opportunities that can contribute to strengthening rural vitality in Europe. Both challenges and opportunities vary depending on rural areas and among different regions of Europe.

While the potential opportunities and benefits are significant, some risks could, for example, lead to the closure of local stores. Similarly, it is essential to create suitable tools to ensure that more people can benefit from the digital transition. Establishing high-speed digital infrastructure in rural areas and digital education and training must

Fig. 5 Smart villages. *Souce* EU.com

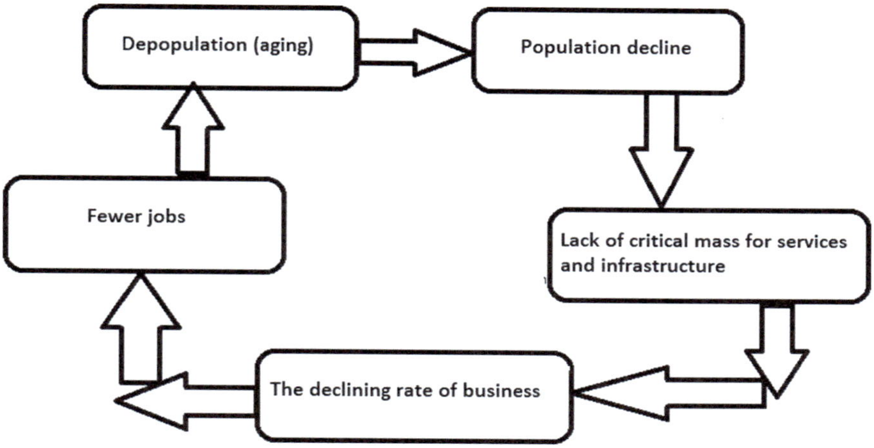

Fig. 6 The circle of decline (Own realization)

be an ongoing investment priority. Both are needed to combat the digital divide and strengthen the ability of stakeholders at the European level and rural communities to harness and utilize digital potential. Smart villages take digitization a step further. Their aim is not only to keep pace with urban areas by narrowing the differences but also to enhance the attractiveness of rural areas and develop new roles in transforming Europe into a digital economy.

Revitalizing rural services through social innovation. The provision of services is a rapidly growing sector worldwide, yet in many rural areas, there are anomalies, especially in regions where services have declined over the past decade. In this context, the communities must take the initiative to bridge the disparities, intervening in areas such as accommodation, transport and mobility, finance, vocational training, energy, recycling, and economic development. The triggers vary, ranging from the closure of a vital service (such as schools, post offices, shops, transportation) to the decline in service quality—like social assistance—the identification of a specific issue or a missed opportunity.

Key among these are leadership (best practice models) and social capital, as well as connections and trust among locals for collective actions.

Bio-energy villages. In our country, they are almost non-existent. However, considering the following steps, we can hope that at least five villages will enjoy increasing bio-energy. These steps are: 1. Initiation: informing about the project, starting the selection procedure, and involving locals; 2. Planning: including a questionnaire on heat consumption, investment requirements and operational costs, feasibility study, and financial plan. 3. Construction: building the biogas plant/factory, wood stove, and the village's heating network.

Digital transformation of rural areas. Regarding the digital rural gap, there are three pillars: broadband infrastructure, the adoption of digital services, and the digital literacy of the locals. The availability of digital services is essential in leveraging the potential of good broadband infrastructure. In urban areas, such services can operate autonomously due to the abundant availability of inhabitants and service providers.

If we **discuss education in smart villages**, we discuss challenges in this field. More than 75,000 teachers teach in village schools in Romania, according to data available from the National Institute of Statistics, a decrease from ten years ago when there were over 82,000 teachers in rural areas. In cities, the number of teachers exceeded 162,000 in 2021, almost similar to a decade ago.

The lack of school facilities, the ageing rural population, and the unattractive salary package are the main problems teachers' face, who prefer cities over rural areas. Thus, if ten years ago, students from villages were half the number of students from cities, now the balance tilts towards urban school children. In 2021, there were just under 1 million students in rural areas; in cities, the number exceeded 2.5 million students, according to INS data. The migration trend from cities to villages, especially among young people looking for green, pollution-free homes, must be reflected in education. In other areas, such as tourism, this population movement has been evident through the emergence of local businesses and accommodation units that have capitalized on the appetite for a return to traditions.

According to INS data, in 2021, 78,000 residents left the countryside for the city, and 116,000 moved from the city to the countryside, with a net difference between the two migratory flows of 38,000 people, double compared to previous years. There are 7,000 educational units in Romania, of which 3,800 are in urban areas and the rest are in rural areas.

3 Research Methodology and Analysis

Questionnaire on the Effectiveness of a Bookstore's Marketing Decisions in Using Information from Its Database

In the context where the decisions made by a company in a rural area are closely linked to a multitude of factors (both internal and external), which influence their success to a greater or lesser extent, the questionnaire aims to provide the information needed to analyze the efficiency with which certain villages use the information from their databases concerning the residents of the villages or communes. More precisely, marketing communications towards the inhabitants are monitored to see if the data held by the villages/communes are used correctly, meaning if they make the best decisions. Following the analyses conducted on the questionnaire data it is intended to obtain relevant information for the administration of the respective villages or communes.

The method used is a survey based on a questionnaire, meaning primary data collection was used. Simple random sampling was conducted, the characteristics of the population having no significant influence on the chosen objective, and the sample size was 41 people (Fig. 7). The questionnaire has a total of 17 questions, all mandatory (Table 1). The first question is a filter to distinguish respondents relevant to the study from the others (these were directed to the last section of the questionnaire), and questions from 1 to 17 were designed to obtain data necessary for the research problem.

The questions used in the questionnaire are as follows:

1. What age range do you fall into?
2. Please indicate your gender.
3. Are you from a village or a commune?
4. Which University do you study at?
5. Please indicate your study cycle.
6. Have you had subjects/courses regarding setting up businesses and/or entrepreneurship in villages?
7. What are your intentions after completing your studies to return to your native village?
8. Are there entrepreneurs in your family?
9. If your answer is "YES," what is your relationship with them?
10. Do you believe that men have more chances of successful business in villages?
11. Have you ever seriously thought about becoming an entrepreneur in a village?

Fig. 7 Interpretive graphics from the database

12. What is the probability of creating your own business in the village?
13. Is there a digital transformation in villages?
14. Do you agree to support digital transformation and innovation in villages?
15. What would be the real motivation?
16. Have you heard of the Smart City Platform/Project?
17. Have you heard of the Smart Villages project?

Table 1 Coding questionnaire data

No. question	Question content	Options to answer	Code attributed to answers
Q 1	What age range do you fall into?	a b c d	1 2 3 4
Q 2	Please indicate your gender;	feminine masculine	1 2
Q 3	Are you from a village or a commune?	village commune	1 2
Q 4	Which University do you study at?	UVT altele	1 2
Q 5	Please indicate your study cycle	bachelor master	1 2
Q 6	Have you had subjects/courses regarding setting up businesses and/or entrepreneurship in villages?	yes no	1 2
Q 7	What are your intentions after completing your studies to return to your native village?	You intend to have your own business You want to work for an employer	1 2
Q 8	Are there entrepreneurs in your family?	yes no	1 2
Q 9	If your answer is "YES," what is your relationship with them?	parents 2nd degree relatives	1 2
Q 10	Do you believe that men have more chances of successful business in villages?	yes no	1 2
Q 11	Have you ever seriously thought about becoming an entrepreneur in a village?	yes no	1 2
Q 12	What is the probability of creating your own business in the village?	probably very probably less probably	1 2 3
Q 13	Is there a digital transformation in villages?	yes no	1 2
Q 14	Do you agree to support digital transformation and innovation in villages?	yes no	1 2
Q 15	What would be the real motivation?	Transformation Renwal Digitalalization Digital transformation Innovation Development	1 2 3 4 5 6

(continued)

Table 1 (continued)

No. question	Question content	Options to answer	Code attributed to answers
Q 16	Have you heard of the Smart City Platform/ Project?	yes no	1 2
Q 17	Have you heard of the Smart Villages project?	yes no	1 2

Column1		*Column2*	
Mean	1.1	Mean	1.1
Standard Error	0.048038446	Standard Error	0.048038446
Median	1	Median	1
Mode	1	Mode	1
Standard Deviation	0.30382181	Standard Deviation	0.30382181
Sample Variance	0.092307692	Sample Variance	0.092307692
Kurtosis	5.979137032	Kurtosis	5.979137032
Skewness	2.771707741	Skewness	2.771707741
Range	1	Range	1
Minimum	1	Minimum	1
Maximum	2	Maximum	2
Sum	44	Sum	44
Count	40	Count	40

Fig. 8 Data analysis (regression) in parallel of questions 16 and 17

The graphical representations were derived from the responses to the survey questions. Notably, for questions Q16/Q17, the results indicate a near parity in responses, with a marginal error of 0.04, as illustrated in Fig. 8.

4 Software Program (a Procedure for Record-Keeping Messages) – Developed in VBA

The software, tailored to address specific requirements, can accommodate both predetermined working variables and those derived from the questionnaire responses across all Romanian villages (Fig. 9). Given the inclusion of data from 40 respondents, it is imperative to articulate and delineate the program's modifications, updates, and deletions, ensuring they align with the distinctive administrative needs and requirements of each village or commune.

```
my Var = 2685 my Var = 13,285
Dim my Var As Integer
my Var = 1000
Dim Villages (1 To 1000) as String
Village (1) = "Village 1"
Village (2) = "Village 2"
Village (3) = "Village 3"
```

Regression Statistics	
Multiple R	0.066776
R Square	0.004459
Adjusted R Square	-0.02174
Standard Error	0.269631
Observations	40

ANOVA

	df	SS	MS	F	Significance F
Regression	1	0.012374	0.012374	0.170201	0.682251568
Residual	38	2.762626	0.072701		
Total	39	2.775			

	Coefficientsandard Err		t Stat	P-value	Lower 95%	Upper 95%ower 95.0%		Upper 95.0%
Intercept	1.020202	0.1395	7.313269	9.3E-09	0.737798773	1.302605	0.737799	1.302605267
X Variable 1	0.035354	0.085694	0.412554	0.682252	-0.1381254	0.208832	-0.13813	0.208832471

Fig. 9 Students' analysis focused on the development of villages through digitalization

```
Village (4) = "Village 4"
Villages (1000) = "Village1000"
Dim Message As String
Message = "Digitalalization"
Message = Message & "Digitalalization Sat1"
Message = "Digitalalization with VBA" ' this line will be executed
'this line will be ignored
        For x = 1 To 1000
                'do something
        Next x
Dim Village
Dim Village (1 To 1000) As String
Village (1) = "Village_1"
'Initialize Village
        For Each Sat in Villages
                'do something with Village
        Next
```

According to the research model and considering the determined variables, we formulated the following equations: A = f (C, PR, SN, Î, NR), where:

Personal attitude (A) of village students is a behavioral characteristic of the individual, which is discerned from their actions.

Control (C) is a psychological trait of the rural individual.

Risk propensity (PR) is a psychological trait of the rural individual.

Self-confidence (Î) is a psychological trait of the rural individual.

Need for achievement (NR) is a psychological trait of the rural individual.

NS = *f* (C, PR, SN, Î, NR), where:

Subjective norms (NS) is a behavioural characteristic referring to an individual's perceptions of the general social pressure to act or not to act.

Ccp = *f* (C, PR, SN, Î, NR), where:

Perceived behavioural control (Ccp) is a behavioural characteristic that allows predicting behaviours that have been relatively direct (i.e., under will control).

IA = *f* (NS)

IA = *f* (Ccp)

1).IA = *f* (A)

	A	B	C	D	E	F	G	H	I	J
A	1									
B	-0.41085	1								
C	0.368125	-0.43454	1							
D	-0.43868	0.18996	-0.39287	1						
E	-0.33225	0.331613	-0.12556	0.165165	1					
F	-0.4071	0.401719	-0.32623	0.302747	0.565239	1				
G	-0.42504	0.432961	-0.28842	0.292234	0.384021	0.539237	1			
H	-0.33978	0.527783	-0.48164	0.563817	0.362994	0.633074	0.367805	1		
I	-0.49888	0.601972	-0.55514	0.483344	0.32796	0.613172	0.595603	0.700243	1	
J	-0.35758	0.399331	-0.40014	0.40515	0.439769	0.471955	0.382691	0.59952	0.692398	1

Fig. 10 Hypothesis verification - entrepreneurial intentions of students in villages (Correlation matrix between variables)

Hypothesis verification was achieved using simple linear regression between psychological traits and behavioural traits on the one hand and between behavioural characteristics and entrepreneurial intentions of village students on the other hand (Fig. 10).

From the analysis of the results, it can be observed that:

1. attitudes (independent variable) directly, positively, and significantly influence the entrepreneurial intentions of rural students (dependent variable);
2. perceived behavioral control (independent variable) directly, positively, and significantly influences the entrepreneurial intentions of rural students (dependent variable);
3. subjective norms (independent variable) directly, positively, and significantly influence the entrepreneurial intentions of rural students (dependent variable).

5 Conclusions

In conclusion, the present is digitized, and the business environment is most aware of this aspect. Every company, whether from the city or the countryside, tries to maintain a high level of technology to gain a competitive advantage in a market, but information is equally important. Daily, city or rural companies can collect, manage, and process vast amounts of data, obtaining raw information from these processes. This will only be valuable when interpreted correctly and when decisions are made considering the acquired information. Even if more data means more information can be obtained, not all will be relevant to the company, and to avoid the illusion surrounding Big Data, it is crucial to determine from the start which data would be helpful and store only the significant ones. As a result of this illusion, companies encounter "*optimization paralysis*". Technologies like Data Mining or data analysis processes are helpful in this context.

In the context of digital transformation and innovation within rural areas of Romania, progress is facilitated through the proactive engagement of the private sector. This advancement is further catalyzed by the emergence of innovative agricultural enterprises and the initiation of public–private collaborative ventures. Additionally, synergistic collaborations between academic researchers and stakeholders such as farmers and rural entrepreneurs play a pivotal role in this transformative journey. In pursuing the goals related to innovation and the sustainable progression of agriculture, it is pertinent to note that the spectrum of innovation within the agricultural domain, particularly in rural development, encompasses technological, non-technological, organizational, and societal dimensions. These innovations may derive from either contemporary methodologies or rooted traditional practices.

An emergent concept, whether it manifests as a novel product, practice, service, production methodology, or organizational restructuring, ascends to the status of an innovation when it garners widespread acceptance and demonstrates tangible efficacy in real-world applications. The trajectory of such concepts, transitioning from mere ideas to bona fide innovations, is contingent not solely on the intrinsic merit of the idea but also on the market dynamics and its receptivity to assimilate and operationalize it.

Also, this aspect depends on the willingness of other operators to take over the idea, which will be influenced by their needs, knowledge, and perceptions and the cost-effectiveness of the new idea in the specific business context. In conclusion, the funds available for rural development, divided into various types of projects within the measures and sub-measures defined in the National Rural Development Program (PNDR 2023–2027), can be used to address some of the issues identified locally. We can call them professional skills. These could be:

C1. Digital transformation and innovation in the development of rural areas in Romania.

C2. Proper use of concepts related to rural development;

C3. Identification and assessment of local resources in diagnostic studies;

C4. Development of development strategies (for revitalizing rural areas) related to local specifics;

C5. Proper use of statistical data series and the development of graphic and cartographic materials allows territorial analysis.

References

Bold I, Buciuman E, Drăghici M (2003) Spațiu rural, Editura Mirton, Timișoara

Dumitrescu L, Fuciu M (2010) The role of database marketing in the operationalization of the services relationship marketing, University of Craiova, Faculty of Economics and Business Administration. Manag Mark J 0(1): 15–24

Eupen VM, Metzger MJ, Pérez-Soba M, Verburg PH, Doorn VA, Bunce RGH (2012) A rural typology for strategic European policies, land use Policy 29:473–482

Hertel M (2015) Risk and return management for the digital economy, Universitatea din Augsburg

Kerekes K, Pakucs B, Szőcs E, Veres E, Vincze M (2010) Dezvoltare rurală. Ocuparea forței de muncă în mediul rural, Editura Accent, ClujNapoca, pp 320

MacKinnon D (2002) Rural governance and local involvement: assessing state community relations in the Scottish Highlands. J Rural Stud 18(3):307–324
Planul National Strategic 2023–2027

The Digital Euro and the Future of the European Project

Jovan Zafiroski and Elena Neshovska Kjoseva

Abstract The creation of the single currency was one of the most important events in the history of the European integration. The single currency is a milestone in this process which should foster the economic integration and contribute to strengthening of the European identity. However, since 2008 and the global financial crisis the European economy is facing permanent challenges while in certain periods the very existence of the single currency was jeopardized. In order to deal with the consequences of the GFC the European monetary authorities have undertaken series of reforms in the European financial and banking system. Also, an unconventional monetary policy which included negative interest rates and massive monetary expansion though the so-called *quantitative easing* was implemented. The COVID 19 pandemic forced the authorities to implement even more aggressive monetary policy which alongside the geopolitical turbulences in the last two years is causing very high inflation rates. Moreover, the monetary union faces another challenge brought by the global process of digitalization of the economy and introduction of new technologies in the realm of money and finance. The process of digitalization of money and the prospect of creation of the central bank digital currencies offers new prospects for the European economic and monetary union. The launch of the digital euro might be a catalyst that will strengthen the monetary union and will foster the economic growth. However, there are some concerns about the possible negative effects deriving from the introduction of the digital euro that could undermine the normal functioning of the monetary union and the European financial system. The text will try to explain the significance of the launch of the digital euro for both the monetary union and for the entire project of European integration.

Keywords Digitalization · Digital economy · Digital euro · European union · Global financial crisis · Inflation

---

J. Zafiroski (✉) · E. N. Kjoseva
Ss. Cyril and Methodius University, Department of Financial Law, Iustinianus Primus Faculty of Law, Skopje, Republic of North Macedonia
e-mail: j.zafiroski@pf.ukim.edu.mk

E. N. Kjoseva
e-mail: e.neshovska@pf.ukim.edu.mk

G. Pellat et al. (eds.), *Data-Centric Business and Applications*, Lecture Notes on Data Engineering and Communications Technologies 258,
https://doi.org/10.1007/978-3-031-95068-1_5

1 Introduction

The world is rapidly changing. As new tendencies emerge the actors present for centuries and the rules forged through generations are simply becoming obsolete and outdated. The monetary sphere is not an exception. As a very refined part of the economy and the law, the monetary system is not a major preoccupation for the broader public while its changes are done very carefully and slowly. Basically, they pass unnoticeable. If the changes are noticeable, then something big has happened. In other words, the conditions for changes in the monetary system are developing gradually through years and then a dramatic transformation suddenly occurs. History of money is full with events that profoundly changed the monetary system and the way how the payments are conducted. Usually, those changes are triggered by certain innovation that alter the traditional notion of money and payments. Nevertheless, the innovation is in the very essence of money. It replaced the barter system and revolutionized the system of commerce.

Nowadays we are facing many events and processes that have potential to dramatically change the face of the monetary system that existed for seventy years and on which we are all accustomed. The changes are triggered by different events and processes (See: Zafiroski 2023). Firstly, it is the return of the inflation which at least in the last twenty years, we forgot that exists. The unconventional monetary policy started as a solution for the global financial crisis and culminated in 2020 during the pandemic. In that period, the central banks were pushed to the limits when by different programs and actions, they were one of the main actors for preserving the stability not only of the economy but of the entire society (Dimitrijevic 2023, pp. 256–263). This unconventional monetary policy that included negative interest rates and different liquidity instruments, such as the *quantitative easing*, opened the Pandora's box of inflation that we are facing in 2023. Secondly, the geopolitical turbulences and the conflicts at the European soil and in the Middle East brought unparalleled tensions between world's major military and economic powers. As a result of the conflict on different levels and fields we have sanctions that included measures targeting the foreign exchange reserves and the international payment infrastructure on a scale never seen before. All these actions are threatening the future of the international monetary system built after the Second World War. Some academics and experts (Pozsar 2022) are already talking about Bretton Woods III. Finally, perhaps the most important challenge in the monetary field comes from the technological breakthroughs and the process of digitalization of money.

All these challenges will affect the functioning of the European monetary union. While the challenges form inflation and the geopolitical shifts can be managed by good and well-timed policies within the current institutional and legal framework of the single currency, the process of digitalization of money is a new phenomenon that requires intervention in the existing structure. The European monetary authorities are facing a difficult situation. There is not only a rise of the private money in terms of cryptocurrencies and other digital monies, but also the part of the payments in cash in the daily payments is constantly decreasing due to various reasons. It means that the central bank money is less and less used by the general public which undermines the monetary sovereignty and the effectiveness of the monetary policy as an important tool of the economic policy. Furthermore, while the level of the payments in cash is decreasing the demand for cash is increasing which is a clear signal that the households and firms are keen to hold central bank money as a store of value, especially in times of crisis. This situation forced the central banks to react and to be in a quest for solutions that will provide sound central bank money in times of highly digitalized economy.

These events bring the project of European monetary integration on a crossroad. There is no doubt that the digitalization of central bank money is a great challenge from an economic, legal and technological point of view. However, if this process is undertaken properly, it would bring many benefits, strengthening the common currency and making it more credible and trusted, a real symbol of European identity.

The text will discuss the current developments in the process of monetary integration at European soil amid inflationary, technological and geopolitical challenges (2). It analyses the proposal for introduction of the digital euro as a new form of single currency that should give a proper answer to the challenges of money and payments in a digitalized economy (3). The conclusion follows.

2 The First 25 years

The European economic and monetary union is a unique example for monetary integration of independent states sharing a common currency. The decision for monetary integration was political. In times of huge geopolitical shifts amid the collapse of the Soviet Union and an ongoing process of unification of Germany the European leaders decided to continue with the process of European unification meaning deeper economic and political integration. From the very beginning, the project was controversial and there were many sceptics about the feasibility of the Eurozone. Also, the economic theory did not provide a clear argument that the introduction of the euro could bring more positive than negative effects. The theory of the *optimal currency area*, usually applied when considering the usefulness and benefits from monetary integration did not provide clear evidences for introduction of the common currency. However, looking from a historical perspective and from a time distance of more than twenty years one may say that the introduction of euro as a common currency for the EU Member States is a successful story.

Today, the euro is the second most important global currency for borrowing, lending and central bank reserves and accounts for close to 40% of international payments, a proportion that is almost equal to the US dollar. The euro is a common currency for more than 340 million Europeans whereas more than 60 countries or territories have directly or indirectly pegged their national currency to the euro. (See: The euro at 20). After couple of waves of enlargement, currently twenty EU members have adopted the euro as a national currency. The enlargement of the monetary union is an ongoing process that will continue in future taking into account the fact that no *opt-out* clause will be provided for the future members of the EU as it was the case with the UK and Denmark at the beginning of the creation of the monetary union.

The euro has contributed for better functioning of the internal market and, more importantly, it has provided a monetary stability even in the countries which historically had high volatility of the exchange rates and where monetary instability and monetary crisis were very frequent, as was the case with Mediterranean countries such as Italy, Spain and Greece. The euro is a success but the journey of twenty-five years was not a smooth experience and straightforward process. It was a bumpy road for the common currency with different tests. Moreover, its future is also uncertain and depends on the ability to adapt to the new circumstances arising primarily form the process of digitalization of money.

2.1 Bold Step in the Path of European Integration

The monetary integration at the European soil is part of a broad process of reconciliation and unification of the European nations. This process should assure a long-lasting peace in Europe creating environment for economic development while providing high living standards of the citizens. The creation of the Eurozone and the euro as a common currency is the last stage of a gradual process of economic integration that went through different phases including free trade area, customs union and common market. However, the final destination of the process of European integration is a political union. As it might be said *the whole European integration experiment, from the Coal and Steel Community on, has been political wolf dressed in economic sheep's clothing* (Buiter 1999). Additionally, it is almost certain that the Eurozone will be the first tier of the possible creation of a multi-speed Europe considering the importance of the common currency for the entire European project.

Monetary integration in Europe is a permanent goal that had various attempts to be realized in different periods. In the past, it was the gold standard that was used as a common value to link the national monies. Historically, in 1929 Gustav Stresemann was the first European politician who publicly raised the question of creation of a common European currency. He discussed the consequences of the financial crises and the significance of monetary unification in Europe in front of the League of nations. After the Second World War the European countries were part of the Bretton Woods system and in 1962 the European commission presented the first document—the Marjolin Memorandum that proposed creation of a common currency for the members of the European Economic Community at that time. The gold standard was suspended in 1971 and the period that followed brought monetary instability all over the world. The European nations tried to create some mechanism, as the so called "snake in the tunnel" in order to provide stability of exchange rates. In 1979 the European monetary system was created with the Exchange rate mechanism at its core that had one simple goal—to provide monetary stability in Europe and avoidance of volatility of the exchange rates. This mechanism is perceived as a harbinger of the monetary union. The most important document for creation of the

European monetary union is the Delors report (1989) from the late'80, period of uncertainties as a consequence of the collapse of the soviet block and time when unification of Germany took place. Based on this report the Maastricht treaty from 1992 created the European and monetary union as a *sui generis* monetary union for member states of the European Union. As it was once said *it is bold step into the unknown, not unlike Alice's leap down the rabbit hole* (Buiter 1999, p.182).

The first test of the monetary union came in 2008 during the global financial crisis that sparked in the US but provoke an avalanche of crisis in different sectors of the global economy. In order to combat the crisis and to mitigate the negative effects for the economy the European governments intervened on the financial markets. The spending of the public money increased the level of the public debt in majority of the Member States which opened the door for debt crisis in the EU. Consequently, some countries, Greece for instance, were under very strong pressure for undertaking austerity measures in order to preserve the credibility of the entire monetary union. At that time, the entire project of the monetary integration was under threat. The crisis made trade imbalances not only between the EU and the rest of the world but also within the EU Member States. In addition, the EU's share of the global GDP continued to shrink, from 26% in 2004 to 22% in 2015 when one of the most important members states decided to leave the Union.

The pandemic crisis from 2020 was another test for the monetary union. It has shown all the weak spots in the construction of the European project and endangered the main pillars of the EU such as the European single market, the currency union and the entire legal and institutional construction (See: The Economist 2020). Thus, the EU rules on subsidies were suspended which opened the door for more than 2 trillion-euro subsidies for European companies in the countries that are able to afford help for the economy which puts in bad position those that could not afford to help their companies. As a consequence, the public debt in some countries in the Eurozone, Italy for example, has skyrocketed that could undermine the stability on the entire monetary union. Moreover, the ECB itself has continued with purchase of assets on financial markets bringing the total value of the securities purchased to more than 2.55 trillion euro. Finally, as a result of different liquidity programs, expansionist fiscal policy and the geopolitical turbulences that increased the prices of commodities, oil in particular, the inflation started to increase and, since the 2022, some of the Eurozone members have a double-digit inflation number. The inflation is well above the ECB's target which undermines the very essence and the purpose of the monetary union, i.e. the price stability as a primary objective.

In the following years, the ECB must maintain its credibility and prove that it can successfully achieve the principal objective of the monetary union and to provide price stability. The digital euro might contribute to this end for the reason that it will bring more solutions for the central bank to implement its monetary policy while the entire credibility of the ECB will be strengthened.

Nevertheless, since the global financial crisis, the EU was working to improve the legal framework and the overall capacities for preventing, dealing with and mitigating the negative effects of the financial crisis. There are several reforms that resulted with creation of mechanisms for better functioning of the monetary union.

The crisis has undoubtedly shown that the EMU has not been completed project yet and profound reforms are necessary in two directions: banking union and fiscal union. The first is achieved through the project for creation of the European banking union which included a Single supervisory mechanism, Single resolution mechanism and Common deposit insurance schemes (see: Zafiroski 2014, pp. 48–55). The second is part of reforms that were undertaken in the field of the public finances. The crisis has unambiguously shown that a supranational monetary policy requires a common or coordinated fiscal policy. Thus, the Fiscal compact[1] was put in place to strengthen the fiscal governance framework in the Eurozone. It contains provisions relating to the economic policy coordination and convergence of the Eurozone. The fiscal compact consists of: a mandatory balanced budget rule, strengthening of the excessive deficit procedure, benchmark for government debt reduction and public debt issuance plans. Nevertheless, the monetary union still lacks permanent mechanism for dealing with the fiscal imbalances in the countries within the Eurozone and should be created in the future reforms of the EU as it is the proposal for a European recovery fund.

2.2 *New Paradigm in the Realm of Money*

The digitalization brings fundamental changes in every aspect of human living and the monetary sphere is not an exception. The incumbent players and the post-war monetary order are facing with challenges deriving from technological breakthroughs that are significant and offer possibilities that were deemed unimaginable in the past with potential to alter the traditional understanding of money and payments. The cryptocurrencies are in the center of this story. They are based on the idea for creation of a decentralized monetary system that would be independent and outside the powers of the state without a third or trusted party in the system of payments and settlements. The core element is the *blockchain* technology that offers a possibility for bridging the double spending problem. The enhanced use for payments of the private digital money in various forms and on different levels, on one hand, and the reduced use of the cash in everyday payments with its limit to be medium for exchange in electronic commerce on the other, forced the central banks around the globe to act and to offer solutions for the changed environment in which there is a need for digital central bank money.

In the current monetary landscape, we have two types of money in circulation: (1) public money issued by the central bank i.e., banknotes and coins and electronic central bank deposits held by banks and other financial market participants and (2) private or book money or deposits held on the accounts of commercial banks. In times when there is a rise in the private digital money while the cash payments are in decline the main objective of the central banks is to find a solution to provide

[1] Treaty on Stability, Coordination and Governance in the Economic and Monetary Union, agreed at the EU summit of 30 January 2012, signed on 2 March by the Heads of State or Government of all EU countries, with the exception of the United Kingdom and the Czech Republic.

a risk-free central bank money in a highly digitalized and cashless economy. The technological breakthroughs in the payment services and the appearance of a new concept of digital money might be useful for central banks in their quest for solutions. Thus, the new form of money should be digital and should also have the power of the state behind it. As declared by the major central banks in the framework of the cooperation in the Bank of International Settlement in their joint report from 2020 (BIS 2020, pp. 5–10) the main objective of the CBDCs should be its use as a means of payment while all others, as monetary policy for example, are secondary. Also, this report lays down the core principles that should underpin the research and the possible design of the CBDCs. Thus, the three key principles that lay the foundations for any work on the CBDCs are: 1. do not harm principle which provides that CBDCs should not hamper the delivery of the central banks' objectives; 2. creation of an ecosystem that would provide coexistence among all types of money; and 3. the introduction of the central bank digital money should encourage innovation and improve efficiency in the payment system.

The idea for introduction of a central bank in digital form seems progressive and adequate in times increased digitalization of the economy. However, the success of this new type of money will primarily depend on its design and key characteristics. There are multiple proposals and ideas how to design the digital central bank money. The choice of the design is related to some important issues that should be decided when introducing the CBDCs. Different choices entail different legal, economic and overall social consequences. Therefore, perhaps the most important question in the design of the central banks digital money is who can have access to this new form of money. There are two possibilities. The first is a *wholesale* CBDC that is something similar to the digital money that are now available to the banks and other market participants and the second option is issue of a *retail* CBDC that will be used by households and companies. The first option will not profoundly affect the current system of payments while the second will make serious changes in the monetary system and will require changes in the legal framework.

Another decision should be made on the identification for access to the CBDCs. Depending on the design they might be *account-based* that requires the identity of the account holder at the central bank for payment authentication which poses great concerns for data privacy and security or *token-based* that is anonymous and, similar to the payments in cash today does not require user's identity. Furthermore, based on the choice of the transfer mechanism or on the ledger structure used the system might be *decentralized* without a central ledger, as it is the case with the cash and various cryptocurrencies providing anonymity, or a *centralized* system based on deposit accounts within the central banks that requires a trusted party or an institution that regulates and manages transactions.

All of the choices in the design of the CBDCs raises different legal question such as the choice of the legal basis for introduction of this new form of digital money, the relations with other types of money, the status of legal tender, monetary sovereignty and the stability of the banking and overall financial system. For these reasons, the launch of the central bank digital currency is a process that should be proceeded very cautiously taking into account the opinion of all stakeholders whereas the final product should be the issue of a new form of money that will bring positive changes for the entire society.

Central banks are proactive and work on introduction of the central bank digital currencies. Different studies suggest that more than a half of the world's central banks are in a process of exploring or developing digital currencies. However, even if the Central banks are conducting research into the CBDC their issuance is unlikely on short or medium term (Stanely 2022; Kosse, Mattei 2022). Thus, we have projects in different stages for creation of the CBDCs. The Sand Dollar in the Bahamas is the first launched central bank digital money while many countries such as Uruguay (E—peso), China (E—CNY), Sweeden (E—krona), Russia (Digital ruble), Ecuador (Dinero Electronico) and others are on very advanced level in this process. As far as the major global central banks are concerned, there are some documents that are giving the direction of their research. The FED as a central bank of the dominant reserve currency—the dollar, is expected to lead the way of digitalization of money. However, at least in the public available information, the prospects of launch of the digital dollar remains remote. The FED presented a document on the future of the U.S. dollar in times of digital transformation and rapid changes in the monetary sphere (Board of Governors of the Federal Reserve System 2022). The authorities in the UK launched a debate for possible issue of a digital pound by presenting a Consultation paper for the possible launch of digital pound (Bank of England and HM Treasury 2023). It seems that, from all major currencies the euro is the closest to have its digital form. The ECB was the first major central bank to launch a debate for the design and key features of the digital euro. The Report on a digital euro (ECB 2020) summarizes the main reasons for the research and the main objectives of this new form of public money in the Eurozone. Also, the Commission proposed a Regulation on the establishment of the digital euro which is the key document for the future of the digital European currency that will have outstanding effects not only on the process of monetary integration, but also on the entire European project. The following part of the text will focus on the proposal for establishment on the digital euro.

3 The Euro and the Digitalization of Money

The wave of digitalization of the economy is present at the European soil and has profound effects on the payment system. The European companies and people are increasing the use of private digital means for payments while banknotes and coins that are the sole central bank money available to the public cannot answer the needs of the highly digitalized economy. Online purchases are increasing at the same time as cashless payments are on rise. The only available central bank money today i.e. banknotes and coins cannot be used in online transactions which may restrict the demand and the use of the central bank money in general. Also, the emergences of digital central bank money issued from other central banks and enhanced role of private digital money in payments may jeopardize the role of the euro not only as a means of payment but also as a sole legal tender in the Eurozone. That is why the European institutions have commenced a long and complex process of debate considering the introduction of a euro in a digital form.

This debate is part of a broader strategy related to digitalization of the economy and of the financial system in particular. Thus, in the framework of the Commission's Digital Finance (European Commission 2020a, p. 10) and Retail Payment Strategies (European commission 2020b, p.14) adopted in September 2020 the exploration of a digital euro was supported and presented as a tool for reducing the fragmentation of the European retail payment market. Also, the digital euro is seen as one of the instruments that should provide efficiency and resilience of the payment market while promoting competition and innovation in the payment industry. The digital euro is expected to foster the process of digitalization of the European economy and to act like a catalyst for innovation in the payment industry.

Moreover, in autumn 2020 the ECB published its Report on the digital euro (ECB 2020) where it presented the idea for the launch of a digital euro that will complement the euro banknotes and coins as a legal tender in the Eurozone. The objective was to consider the views from all the parties concerned from the introduction of this CBDC and to analyze the possible benefits and risks deriving from its introduction. Thus, the ECB launched a "Public consultation on the digital euro" that included 18 questions aiming to collect opinions from different stakeholders on the idea for launching the digital euro, its design and key characteristics. It ran until January 2021 and summarized point of views of individuals and professionals regarding the possibility for launch of a digital central bank money in the Eurozone. In addition, the Commission's consultation strategy included targeted consultation including payment industry specialists, National central banks, civil society and academia, different national and EU authorities, private business representatives etc.

In November 2023, taking into account the findings of the investigation phase, the Governing Council of the ECB has decided to move to the preparatory phase as a next one of the digital euro project, without prejudice of a final decision on its issuance. In the first stage of the preparation phase that will last two years the focus will be on testing, experimenting and consultation of all the parties involved in the process. After this period, the Governing council of the ECB will decide about the future steps in the introduction of the digital euro, depending on the legislative process and the results of the first stage (ECB 2023).

As a result of the consultation process, the Commission made a proposal for a Regulation on the establishment of the digital euro (European commission 2023) based on the Article 133 TFEU the European Commission that provides a legal framework for issuance and effective use of the digital euro. The regulation should establish and regulate the digital euro as a new form of central bank money that alongside the banknotes and coins will be available to the general public. This Regulation is the main document that governs the digital euro but it may be supplemented by the delegated acts that the Commission and the ECB are empowered to adopt.

3.1 The Proposal for Digital Euro

The European commission made a bold step in the process of introduction of the digital euro and proposed a Regulation on the establishment of the digital euro which makes the European currency dominant global currency as the most advanced one in the process of introduction of its digital form. The organizational structure of the Eurozone, and the complicated legislative procedure that includes several instances are the main reasons why the digital euro started this process before the other currencies, such as the dollar for instance. Nevertheless, this proposal gives a solid base for an analysis of the future design of the digital euro which is essential for its success.

The main **objective** of the Regulation is to establish the digital euro as one of the forms of the single currency in the Eurozone and to lay down the principles and key characteristics of its design. Thus, the Regulation introduces the rules concerning legal tender status, distribution, use and important technical characteristics of the digital euro. The **digital euro is defined** as form of the single currency available to natural and legal persons. It is clear that the European institutions opted for a *retail* CBDCs which will be a novelty in the European monetary system and could potentially bring profound changes. As it is the case with euro banknotes and coins the digital euro is a direct liability of the ECB or of the NCBs. The ECB decides or authorizes the issue of the digital euro.

The idea behind the launch of the digital euro is to offer central bank money in a digitalized economy that will complement, but not to replace cash. That is the rationality why convertibility of the digital euro and the euro cash at par is assured. The purpose of the digital euro is to serve primarily as a means of payment while limiting its use as a store of value. As a result, the Regulation stipulates that the **digital euro should not bear interest**. In this sense, the proposal tries to make the digital euro equally attractive as the cash, as far as the interest is concerned. The digital money held on the accounts at the commercial banks bear interest depending of the type of deposit, so one may expect that the digital euro will have same characteristics. However, in respect to the interest the digital euro will be identical as euro cash.

Legal tender status is a key characteristic of the central bank money. Currently, by virtue of the Article 128 (1) of the TFEU and Article 10 and 11 of the Regulation No 974/98 on the introduction of the euro only euro banknotes and coins have status of legal tender in the Eurozone. The digital euro should not replace the euro banknotes and coins but should complement it. Euro cash remains to be legal tender in the euro area. Moreover, after the cases in front of the Court of Justice of the EU (CJEU 2021) relating to the euro as a legal tender the Commission prepared a proposal for a Regulation on the legal tender of euro banknotes and coins (European Commission 2023a) that should harmonize the rules governing the legal tender status of cash and to provide that euro banknotes and coins are widely distributed and used by the economic agents and individuals. The Regulation on the establishment of the digital euro provides that the digital euro will have a **legal tender status** which entails mandatory acceptance by payees. The status of legal tender implies that in settlement of a debt denominated in euro the digital euro cannot be refused as a means of payment for both offline and online payments. Also, payees cannot unilaterally or though contractual terms that have not been individually agreed exclude payments in digital euro. Taking into consideration the fact that the obligation for mandatory acceptance of the digital euro might be costly for some economic operators there are some exceptions from the general rule of mandatory acceptance of the digital euro. Thus, exceptions are provided for natural persons acting in the course of purely personal household activity, microenterprises and to non-profit legal entities. This is the key characteristic of the digital euro which makes it equal to the euro cash. Moreover, the technical features of the digital euro that will enable the use of the digital euro for both online and offline payments will make this new form of money attractive for payments without any disadvantages compared to the payments in cash.

The Regulation creates mechanisms for functioning of the payments with the digital euro. The choice in the design is neither an account based nor token based system but the current payment system with payment system providers will be used for this purpose. Thus, the **payment service providers** are responsible for the digital euro accounts and provide digital euro payment services. Digital euro payment accounts are particular payment accounts denominated in euro for caring out different transactions such as place fund, withdraw cash and payment transitions. In order to hold and to use digital euro the users need to establish contractual relationship with the payment service provider. In this context, in order to assure wide use of the digital euro and to promote financial inclusion of certain groups of the society, the digital euro payment services include basic and additional euro payment services. Thus, the former are considered essential for the use of this new form of money for natural persons. The latter include additional payment services such as conditional digital euro payment transactions.

The Regulation gives a possibility for the ECB to set **limits on the use of the digital euro as a store of value**. As discussed earlier, one of the risks deriving from the introduction of the CBDCs is the possibility of disintermediation of the financial system and most notably digital bank runs in times of financial crisis. For this reason, respecting the principle of an open market economy with free competition, in line with the principle of proportionality, the ECB may introduce limits on the digital euro's use as a store of value through different tools including, but not restricted to quantitative limits to individual digital euro holdings and limits to conversion of other categories of funds to digital euro in a specified timeframe. In order to enable proper implementation of these limits and to provide a smooth functioning of the payment system automated mechanisms that link digital with non-digital euro payment accounts will be introduced. Thus, via "waterfall functionality" users will be able to receive digital euro payment transactions even in situation when the amount of transaction exceeds the limit on digital euro holdings by transferring funds to non-digital euro payment account and conversely by "reverse waterfall functionality" digital euro users can make payment transitions even in situations when their digital euro holdings in inferior to the amount for transfer by automatically mobilizing funds from non-digital payment account.

The digital nature of the CBDCs brings limitless opportunity for programing and designing of the money to fulfill tasks of some other policies that the State deems important in certain periods of time and may decide to use the money as a tool for achieving these goals. On first glance, this feature of digital money seems attractive and useful, but on long run term it may undermine the principle of neutrality of money and make distortions of the uniformity of the monetary system by creating various types of money with different market value. The digital euro, at least in the period of its introduction will not offer this possibility. The proposal clearly declares that the digital euro will **not be programmable** money. It means that it cannot be used in a way that limits the spending or direct spending towards certain goods and services, on a particular territory or during a certain period. Instead, the digital euro will be fully fungible for other types of money. However, the digital euro will support the programing of conditional digital euro payment transactions as payments that are automatically triggered by software. The payment is undertaken on predefined and agreed conditions. Thus, the digital euro is going to be one of the instruments for promotion of innovation in the field of payment services.

3.2 Effects of the Digital Euro

The introduction of the digital euro will inevitability have series of effects on the financial and monetary system of the EU and on the payment industry in particular.

Today, online payments relay exclusively on commercial money which on long term could jeopardize the status of the central bank money as a monetary anchor in the system. The digital euro should fill this gap and provide a central bank money that properly meet the **needs of the digitalized economy** and facilitate online payments. Furthermore, the design of the digital euro should enable the so-called machine-to-machine payments that will answer the future payments needs in the context of the industry 4.0 and payments in decentralized internet. In other words, the principal effects of the introduction of the digital euro are related to the system of payment and the status of the central bank money as an attractive instrument for payments in a digitalized economy.

Financial inclusion is another advantage that the digital euro might bring. Digitalization of money and enhanced use of private money for digital payments may exclude certain groups of people in rural and remote areas. The easy access to the digital euro may encourage vulnerable groups of the society to use this new form of money and to expend the level of people that are using regular systems of payments in everyday use. This will largely depend on the final version of the design of the digital euro and the rules for its use.

The issue of **monetary sovereignty** is one of the key questions in the debate on the possible effects of the introduction of the digital euro. There are two aspects. First, the European monetary sovereignty should be strengthened by the introduction of the digital euro due to the fact that the public is offered with central bank money that are adequate for the modern times and highly digitalized economy. This keeps the door closed for possible third county CBDCs to interfere in the European monetary space. Additionally, the use of the private digital means of payments will have reduced role in the payments because that there is an alternative of a digital central bank money that enjoys credibility as it is the case with the euro as a single currency. Second, as one of the most important global currencies, the digital form of the euro will have important effects on countries outside the euro area. There might be negative impact on monetary sovereignty in countries where the digital euro replaces the local currency, not only as a store of value but also as a means of payment in domestic transactions. Also, in EU members states that have not yet adopted the euro as a national currency the enhanced use of the digital euro may interfere in the process of the introduction of the euro and fulfilment of the criteria prescribed by the treaties and most notably by the Article 140 TFEU.

The European institutions are actively working on creating conditions for enhanced **international role of the euro**. More prominent role of the euro on international stage will contribute to lower cost and lowers risk for European companies when trading internationally due to the fact that the exchange risk will be removed making the entire system more resilient on exchange rate shock on international marked. Furthermore, the internationally accepted euro will lower the interest rates on the European market while making the European financial markets deeper, liquid and integrated where European business and governments will have easy access even in time of financial turmoil (See: European commission 2018). The introduction of the digital euro is expected to strengthen the international role of the euro. The proposal for the Regulation includes a possibility for arrangements concluded between the ECB and the central banks in member states whose currency is not the euro and with central banks from third countries that will regulate cross currency payments between digital euro and the local currency. Depending on the final design of the digital euro and the possibility of non-residents to have access to the digital euro the international role of the euro, might be promoted.

4 Conclusion

The process of digitalization of the economy requires appropriate forms of money which will provide fast and secure payments and will support online payments. The current monetary system does not provide central bank money that may fulfil this task. The banknotes and coins, as a sole available form of central bank money for the households and businesses, cannot answer the need of the digitalized economy. Private money is used in everyday transactions while some of them, as the cryptocurrencies, are even outside the official monetary system. For this reason, the central banks started to work on developing the idea for introduction of a digital form of the national currency. Today, the vast majority of the central banks have projects for introduction of a digital currency unit that in different countries are in different stage. The choice of the design of the CBDCs is the key for success of this new form of money and should be done taking into account the interest of the entire society.

The EMU is an example for profound research and advanced stage for introduction of a digital euro. The creation of a national currency was a demanding endeavor and that will also be the case with the digital euro. The proposal for introduction of the digital euro lays down the foundations for introduction of the digital euro.

The digital euro is expected to bring benefit in numerous fields It should provide a central bank money that will facilitate online payments and foster the digital trade and work as a catalyst in the digital economy. The new form of the single currency should encourage the financial inclusion by expanding the level of people that are using the regular system of payments. Also, the introduction of the digital euro should strengthen the European monetary sovereignty by providing sound money for digital payments and thus limiting the space for other foreign digital currencies to interfere on the European monetary space. Finally, the euro in a digital form should increase the international role of the euro and make it more attractive for international trade, international markets and investments. However, the process is not without risks. Depending on its final design the digital euro may jeopardize the functioning of the financial and banking system while having negative effects on other countries with small economies that are linked to the European market.

References

Bank of England and HM Treasury (2023) The digital pound: a new form of money for households and business?, Consultation paper, February. Available at: https://www.bankofengland.co.uk/-/media/boe/files/paper/2023/the-digital-pound-consultation-working-paper.pdf

BIS (2020) Central bank digital currencies: foundational principles and core features. Available at: https://www.bis.org/publ/othp33.pdf

Board of Governors of the Federal Reserve System (2022) Money and Payments: The U.S. Dollar in the Age of Digital Transformation, January, Available at: https://www.federalreserve.gov/publications/files/money-and-payments-20220120.pdf

Buiter, Willem H (1999) Alice in Euroland. J Common Mark Stud, 37(2) p 200

Committee for the study of Economic and Monetary Union, Report on economic and monetary union in the European Community (1989) достапно на: http://ec.europa.eu/economy_finance/publications/pages/publication6161_en.pdf

CJEU, Judgment of 26 January 2021 in Joined Cases C-422/19 and C-423/19, Hessischer Rundfunk, EU:C:2021

Dimitrievic M (2023) Pravo Evropske Centralne Banke, Pravni Fakultet Univerziteta u Nisu

European commission (2023) Proposal for a regulation of the european parliament and of the council on the establishment of the digital euro, Brussels, 28.6.2023 COM(2023) 369 final 2023/0212 (COD). Available at: https://eur-lex.europa.eu/resource.html?uri=cellar:6f2f669f-1686-11ee-806b-01aa75ed71a1.0001.02/DOC_1&format=PDF

European commission (2018) Communication from the commission to the european parliament, the european council, the european central bank, the european economic and social committee and the committee of the regions, towards a stronger international role of the euro, Towards a stronger international role of the euro, Brussels, 5.12.2018 COM(2018) 796/4. Available at: https://commission.europa.eu/system/files/2018-12/communication_-_towards_a_stronger_international_role_of_the_euro.pdf

European commission (2020a) Communication from the commission to the european parliament, the council, the European economic and social committee and the committee of the regions on a digital finance strategy for the EU, Brussels, 24.9.2020 COM(2020) 591 final, Available at: eur-lex.europa.eu/legal-content/EN/TXT/PDF/?uri=CELEX:52020DC0591

European commission (2020b). Communication from the commission to the european parliament, the council, the European economic and social committee and the committee of the regions on a retail payments strategy for the EU, Brussels, 24.9.2020 COM(2020) 592 final, available at: eur-lex.europa.eu/legal-content/EN/TXT/PDF/?uri=CELEX:52020DC0592

European commission (2023a) Proposal for a Regulation of the European Parliament and of the Council on the legal tender of the euro banknotes and coins, Brussels, 28.6.2023 COM(2023) 364 final 2023/0208 (COD), available at: eur-lex.europa.eu/legal-content/EN/TXT/PDF/?uri=CELEX:52023PC0364

ECB (2020) Report on a digital euro, October, available at: https://www.ecb.europa.eu/pub/pdf/other/Report_on_a_digital_euro~4d7268b458.en.pdf

ECB (2023) A stocktake on the digital euro, Summary report on the investigation phase and outlook on the next phase. Available at: https://www.ecb.europa.eu/paym/digital_euro/investigation/profuse/shared/files/dedocs/ecb.dedocs231018.en.pdf

European Council, The euro at 20. Available at: https://www.consilium.europa.eu/en/euro-at-20/#group-section-Would-you-like-to-know-more-WzGKpKtUa7

Pozsar Z (2018) War and interest rates, Credit Suisse

Kosse A, Mattei I (2022) Gaining momentum—Results of the 2021 BIS survey on central bank digital currencies, BIS Papers No 125, May, available at: https://www.bis.org/publ/bppdf/bispap125.pdf

Stanley A (2022) The Ascent of CBDCs, IMF, September, available at: https://www.imf.org/en/Publications/fandd/issues/2022/09/Picture-this-The-ascent-of-CBDCs

The Economist (2020) On the blink, The European Union is having bad crisis, May 14th 2020

Zafiroski. Jovan (2023) CBDCs—a (r) evolution of money, Collection of papers International scientific conference "The influence of European integration on the development of monetary law", Jean Monnet Module for European Monetary Law Srđan Golubović Marko Dimitrijević Editors, Nis, pp 9–19 ISBN: 978–86–7148–309–4

Zafiroski, Jovan (2014) Uber die europaische Bankenunion, European law

Artificial Intelligence: A Dangerous Obsession?

Juliana Vassileva and Roger Tsafack Nanfosso

Abstract The discussion concerning the role that artificial intelligence (AI) is expected to assume in modern society, the economy, and business is becoming increasingly intense and comprehensive. (Krugman, Foreign Aff 73:28–44, 1994) has, at times, expressed scepticism about how the concept of "competitiveness" is used in policy discourse, particularly when it's narrowly defined in a way that may not fully capture the complexities of a nation's economic well-being, stamping it with "dangerous obsession" and arguing that it is often misunderstood or oversimplified. Building on the parallel between the hype in the 1990s around the developing theory of competitiveness and nowadays hype over the AI progress, its impact on society and economy, and the expected near future, we try to discuss how dangerous can the obsession with the AI be? Especially in times of crises and instabilities, and the acute need to manage them. However, even that concerns are expressed by various authors, scholars, and experts in the field of AI's ethics and safety, aspects related to crises economic policy and management in the light of AI are still conspicuously not sufficiently present into the scientific discourse. The aim of this study is to systematically investigate the role of AI in economic crises policy and management, considering various aspects of its impact, including its potential to enhance efficiency and productivity, its capacity to establish a new prerequisite for growth, its potential as a threat bordering on the extreme, and its inherent role as a stabilizing factor in the interplay between competitiveness, cooperativeness, and the preservation of the social contract during times of crisis. The research methodology primarily relies on a review of existing scientific knowledge, societal and policy debates. Additionally, attempts are made to accumulate empirical knowledge, although it cannot be claimed to be representative due to the scale and scope of the task. Nonetheless, the results reveal the ambiguity of expectations regarding the level of influence and

J. Vassileva (✉)
Department of Administration and Management, New Bulgarian University, Sofia, Bulgaria
e-mail: jvassileva@nbu.bg

R. T. Nanfosso
Faculty of Economic Sciences and Management, University of Dschang, Dschang, Cameroon
e-mail: roger.tsafack-nanfosso@univ-dschang.org

G. Pellat et al. (eds.), *Data-Centric Business and Applications*, Lecture Notes on Data Engineering and Communications Technologies 258,
https://doi.org/10.1007/978-3-031-95068-1_6

control within the relatively irreversible stage attained in the process of technological progress through artificial intelligence technologies.

Keywords Economic policy and management · Crisis · Technological progress · AI potential and threat

1 Introduction

The alert of 'dangerous obsession' connects to Krugman's critiques in the 1990s of competitiveness concepts and, more specifically, its limited understanding of nations competing as corporations in the global economy. Krugman (1994) draws attention to the possible unproductiveness or incorrect nature of discussions about a country's economic performance or challenges, and thus, the possible negative implications for its domestic and international economic policies, highlighting consequences such as misallocation of resources, trade tensions, and the formulation of bad national economic policies.

Just as the concept of competitiveness is criticized for being a potentially dangerous obsession, the role of artificial intelligence (AI) in modern society, economies, and crises could also be characterized by ambiguity and uncertainty. There are various expectations and concerns regarding AI's impact, including its potential to boost efficiency and productivity, as well as its potential to pose extreme threats. This ambiguity may result in potentially negative consequences if not carefully managed. An uncritical and heated obsession with AI's role may also lead to unclear policy directions and tensions. In both cases, the underlying message is the importance of responsible and nuanced analysis.

We emphasize the necessity of capturing and systematizing the role of AI from various perspectives. This means considering both the potential benefits and the potential threats that AI can pose in times of crises. Potential threats bordering on the extreme highlight the shared concern that unchecked obsessions or unbalanced views can lead to negative consequences. Establishing a link between AI and crises involves examining how artificial intelligence technologies, their deployment, and societal reactions can influence and be influenced by economic, political, and societal downturns or crises.

While none explicitly defines AI as a 'dangerous obsession' in those exact terms, the idea that AI could become a dangerous obsession is a viewpoint expressed, in various ways, by authors, scholars, and experts in the field of AI ethics and safety. It serves as a characterization used to underscore concerns associated with the development and use of AI. Extensive discussions by authors and experts highlight potential dangers and ethical considerations related to AI. In this light, the term 'dangerous obsession' succinctly expresses concerns about the risks, ethical dilemmas, and societal implications of AI when its development and deployment are not carefully managed or when certain objectives, such as safety and ethics, are prioritized over others.

These concerns become even more pronounced in times of crisis and, further, in times of ’polycrisis’ (Janzwood & Homer-Dixon 2022).

Numerous notable figures have extensively spoken about the potential risks and ethical considerations associated with AI. Concerns regarding the future of AI have been raised, with some warning about the potential dangers of superintelligent AI, summoning the demon (Elon Musk, CEO of Tesla and SpaceX), and the potential existential risks that require careful planning for safe development (Stephen Hawking, theoretical physicist). Others emphasize the importance of aligning AI with human values (Max Tegmark, physicist and author), addressing biases and unintended consequences of AI and algorithms (Cathy O’Neil, mathematician and author), and even questioning the terminology itself, such as Noam Chomsky, who recently called to stop referring to it as ’artificial intelligence,’ emphasizing the need for responsible development.

The recent petition to halt AI development for at least six months by the Future of Life Institute (2023) also underscores the need for ethical guidelines, safety measures, and responsible development to ensure that AI technology benefits society while minimizing potential risks. Regardless of the petition’s purpose, it contributes to the ongoing debate on the role of AI in our future and highlights the active stakeholders in this critical discussion.

The aim of this study is to systematically investigate the role of AI in economic crises policy and management, considering various aspects of its impact. These aspects include its potential to enhance efficiency and productivity, its capacity to establish a new prerequisite for growth, its potential as a threat bordering on the extreme, and its inherent role as a stabilizing factor in the interplay between competitiveness, cooperativeness, and the preservation of the social contract during times of crisis and polycrisis. The research methodology primarily relies on a review of existing scientific knowledge, societal and policy debates. Additionally, attempts are made to accumulate empirical knowledge, although it cannot be claimed to be representative due to the scale and scope of the task. Nonetheless, the results reveal the ambiguity of expectations regarding the level of influence and control within the relatively irreversible stage attained in the process of technological progress through artificial intelligence technologies.

2 AI as Prerequisite for Growth

The benefits of implementing AI technology across various industries and businesses have increasingly become the focus of scholarly research since its gradual widespread application approximately 10 years ago. Evidence for positive economic effects is discernible throughout the entire business value transformation system.

In the fintech sector, for example, AI has rapidly transformed how financial agents operate in fields such as customer service, communication, risk management, and other core functions and business processes in finance. Cumulatively, within the context of dynamic market expansion, the improvements in AI automated decision-making serve as key high-efficiency indicators. These indicators include reduced decision-making time from 15 min or more to instant decision-making, a decreased nonperforming loans rate, an increased reliance on completely non-human-based decision-making to 70% or more, reduced operational costs, a multiplied processing number of credit applications, and the retention or maintenance of customer satisfaction levels exceeding 90% (Hadjitchoneva 2019). The utilization of AI extends the knowledge base, secures the rationality of decision-making, outperforms programmed decision-making by transitioning from risky towards certain contexts, and avoids irrationality in information asymmetry. It institutes confidence, ensures velocity, adaptability, focus, and flexibility in a demanding competitive environment.

In the retail sector, the advantages of applied AI are numerous and impactful. These include enhanced accuracy and reliability, swift real-time decision-making, significant cost savings, ease of use, and improved accessibility. AI facilitates the expeditious expansion into new markets without relying on specific human skills. One notable benefit is the substantial time savings, amounting to at least 5,000 h per year for daily management decision-making at an average level. In contrast, the AI tool requires just a few minutes of calculation per day for monitoring all points of sale. The time saved through AI implementation is strategically redirected towards more business-friendly activities where human resources excel and prove more productive than machines. This approach not only optimizes efficiency but also has a 'hygienic' effect, positively influencing the organizational culture, dynamics, and overall environment of the company.

Additionally, the reduction in waste levels at the point of sale is noteworthy, plummeting from 7% to less than 2%. By the third year of AI introduction, the return on investment (ROI) stands at an impressive 5–6 times higher, with EBIT showing a calculated growth of 2% (Hadjitchoneva 2020). Management attests to specific benefits such as a 50% reduction in waste in outlets without AI, a 35% increase in instances of assortment unavailability in outlets without AI, while outlets applying AI experience no 'catastrophic events'—avoiding scenarios of overstocking or understocking that lead to higher wastage or lost profits. Moreover, since the introduction of AI, there has been no instance of monstrous wastage or shortages where decisions are made entirely without human intervention.

In the realm of e-commerce, studied in crisis period, where more than 12,000 units of fast-moving consumer goods are handled, the strategic business challenges identified by the management align seamlessly with the outcomes generated by AI (Hadjitchoneva 2021, 2022). These challenges include the delicate task of balancing the flow of goods to ensure consistent availability without leading to unnecessary waste. The management recognizes that while minimizing waste is crucial, it's equally essential to ensure that products are consistently available to prevent revenue loss and customer dissatisfaction. Another important issue is maintaining stock rotation—ensuring a steady flow of products while preventing capital freezes. Every product holds importance, but a missing item can result in the loss of an entire consumer basket. As sales volume increases, the absolute value of goods wasted may rise, but the relative percentage decreases. The management's objective is to achieve a waste rate of 0.6%, significantly reduced from the initial 3%, while concurrently achieving a remarkable revenue growth of over 235% year-over-year. It's noteworthy that these impressive results are achieved despite the significant complexity involved in forecasting tasks due to ever-changing sales dynamics and a growth rate exceeding 100% per annum.

AI applied to meet concrete business needs enhances efficiency and productivity by automating and streamlining processes, replacing repetitive and time-consuming tasks. This improves operational efficiency and empowers human workers to focus on more complex and creative aspects of their jobs. The integration of AI leads to cost savings by reducing the reliance on manual labour and increasing overall productivity. Additionally, advanced analytics, utilizing AI algorithms like machine learning, enable the rapid analysis of vast amounts of data. This facilitates the identification of patterns, trends, and insights that might be challenging for humans to discover, ultimately providing data-driven insights, and contributing to better decision-making. Another significant benefit lies in communication and engagement, as AI enables personalized experiences for customers by analysing their preferences and behaviours. This results in more targeted and relevant interactions through AI-powered chatbots and virtual assistants, enhancing customer service by providing instant responses, handling routine queries, and ultimately improving overall customer satisfaction.

The fastest-growing category in AI, Generative AI technology, is poised to not only solve real-world problems but also revolutionize various sectors, from healthcare and education to entertainment, according to Trenker et al. (2023, 2). In a conservative scenario, the total addressable AI market in 2030 is projected to be US$12.14 trillion, while in an optimistic scenario, it could reach US$15.72 trillion. The Generative AI category alone is expected to contribute between US$3.33 and US$4.31 trillion to the US$15.72 trillion estimate.

AI is anticipated to reshape the business landscape, leading to increased productivity, and reaching a market value of US$755 billion by 2030. Furthermore, it has the potential to significantly boost economic performance, with the share of the AI market projected to range from 7.3% to 9.4% of the total GDP. The global AI funding landscape reflects this growth, with investments reaching US$31 billion in 2019, US$72 billion in 2021, and US$46 billion in 2022. Figure 1 illustrates the size of the AI economy and investments by leading industries in AI, along with the AI ecosystems size in the UK, US, and EU.

After the assimilation of AI impact into the economy, anticipations suggest a subsequent economic growth. According to Statista (2023a, b), sectors such as manufacturing, wholesale and retail, agriculture, forestry and fishing, accommodation and food services, utilities, public services, and other services are projected to experience equal or higher GVA growth rates in 2035, attributed to the implementation of AI atop the anticipated baseline growth rates (Fig. 2). The sectors of information and communication, manufacturing, financial services, and wholesale and retail are expected to achieve growth rates of 4% and higher by 2035, driven by the added value contributed by AI.

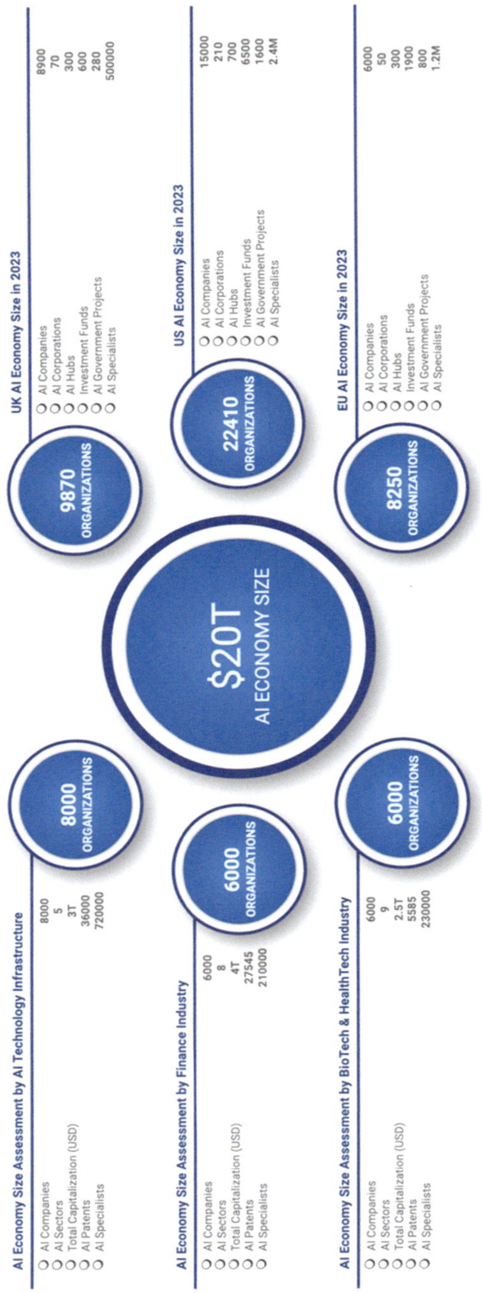

Fig. 1 AI Economy size investment. *Source* AI-Ecosystem (2023)

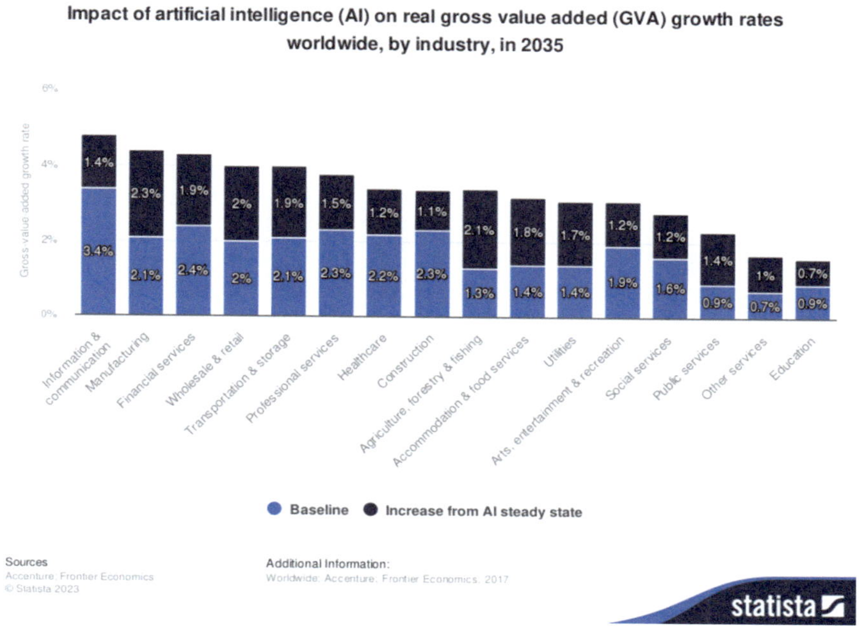

Fig. 2 Impact of AI on GVA growth rates in 2035. *Source* Statista (2023a)

3 AI as Prerequisite for Crisis

The frequency and scale of cyberthreats are anticipated to grow significantly, running in parallel with the expanding scope of digitalization, the intensification of digital transformation during the Fourth and Fifth Industrial Revolutions, the crisis and post-crisis period since 2020, and the ongoing polycrisis. The possibility of extreme cybersecurity threats looms large. Technological advances, particularly in AI, have undoubtedly contributed significantly to increasing productivity, efficiency, and optimizing processes, thereby enhancing business value (Hadjitchoneva 2022). However, these advancements have also led to an increase in potential risks and challenges associated with information security.

Cyberattacks have evolved into sophisticated tactics aimed at exploiting vulnerabilities in information systems and gaining access to sensitive data. In just the last year, global cyberattacks have surged by over 125%, doubling in recent years and resulting in multi-billion annual losses. The annual cost to the global economy from cybercrime was estimated at €5.5 trillion as early as 2020, soaring to over $10 billion in the US in 2022 and an anticipated $8 trillion in global annual cybercrime costs in 2023. The growing cost of cybercrime damages is projected to reach a staggering $10.5 trillion by 2025 (EP 2022; Cybersecurity Ventures 2023; Griffiths 2023).

Due to the globalization of supply chains, SMEs are increasingly vulnerable to cyber-attacks, accounting for 28% of attacks in 2021 (ER 2022). The service, finance, banking sectors, and software supply chains are highly affected, along with the public sector, including healthcare, and virtually all other sectors (ENISA 2022). E-commerce fraud is projected to cost $48 billion globally in 2023, with online payments incurring $343 billion in costs to businesses between 2023 and 2027. Supply chain vulnerabilities rank first among cyber threats, according to executive management (Griffiths 2023).

The relationship between cybersecurity and artificial intelligence (AI) is twofold. On one hand, the potential risks and challenges associated with the use of AI, especially with advancements like generative AI (ChatGPT, etc.), are significant topics of discussion (Funk 2022). Companies need to understand these risks to mitigate them, as these technologies enable more sophisticated and scalable cyberattacks, often avoiding detection. Cybercriminals continuously exploit vulnerabilities in systems, with machine learning models increasingly becoming targets. Misinformation and 'deep' fakes using AI, along with persona-modelling bots, pose threats to communication processes and community interactions (ENISA 2022, 12). On the other hand, AI technologies, particularly ML algorithms, find applications in qualitative analysis and forecasting of potential risks, aiding in risk assessment and cybersecurity planning (e.g., using tools like SecRiskAI). Funk (2022, 252) notes that cybersecurity teams equipped with AI technologies can make faster decisions, outpacing attackers.

According to a recent study on managing cybersecurity by Hadjitchoneva & Mandadjiev (2023), 75% of surveyed experts believe there is insufficient awareness, despite public interest, regarding the need for organizational focus and investment in cybersecurity. Half of the respondents noted some efforts but no perceived organizational priority. Additionally, three-quarters strongly believe there is a growing societal problem due to low-risk awareness. The experts highlighted key priority activities, including building an internal company policy and culture on cybersecurity, implementing appropriate technical tools for proactive prevention, and regularly training staff directly or indirectly involved in information technology, including vendors. The main challenges faced by companies in cybersecurity encompass a lack of a clear vision and strategy, especially at senior management levels, insufficient skilled personnel to cope with evolving trends, and difficulties in securing adequate corporate funding for cybersecurity.

The AI market is experiencing a remarkable expansion, poised to transition from a valuation in the billions to trillions of U.S. dollars in the coming years. As of 2023, the global AI market is valued at 142.3 billion U.S. dollars (Statista 2023b). Investments in AI start-ups surged by five billion U.S. dollars from 2020 to 2022, nearly doubling previous figures. Much of this influx is credited to private capital from U.S. companies, with the latest top-funded AI businesses focusing predominantly on machine learning and chatbot technologies. Corporate investments in AI experienced remarkable growth, surpassing a sevenfold increase from 2015 to 2021 and reaching over 90 billion U.S. dollars after 2020 (Fig. 3).

The question arises: could this significant concentration of investments in AI disturb the equilibrium of investments in various economic sectors, activities, and

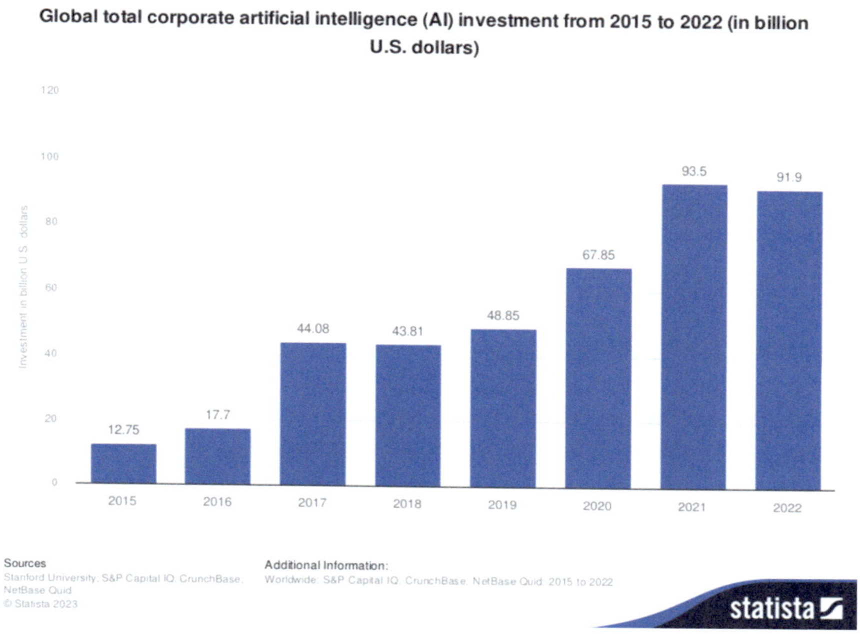

Fig. 3 Global total corporate AI investment (2015–2022). *Source* Statista (2023b)

the delicate balance in society? This inquiry gains particular relevance in the context of crises, especially during times of polycrisis, where the allocation of resources and investments plays a pivotal role in shaping economic and social outcomes.

Beyond the potential security risks, the swift development and deployment of AI technologies give rise to profound ethical concerns. Matters such as biased algorithms, privacy erosion, and the malevolent use of AI not only pose immediate threats but also have the potential to foster societal distrust. In extreme cases, these ethical quandaries may culminate in crises, underscoring the critical need for a balanced and conscientious approach to AI advancement.

The widespread adoption of AI, particularly in automation-heavy industries, may lead to job displacement for certain sectors of the workforce that will necessitates different active and passive employment and labour market policies (Hadjitchoneva 2023). If not managed properly, this could result in economic inequalities, social unrest, and potential crises. It is estimated that 14% of jobs could be automated in OECD countries, while another 32% could undergo significant changes. The Goldman Sachs (2023) review of the potential macroeconomic implications of AI technologies shows that they clearly have the potential to significantly shake up labor markets. Projections predict that nearly 1/5 of jobs globally will be automated with AI technologies, about 2/3 of jobs will be exposed to some degree of automation in Europe and the US, and generative AI could displace up to 1/4 of current job tasks (with up to 300 million full-time jobs globally when extrapolated). Administrative

(46%) and legal (44%) professions are particularly exposed. Conversely, physically intensive occupations such as construction (6%) and maintenance (4%) are rated low in terms of exposure to technology. Analyses by OpenAI (the company that developed ChatGPT) indicate that 15% of all tasks for U.S. workers could be completed significantly faster at the same level of quality. However, this share increases to 47–56% when LLM is upgraded with software and other tools. One in 6 American workers would have at least a 10% impact of GPT technologies on their work, and one in 10 would have a 50% impact.

4 AI as Prerequisite for Stability

Stability requires proper administrative regulation and administrative control. The strategy and regulatory landscape for AI and its transparency is in a constant state of evolution as governments worldwide grapple with the intricate challenges and opportunities presented by AI. Across the globe, governments are introducing a diverse array of policies, strategies, and regulations to guide the development and deployment of AI technologies. Table 1 offers detailed examples of government initiatives from key regions, including Australia, Canada, China, the European Union (EU), Singapore, the United Kingdom (UK), and the United States of America (USA).

Table 1 systematise and communicates the global response to AI challenges. These initiatives shed light on the varied approaches and priorities in AI governance. Governments have also taken proactive measures by introducing multiple regulations on AI that address diverse sectors or specific aspects of AI deployment, e.g. the U.S. has various sector-specific regulations and guidelines that impact AI, especially in sectors like healthcare, finance, and transportation, China's plan for AI development covers various sectors, including technology, industry, agriculture, national defense, and more, the EU is building on implemented regulation, namely the General Data Protection Regulation (GDPR). This includes a multitude of AI-related non-regulatory initiatives such as ethical principles, guidelines, surveys, expert groups, high decision-makers' groups, projects, models, registers, case studies, and more, showcasing the comprehensive nature of global efforts in shaping responsible AI practices.

The effective implementation of AI regulations and policies often hinges on substantial financial support from the public sector. This support is vital for ensuring robust enforcement, compliance mechanisms, ongoing monitoring, and the overall success of AI governance initiatives. Public finances play a crucial role in various aspects, including implementation and enforcement, infrastructure development, research and development (R&D), education and training programs, public awareness and engagement efforts, innovation funds, incident response, crisis management, grants and support for SMEs, and ethical oversight.

While it remains challenging to provide a comprehensive assessment of global public spending on AI policies, specific cases highlight the commitment of governments. For instance, the Australian Government has invested approximately half

Table 1 AI policies, regulations and other initiatives worldwide

Country/ Union	AI Initiative	Main Scope	Main Addressed Field
Australia	Australia's AI Ethics Framework	• Guiding businesses and governments to responsibly design, develop and implement AI	Collaboration International Collaboration
	Australia's AI Ethics Principles	• Ensuring AI is safe, secure, and reliable by human-centred values, fairness, transparency and explainability, contestability, accountability, reliability and safety, and human, societal and environmental wellbeing	Non-regulatory Ethical
	Australia's AI action plan	• Making Australia a global leader in responsible and inclusive AI, • Lifting development and adoption of AI to create jobs and boost productivity, • Growing and attracting world-class talent and expertise, • Harnessing world-leading AI capabilities to solve national challenges and benefit all Australians, • Ensuring AI technologies are responsible, inclusive and reflect Australian values	Politic Economic
Canada	Pan-Canadian AI Strategy	• Aiming to position Canada as a global leader in AI research and development, • Focusing on research excellence and talent development, • Fostering innovation ecosystems	Economic R&D
China	New Generation AI Development Plan	• Introducing strategic objectives for competitiveness, • Planning to become a world leader in AI by 2030, • Emphasizing the development of key AI technologies, fostering innovation, and integrating AI across various sectors	Political Regulatory Economic
	Beijing AI Principles	• Guiding the research and development, implementation, and governance of AI • Emphasizing fairness, transparency, and accountability as main principles for the ethical use of AI	Societal Ethical

(continued)

Table 1 (continued)

Country/ Union	AI Initiative	Main Scope	Main Addressed Field
EU	EU AI Act (still debated)	• Regulating AI systems, • Introducing AI Risk Categories Classification (Unacceptable Risk, High Risk, Limited Risk and Minimal Risk) corresponding to AI Management (Prohibited, Conformity Assessment, Transparency and Code of Conduct), • Ensuring Ethical Use of AI technologies	Political Regulatory Ethical
	EU Coordinated Plan on AI	• Driving the development and deployment of AI across the EU, • Fostering collaboration between member states, • Promoting investments, • Ensuring adherence to shared ethical principles	Economical Ethical Ecologic Environmental International Collaboration
	EU Member States National AI Strategies	• Supporting AI development, • Increasing financial commitment to AI, • Introducing AI Public–Private venture capital, • Implementing different approaches, e.g. horizontal or sectors' oriented	From highly political umbrella strategies to operational strategies with concrete measures and budgets
Singapore	Model AI Governance Framework	• Guiding organizations in the responsible and ethical deployment of AI technologies, • Translating ethical principles into practical recommendations that organisations could readily adopt to deploy AI responsibly, • Guiding organisations to assess the alignment of their AI governance practices	Economic Ethical
	A Guide to Job Redesign in the Age of AI	• Guiding to job redesign by transforming jobs, charting clear pathways between jobs, clearing barriers to digital transformation, and enabling effective communication between employers and employees	Collaboration
UK	AI Sector Deal	• Driving growth in the AI sector by investing in research and development, • Promoting skills development, • Fostering collaboration between industry and academia	Political Regulatory Economic R&D

(continued)

Table 1 (continued)

Country/ Union	AI Initiative	Main Scope	Main Addressed Field
	Centre for Data Ethics and Innovation	• Providing independent advice to the UK government on the ethical use of AI and data-driven technologies, • Developing tools to give organisations confidence in AI and data-driven technologies, • Developing governance frameworks, guidance, and standards to ensure public trustfulness, • Demonstrating positive uses of data and AI to tackle global problems	Ethical
USA	Executive Order on AI	• Prioritizing AI development, • Maintaining U.S. competitiveness in AI technologies, • Focusing on research and development investments, • Caring for workforce development, • Establishing international collaboration	Political Regulatory Economic Social Ecologic Environmental R&D
	National AI Research and Development Strategic Plan	• Advancing AI research and development, • Emphasizing partnerships between government, academia, and industry	Economic R&D

Source Authors' elaboration based on Australian Government (2023a; 2023b), Government of Canada (2023), Stanford University (2023), Springer (2023), EC (2021a; 2021b), IAPP (2023), PDPC (2023), Government of UK (2023a; 2023b), White House (2023a; 2023b)

a billion dollars in AI since 2018 (Australian Government 2023b). This substantial investment aims to bolster Australia's AI capability through initiatives such as the creation of the National AI Centre and four AI and Digital Capability Centres. It also includes support for Australian businesses to collaborate with the government on piloting AI projects, the establishment of the Next Generation AI Graduates Program to train home-grown AI specialists, and the AI Opportunity initiative, offering co-funded competitive grants to incentivize AI practitioners to develop solutions for regional challenges in collaboration with businesses.

AI investments can stimulate economic growth by fostering innovation and creating job opportunities in the AI industry. As AI technologies advance, new roles and industries may emerge, contributing to overall employment and economic stability. AI implementation in businesses can enhance efficiency, automate processes, and improve productivity. This, in turn, may lead to increased competitiveness, economic output, and overall stability (Hadjitchoneva 2022; 2023).

Chui et al. (2023) estimate that the productivity impact of generative AI could add \$2.6 to \$4.4 trillion per year (in partial usage analyses) to the global economy, which would increase the impact of all AI by 15 to 40%. This estimate doubles if the effect of embedding generative AI in software is included. Labour productivity growth is expected to jump from 0.1 to 0.6% per year by 2040. When combining generative AI with all other technologies, labour automation could contribute 0.2 to 3.3 percentage points per year to productivity. These projections are conditional - they depend on the extent of technology adoption, the reallocation of workers' time to other activities, and support in learning new skills or occupational change.

5 Conclusion

In conclusion, our discussion emphasizes the importance of maintaining a cautious perspective when dealing with concepts such as competitiveness and AI. A careful and responsible analysis is crucial to avoid unintended and potentially harmful outcomes associated with an obsessive focus on these concepts.

The substantial investments in AI, whether by corporations or in start-ups, bring about a complex interplay between economic growth and crisis scenarios. On one hand, AI investments can serve as a catalyst for economic growth and stability, enhancing efficiency, productivity, and innovation. On the other hand, these investments introduce challenges related to job displacement, ethical concerns, and security issues, which, if not managed properly, could contribute to crises.

In summary, while AI investments hold the potential to drive economic growth and stability, a balanced approach is necessary. Proactive measures, including robust regulatory frameworks and ethical guidelines, are crucial to mitigate the associated risks. This balanced approach aims to maximize the positive impact of AI investments while minimizing the potential for crises related to job displacement, ethical considerations, and security challenges.

Furthermore, the examples discussed highlight that various countries are adopting comprehensive strategies for AI regulation, encompassing diverse sectors and applications. The overarching goal is to ensure the responsible and ethical development and deployment of AI technologies, aligning with societal values. The cautionary perspective, reminiscent of concerns raised about discussions on a country's economic performance, underlines the need for nuanced analysis to avoid misallocation of resources, trade tensions, and the formulation of misguided economic policies. This cautionary approach is integral to harnessing the benefits of AI while minimizing potential negative consequences.

References

AI-Ecosystem (2023) AI Economy size investment. https://www.ai-ecosystem.org/. Accessed 20 Nov 2023

Australian Government (2023a) Australia's artificial intelligence ethics framework, https://www.industry.gov.au/publications/australias-artificial-intelligence-ethics-framework. Accessed 24 Nov 2023

Australian Government (2023b) Australia's AI action plan, https://www.industry.gov.au/news/action-plan-artificial-intelligence-australia. Accessed 24 Nov 2023

Bucknall BS Dori-Hacohen S (2022) Current and Near-Term AI as a potential existential risk factor. In: Proceedings of the 2022 AAAI/ACM conference on AI, ethics, and society (AIES'22). https://doi.org/10.1145/3514094.3534146

Chomsky N (2023) Les fausses promesses de ChatGPT. Philosophie Magazine, Hors-Série IA, le mythe du XXIe siècle. https://www.philomag.com/articles/les-fausses-promesses-de-chatgpt. Accessed 20 Nov 2023

Cybersecurity Ventures (2023) 2022 official cybercrime. https://www.esentire.com/resources/library/2022-official-cybercrime-report. Accessed 8 Sept 2023

ENISA (2022) ENISA Threat landscape 2022. https://www.enisa.europa.eu/publications/enisa-threat-landscape-2022. Accessed 8 Sept 2023

European Commission (EC) (2021a) COM(2021) 206 final. Laying down harmonised rules on artificial intelligence (artificial intelligence act) and amending certain union legislative acts. https://eur-lex.europa.eu/legal-content/EN/TXT/?uri=celex%3A52021PC0206. Accessed 22 Nov 2023

EC (2021b) COM(2021) 205 final. Fostering a European approach to artificial intelligence, https://digital-strategy.ec.europa.eu/en/library/coordinated-plan-artificial-intelligence-2021-review. Accessed 22 Nov 2023

European Parliament (EP) (2022). Cybersecurity: why reducing the cost of cyberattacks matters. https://www.europarl.europa.eu/news/en/headlines/society/20211008STO14521/cybersecurity-why-reducing-the-cost-of-cyberattacks-matters. Accessed 8 Sept 2023

Funk P (2022) Artificial intelligence and cybersecurity implications for business management. J Int Sci Publ 16:252–261

Future of Life Institute (2023) Pause giant ai experiments: an open letter. https://futureoflife.org/open-letter/pause-giant-ai-experiments/#:~:text=We%20call%20on%20all%20AI,more%20powerful%20than%20GPT%2D4.&text=AI%20systems%20with%20human%2Dcompetitive,acknowledged%20by%20top%20AI%20labs. Accessed 21 Nov 2023

Government of Canada (2023) Pan-Canadian AI strategy. https://ised-isde.canada.ca/site/ai-strategy/en. Accessed 25 Nov 2023

Government of UK (2023a) AI sector deal. https://www.gov.uk/government/publications/artificial-intelligence-sector-deal/ai-sector-deal. Accessed 25 Nov 2023

Government of UK (2023b) Centre for data ethics and innovation, https://www.gov.uk/government/organisations/centre-for-data-ethics-and-innovation. Accessed 25 Nov 2023

Griffiths C (2023) The latest 2023 cyber crime statistics. https://aag-it.com/the-latest-cyber-crime-statistics/#:~:text=Potential%20losses%20to%20cyber%20crime,lost%20an%20estimated%20%246.9%20billion. Accessed 8 Sept 2023

IAPP (2023) PDPC Model AI governance framework, https://iapp.org/resources/article/pdpc-model-ai-governance-framework-second-edition/. Accessed 25 Nov 2023

Janzwood S, Homer-Dixon T (2022) What is a global polycrisis?. discussion paper 2022–4. Cascade Institute, https://cascadeinstitute.org/technical-paper/what-is-a-global-polycrisis/. Accessed 25 Nov 2023

Hadjitchoneva J (2023) The future of work: transformations and implications for economic growth in the 21st century. https://doi.org/10.13140/RG.2.2.28961.58728

Hadjitchoneva J, Mandadjiev P (2023). Managing Cybersecurity : the Mission Becomes Urgent. https://doi.org/10.13140/RG.2.2.31478.16961

Hadjitchoneva J (2022) Entrepreneurship, innovations and artificial intelligence in bulgaria—dynamics and management. Sofia: UNWE
Hadjitchoneva J, Ruff C, Ruiz M, Matheu A (2021) El uso de la inteligencia artificial en el sector minorista online: el caso del supermercado eBag. 2021 16th Iberian Conference on Information Systems and Technologies (CISTI), https://doi.org/10.23919/CISTI52073.2021.9476484
Hadjitchoneva J (2020a) L'utilisation de l'intelligence artificielle dans le secteur des ventes au détail en ligne : une étude de cas. Dans : L'Intelligence artificielle et l'Humain. La Rochelle : Excelia. 140–160
Hadjitchoneva J (2020b) L'intelligence artificielle au service de la prise de décisions plus efficace. In Escobar C. R., Redslob A. and Malaga K. (Sous la direction de). (2020). Pour une recherche économique efficace. Santiago: Ediciones UBO. 149–170
Hadjitchoneva J (2019) Efficient Automation of Decision-making Processes in Financial Industry: Case Study and Generalised Model. Enterprise Engineering and Knowledge Management 2019. Selected Papers of the XXII International Conference "Enterprise Engineering and Knowledge Management (EEKM 2019), 2413, 42–52
Krugman P (1994) Competitiveness: a dangerous obsession. Foreign Aff 73(2):28–44. https://doi.org/10.2307/20045917
PDPC (2023) Singapore's approach to AI governance, https://www.pdpc.gov.sg/ai. Accessed 25 Nov 2023
Springer (2023) Beijing AI Principles, https://link.springer.com/content/pdf/https://doi.org/10.1007/s11623-019-1183-6.pdf. Accessed 24 Nov 2023
Stanford University (2023) DIGIChina: China's 'new generation artificial intelligence development plan' 2017, https://digichina.stanford.edu/work/full-translation-chinas-new-generation-artificial-intelligence-development-plan-2017/. Accessed 24 Nov 2023
Statista (2023a) Impact of AI on GVA growth rates worldwide by industry in 2035. https://www.statista.com/statistics/757763/worldwide-artificial-intelligence-impact-on-gross-value-added-by-industry/. Accessed 25 Nov 2023
Statista (2023b) Global total corporate AI investment from 2015 to 2022. https://www.statista.com/statistics/941137/ai-investment-and-funding-worldwide/. Accessed 25 Nov 2023
Trenker J, Menon SS, Tavva, N, Owens T., Tas O, Blumtritt C (2023) Unleashing artificial intelligence's true potential: How generative AI could empower innovation, redefine productivity, and transform the workforce. Statista, Insights Compass 2023
White House (2023a) Executive order on the safe, secure, and trustworthy development and use of artificial intelligence. https://www.whitehouse.gov/briefing-room/presidential-actions/2023/10/30/executive-order-on-the-safe-secure-and-trustworthy-development-and-use-of-artificial-intelligence/. Accessed 23 Nov 2023
White House (2023b) National artificial intelligence research and development strategic plan 2023 update. https://www.whitehouse.gov/wp-content/uploads/2023/05/National-Artificial-Intelligence-Research-and-Development-Strategic-Plan-2023-Update.pdf. Accessed 23 Nov 2023

The Covid Pandemic as a Paradigm Shift and the Emergence of a New VUCA Situation

Olivier Dumontel, Marian Šuplata, Frédéric Delaneuville, and Alexandra Sekáčová

Abstract According to many economists, the coronavirus crisis, which appeared at the start of 2020, seems to belong to the famous "Black Swan" evoked in Nassim Taleb's book (The Black Swan 2007) (Taleb 2020). Author Nassim Taleb preferred to evoke a situation closer to the theory developed by Michele Wuckeren (Wucker 2016). Evoking not an improbable event, but rather a "Gray Rhino" event relating to a highly probable threat, also having a considerable impact, due to a poor assessment of the risks involved. Whatever the case, unpredictable or predictable, this historic event, like the Spanish flu between the two world wars, called into question the world order inherited from 1945 (Dalio 2021). Human history has always been marked by other epidemics, such as the Spanish flu at the beginning of the twentieth century, or the Great Black Death which severely affected and scarred Europe in the fourteenth century (Pascal Boniface). Many stories, such as Albert Camus' "The Plague" (Boniface 2020), tell of events that plunged humanity into confusion, denial, panic, disarray and a sense of powerlessness. It is therefore through a "weakened" system (Pascal Boniface) (Dumontel 2022). that we can focus our analysis on the questioning of a global model that operates on a just-in-time basis, but has become VUCA (Chaminade 2022, i.e. highly volatile due to the many changes, uncertain due to the level of unpredictability, complex due to the interconnection and interdependence of countries, and finally ambiguous because the superstructure is no longer in line with the infrastructure (Marx 1999). The big question is whether this event will lead our Western societies, particularly France and Slovakia, towards a more local,

O. Dumontel · M. Šuplata · F. Delaneuville (✉) · A. Sekáčová
Faculty of Management, Comenius University Bratislava, Bratislava, Slovakia
e-mail: frederic.delaneuville@fm.uniba.sk

O. Dumontel
e-mail: olivier.dumontel@fm.uniba.sk

M. Šuplata
e-mail: marian.suplata@fm.uniba.sk

A. Sekáčová
e-mail: sekacova18@uniba.sk

G. Pellat et al. (eds.), *Data-Centric Business and Applications*, Lecture Notes on Data Engineering and Communications Technologies 258,
https://doi.org/10.1007/978-3-031-95068-1_7

fairer, more ethical and, of course, more resilient model, i.e. from a "just in time" model to a "just in case" model.

1 Introduction

1.1 Relevance

In 1947, Albert Camus wrote that "the only way to bring people together is to send them the plague" (Camus 1972). This posthumous success was given a real boost by the health crisis, which saw sales of Camus's La Peste regain a certain readership in several European countries, notably France (Edistat 2025) and Italy (Repubblica 2020), as shown by Edistat's March 2020 study.

In fact, since the start of 2020, it's not plague that's invading the world, but a new coronavirus (MEYER, Fever 2017), originating in China, bringing with it a health crisis of exceptional proportions, affecting all countries across continents.

In 1816, after reading the book written by the English ambassador Lord Macartney (Macartney 1792), Napoleon had, according to Alain Peyrefitte, the famous prophetic phrase "Quand la Chine s'éveillera, le monde tremblera" ("When China awakens, the world will tremble") (Peyrefitte 1980). Since the fall of the USSR in the late 80 s and early 90 s, China has grown considerably with the phenomenon of globalization, and now occupies a dominant position in international trade. China has become the world's factory, and you only have to follow certain indicators such as the BDI (Baltic Dry Index) (Economics and Index 2023) to see that China produces and exports in huge quantities everything Western countries need.

The world has become, as Marshall Mc Luhan put it, a global village (Mc Luhan 1992), and our economies have become highly interdependent and interconnected.

Faced with this global pandemic, which has upset the balance of power and relations between countries, each country has had to put in place an appropriate response strategy, bearing in mind that the constant evolution of the pandemic has necessitated adjustments throughout the crisis. This situation could not do without strong international cooperation (Commission and Coronavirus 2020).

Public authorities at all levels have therefore been called upon to provide relevant emergency solutions on an unprecedented scale, which have given rise to new long-term challenges (SCHWAB - MALLERET, The Great Reset, Covid-19, July 2020) (Schwab and Malleret 2020). The pandemic has truly changed the way we see the world, and its repercussions affect all areas of our shared daily reality (TOUATI 2020; Touati 2020a).

Marc Touati points out in his book that, by the end of 2019, some countries had still not recovered their early 2008 GDP levels. He proposes three hypotheses varying in the recovery of French GDP in the second half of 2020 and the average annual growth of French GDP from 2021 onwards. In the pessimist or realistic hypothesis (with 12% recovery and 1,5% average annual growth of French GDP) he estimates

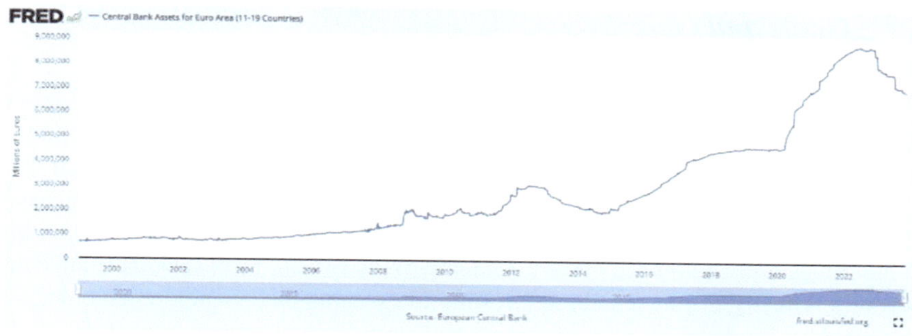

Fig. 1 ECB money creation between 2000 and 2022. *Source* European Central Bank, Euro Reference Exchange Rates (2025)

that the French GDP will retourn to its level of the fourth quarter of 2019 in the first quarter of 2030. In the optimistic hypothesis (with 12% recovery and 2,5% average annual growth of French GDP) the GDP should retourn to its level in the third quarter of 2025 (Touati 2020b). In the case of French GDP, given that its level in the fourth quarter of 2020 was only 10.5% above that of 2008, the fall in the first half of 2020 quickly wiped out this small recovery.

In addition to this, the 2009 crisis, which predates the health crisis, could have serious consequences for central banks such as the ECB, which saw its money creation increase considerably between 2006 and 2022, by a factor of 9 (see Fig. 1). Moreover, we can see from the recent statement by Philip Lan (Lane 2023), the ECB's chief economist, that the latter is currently seeking to normalize its balance sheet and thus reduce its size by withdrawing money from the economic system.

This is what they call "monetary sterilization" (Economics 2025), which could cause tension and stress on the financial market in the months ahead, adding to other difficulties, as Andrea Enria, Chairwoman of the ECB's Prudential Supervisory Council, also pointed out in November 2023 (Reuters 2023a).

From a global perspective, but also from an economic, geopolitical, social and cultural one, we're going to look at how we perceive and manage the situation in a world that has become volatile, uncertain, complex and ambiguous, confronting us with the greatest societal challenge of all, and one that will most certainly leave deep scars on the history of mankind.

This health crisis should enable us to draw lessons in terms of management, so as to better perceive the positive aspects of coordination and communication, but also to better apprehend the risk of a rise in national egoism (Voy-Gillis 2020a) or a possible deep social divide that could be expressed at the next elections, as we saw for example during the last legislative elections in France with the rise of extremist political parties, and in Slovakia with the return to power of Robert Fico, who came to power at the head of a coalition with the Slovak National Party in October 2023.

1.2 Goals and Objectives

The main aim of this document is threefold: firstly, to provide a brief overview of the global effects and impact of the Covid pandemic, which has spared no country, bearing in mind that this health crisis has had consequences at all levels of our societies, in both developing and developed countries; secondly, to make a comparison with previous health and economic crises, in order to gain a better understanding of the issues and challenges currently facing Western countries such as France and Slovakia. Thirdly, it may also be interesting to examine the new approach emerging for tomorrow's world, namely to steer our economies and societies towards greater sobriety and resilience.

2 The Covid Crisis: A Global Impact

As Figs. 2 and 3, below, show, the Covid pandemic in 2020 has highlighted major threads in the global context, with a heightened risk of intensified national and international confrontation, as well as a heightened sense of urgency in the face of some critical global issues (Forum économique mondial, Rapport sur les risques mondiaux 2021).

This summary table clearly shows the global impact of the health crisis on health, psychology, behavior, society, politics, economics and finance, as well as its impact on globalization and its transformation, on the job market and its transformation (Valušová and Lacová 2025) (online work, "uberization" (Marquet 2016) of the

Fig. 2 Overall consequences of the Covid health crisis. *Source* own elaboration, global impact of the pandemic, Dumontel, 2022 (2021)

Map of global Risk Interconnections in 2020:

Environmental Risks
- Natural Disasters
- Extreme Weather Conditions
- Human-made Environmental Disasters
- Loss of Biodiversity
- Climate Action Failure

Societal Risks
- Social Instability
- Involuntary Migration
- Food crises
- Water Crises
- Urban Planning Failure

Geopolitical Risks
- National Governance Failure
- Global Governance Failure
- Weapons of Mass destruction
- Interstate Conflicts
- Terrorist Attacks
- State Collapse

Economic Risks
- Unemployment
- Financial Failures
- Budget Crises
- Unmanageable Inflation
- Energy Price Shocks
- Illicit Trade
- Critical Infrastructure Failures
- Speculative Bubbles
- Deflation

Technological Risks
- National Governance Failure
- Cyberattacks
- Data Fraud or Theft
- Information Infrastructure Failures

Fig. 3 Overall consequences of the Covid health crisis. *Source* Compilation based on the World Economic Forum's Global Risks 2020 Report 2021, Olivier Dumontel, 2022 (2021)

world), and also on the environment (Peya 2020) (as an indicator of inequalities), with a growing awareness of the excesses of globalization and mass consumption.

Many questions are being raised by the historic change initiated by the Covid pandemic. To a certain extent, we are faced with a major change of a magnitude that will have to be considered in the years to come, but which is reminiscent in many respects of other "black swans" or upheavals that have arisen due to a context that is at once political, strategic, economic and financial, but also sometimes social or religious. Didn't France undergo major societal change in the late 60 s and early 70 s, which was also marked by the end of the "trente glorieuses"? Didn't the Middle East experience a serious crisis that turned into an armed conflict known as the Yom Kippur War, the economic consequence of which was an embargo on petroleum products?

Unlike previous crises, whether or not they were "black swans", "gray swans" or "gray rhinos", the Covid pandemic truly affected a very wide range of areas relating to our societies and organizational systems, plunging us into a single certainty: that our world had changed (Boniface 2020), and that we were only at the beginning of a new world order.

As the report by the World Economic Forum (Fig. 3) and Klaus Schwab's book "The Great Reinitialization" clearly show, we are facing a cyclical event with structural consequences that will lead us to a new normal on a macroeconomic level, linked to economic, societal, geopolitical, environmental and technological factors.

But also, a second aspect of this new normal at the micro level, on specific industries and companies but also at the individual level (Schwab and Malleret 2020).

Indeed, the question is: are we ready for the post-Covid world? Will the citizens of the global village be more resilient, as French President Emmanuel Macron hopes? Will the French and Slovaks who belong to this global village via the European Union accept the sobriety that the French president (Monde and Appel 2022) believes is needed in the face of the changes we are experiencing?

Epidemics thus provoke a form of terror that is deeply inscribed in the collective memory by a kind of "atavism". The definition of an epidemic given by the World Health Organization (WHO) (Définition 2022) is both broad and vague, since it states that an epidemic is a disease acquired by a relatively large number of people in a given region over a relatively short period of time. According to this definition, an epidemic is a phenomenon linked to diseases affecting a large number of people. This means that epidemics are not limited to infectious diseases. In fact, according to scientists such as Laurent Toubiana, Professor Christian Perronne and Professor Delfraissy, who chaired the Covid-19 scientific advisory board during the early days of the health crisis, and many others who have taken an interest in the Covid-19 pandemic, when an epidemic affects a very large area, or even the whole world, we speak of a "pandemic".

This pandemic has had a global impact in particular because of the history of our humanity which has of course already experienced numerous crises, but which is currently very globalized, interconnected and also interdependent. We can note in particular the post-covid phenomenon that certain economists like Jacques Sapir evoke when speaking of "deglobalization" which can be seen in particular with the latest events in the Sahel where France has started to lose its influence (Voxafrica 2022), but also with the strengthening of the BRICS since the accession of new countries (Information Portal and News—Membership 2022) which shows in this new reorganization of the world a break with the Western unipolar approach and the birth of a multipolar structure where China, India but also African countries play an essential role. The crisis in Ukraine accompanied by that of energy highlights, in the same way as during the pandemic with masks manufactured in China, the difficulties of countries, particularly Western ones, regarding supplies and the high risks of shortages on certain products such as medicines. The European Union is hit by high and lasting inflation having multiple causes but also by a shock in energy prices without forgetting its exposure due to its strong dependence on rare materials necessary for its industry, or even critical metals such as titanium, aluminum or nickel. We know, for example, that the European Union (Euractiv 2023) continues to buy these critical metals directly from Russia despite Western sanctions linked to the war on the European aeronautical sector. Whether it is the polarization on the management of the pandemic crisis or that linked to energy, as Klaus Schwab clarified in the most recent report of 2023 (Forum économique mondial 2023), the risk of fracture of society is very real due in particular to a loss of confidence of the population with its national or supranational leaders.

A recent study carried out by Ipsos Sopra/Steria in September 2023, for Le Monde, Cvipof, the Jean Jaurès Foundation and the Institut Montaigne, allows us to better perceive, for example, the perception of the French who have been exposed since the start of the pandemic until today to a certain number of anxiety-provoking events such as of course the pandemic but also the conflict in Ukraine, inflation, recession, urban riots. According to the survey, 91% of French people find France violent (82 %*in*2022) believe that France needs a "real leader" to restore order (7 9% in 2022) (Montaigne and Fractures françaises 2023). The results of the study corroborate with the societal risks (social instability) presented in the 2020 World Economic Forum report.

As Ivan Krastev (Krastev 2020a) points out in his book which deals with the paradoxical post-covid-19 world, the pandemic is a phenomenon, whether predictable or not, which is fundamentally different from the previous crises which affected the beginning of the twenty-first century, namely that of September 11, 2001, the Great Recession of 2008 or even the refugee crisis of 2015. For Ivan Krastev, only one thing is certain in this world which has now become very volatile, uncertain, complex and ambiguous, the pandemic has profoundly transformed the world in which we live, particularly in Europe. Its analysis is currently taking on a completely different dimension with the high inflation which has mainly hit Western countries including Europe, the war in Ukraine which hides a terrible disaster in terms of human losses and also on an energy and economic level, without forgetting in certain European countries such as France which are shaken by strong opposition demonstrations against the leaders and even urban violence suggesting large-scale social fractures as explained by criminology professor Alain Bauer (Bauer 2023).

We can actually see in the 2023 report, that the risk of erosion of social cohesion and societal polarization is very high since it is located in 5th place according to the study carried out. Likewise, the geoeconomic confrontation or the cost of living which occupies first place are all factors that can accentuate major fractures in our societies. Likewise, the figures in the report allow us to compare the evolution of risks and show the interconnection of global and therefore global risks that are much greater than those we had in the 2020 report. The main risks include collapse of a systemically important supply chain, erosion of social cohesion and state collapse (World Economic Forum 2023).

We can indeed note a strong progression and deterioration in terms of social cohesion and social instability which can result in particular in the eruption of major protests such as that of the "Yellow Vests" in France who managed to constitute a group and a list for the European elections (Toute l'Europe 2019).

This movement of "Yellow Vests" has become, according to social sciences professor Michel Maffesoli (Maffesoli 2021a), an international phenomenon which has developed in particular in opposition to the management of the health crisis. Furthermore, one of the etymologies of the term crisis "krisis" is also the judgment passed by what is being born on what is dying. Maffesoli therefore associates the health crisis as being a modality of an ongoing societal crisis (Lacová 2022) and therefore of a much more profound paradigm shift.

Klaus Schwab in his book "The Great Reset"(Schwab 2020) explains that like most transformative events, crises have the ability to bring out the best and worst in us. Angel or demon: which will win? This is the question that the director of the World Economic Forum asked himself in 2020. According to him, if we trust history, natural disasters, like hurricanes and earthquakes, bring people together, while pandemics do the opposite, they push them apart. Indeed, according to Klaus Schwab, pandemics are events of longer duration, which necessarily arouses feelings of permanent distrust towards each other due to being rooted in a primitive fear of dying. It is also this rooting that the professor emeritus in sociology at the University of Paris-Descartes, Michel Maffesoli, speaks about, since he mentions in his book an increased risk of a revolt of people who are rooted in a tradition. According to the latter, the episode of the Covid epidemic shows that we are at a turning point, and that the progressive model is being called into question, leaving room for the emergence of new values.

Anaïs Voy-Gillis (Voy-Gillis 2020b) speaks in her book of an increased risk of social fractures and the return of nationalism in several member states due to a global context favorable to the coming to power of nationalist identity parties. This trend could accelerate with a certain number of post-covid scandals concerning in particular the Qatargate affair (Toute l´Europe 2020) or that concerning the supposed links between the president of the European Commission and the director of Pfizer, Albert Bourla, who twice refused to appear before the commission of inquiry of the European Parliament. This latest affair directly involving Ursula Van der Leyen (Monde and Affaire 2022) could further discredit the functioning of the institutions and could create an institutional crisis between the European Parliament and the European Commission in particular. All these post-pandemic Covid events, associated with the conflict on European soil opposing Russia to Ukraine and therefore to the West supporting the latter, to which is added the energy crisis which has crystallized due to this same conflict having led to a polarization of the world, all this could have repercussions in the next European elections which must take place in June 2024. We therefore see that the pandemic has not simply had an impact on health, but also on the level politics in all its meaning, national politics, European politics and even at the global level with a risk of change in the world order, as underlined by French President Emmanuel Macron in September 2022 (Libre and Guerre 2022). This phenomenon has of course been accentuated with the conflict in Ukraine and the Russian military intervention. We are also seeing an acceleration of this change with the strengthening of the BRICS or even more recently the polarization of the world around the conflict between Israel and the Hamas forces in Palestine.

For Michel Maffesoli, we are generally entering an era of uprisings which, according to his study (Maffesoli 2021b), could affect not only Europe but also the globalized world with the Covid pandemic as a trigger or catalyst. According to the professor emeritus of the Sorbonne but also a member of the Institut universitaire de France, "the shadow of Spartacus" is extending over the whole of society, since we are already witnessing a multiplication of uprisings such as for example in France that of the "Yellow Vests" which resurfaces while the restrictive measures linked to the pandemic have been removed in Europe, as China relaxes its "zero Covid" policy, and all this in a context of geopolitical tensions and of conflict escalations that directly affect European soil.

This "shadow of Spartacus" is in reality nothing other than the expression of popular anger in all its forms and modulations, that is to say a "servile uprising" (Maffesoli 2021b) when the people no longer feel in agreement with the elites in power. It is therefore a question of popular power which confronts the institutional power represented by a system that Michel Maffesoli describes as "overadministered". According to his analysis, Covid has accelerated a civilizational crisis, that of mass consumption linked to the commodification of society thanks to modern progressivism and globalization which accelerated after the fall of the Berlin Wall in the 90 s. History has many examples of crises which end a civilization, such as for example the "Antonine Plague" which is associated by historians with the decadence of Rome (Harper 2021) or the "Black Death" (Duhoux 2015) which accompanied the end of the Middle Ages and the beginning of the Renaissance.

Many historians, such as Henri Deleersnijder (Deleersnijder 2021), professor of history at the University of Liège, affirm that epidemics have of course marked Western history and marked our societies as well as their developments. As a result, we can almost be certain that the coronavirus pandemic will give researchers (historians, sociologists, journalists, etc.) the opportunity to collect life stories from the historical event that it triggered.

3 Covid, an Unprecedented Health, Economic and Social Crisis?

In November 2021, INSEE (INSEE 2022) (National Institute of Statistics and Economic Studies) affirmed that several epidemic waves of Covid would have caused heavy losses in terms of mortality on the French national territory and to another extent also at the global level. Some official sites (Our World in Data 2025) on the Internet have even made it possible to precisely follow the evolution of the pandemic and its impact on the world population. Among these sites we also note that of John Hopkins University (Center for systems Sciences and Engineering (CSSE) 2025) which notably made it possible to monitor in real time the evolution of the pandemic as well as the number of cases and then subsequently the number of people who will be vaccinated. "Sick", "hospitalized", "transition to critical care", "positive cases", "dead from Covid" and among the dead those who died in hospital or in retirement homes, EHPADs (accommodation establishments for dependent elderly people) or even at home, many indicators have been used in the media, disseminated to the general public, which can subsequently generate pathologies linked to the anxiety-provoking nature of the epidemic. Every day the French, the Slovaks and a large number of other countries could have access to the "epidemic figures" in their country.

The healthcare system has been hit hard due to spikes in hospitalizations linked to the disease. The epidemic has indeed had a direct impact on the population but also an indirect one due to the consequences for the health system, including the deprogramming of non-urgent care, a probable non-recourse for certain diseases such as cancer, etc. Healthcare activity in the city has also decreased considerably and at the same time mental health (UNICEF 2021) has deteriorated, particularly among young adults.

Covid has made it possible to test but above all to reveal the flaws in the health systems (Perronne 2021) of many countries, some of which acted in imitation (Bergeron 2020) of China when the latter implemented its policy of containment and then of "zero Covid" which had significant repercussions both on a human level but also on the country's economy and the entire global production chain. Yves Mény (Mény 2000) speaks on this subject of institutional mimicry which are ultimately more or less artificial grafts into foreign systems which can, however, lead to strong reactions of rejection.

According to Laurent Toubiana, doctor in physics but also epidemiologist and researcher at the National Institute of Health and Medical Research (INSERM), history repeats itself "bis repetita" (Toubiana 2022), epidemics have always marked the history of humanity. "Epidemics act directly on individuals by modifying their state of health [...] and are felt as a tragedy within the affected populations" specifies Laurent Toubiana in his book.

4 Health Crisis: Towards a New Normal with More Sobriety and Resilience "just in Case"?

4.1 A Questioning of the "just in Time" System?

As we have seen, the model of globalization developed at the end of the last century, designed and built by global manufacturing companies in search of cheap labor, products and components, has certainly found its limits in view of the pandemic situation.

Indeed, it has fragmented international production into increasingly complex parts and resulted in a system operating on the basis of "just in time", or "just in time", which has proven to be extremely lean and efficient, but also extremely complex and, therefore, very vulnerable.

Simplification could therefore be the antidote, according to some experts such as Patrick Artus (Artus 2020a) or Jacques Sapir (Sapir 2015), and should in turn generate greater resilience. This means that "global value chains", which represent around ¾ of all world trade, will inevitably decline. This decline will be compounded by the new reality that businesses that rely on complex, just-in-time supply chains can no longer assume that WTO tariff commitments will protect them from a sudden surge of protectionism somewhere.

They will be forced to prepare accordingly by reducing or localizing their supply chain and developing alternative production or supply plans to guard against prolonged interruption.

Any company whose profitability depends on the principle of a just-in-time global supply chain will have to rethink the way it operates and probably sacrifice the idea of maximizing efficiency and profits in the name of "security of supply" and resilience. Resilience will therefore become the main consideration for any business that seriously wants to protect itself against disruption. Production costs will inevitably increase, but this will be the price to pay for building resilience (Schwab and Malleret 2020).

Across the world, pressure to improve social protection and the salary levels of low-paid employees will increase.

In our post-pandemic world, increasing the minimum wage will most likely become a central issue that will be addressed through stricter regulation of minimum standards and more rigorous enforcement of existing rules.

There is therefore a good chance that companies will have to pay higher taxes with the "whatever it costs" policy such as that implemented in France for example.

The pandemic hit at a time when many diverse issues, from climate change activism to rising inequality to gender diversity and the #MeToo scandals, had already begun to raise awareness and reinforce the essential nature of stakeholder capitalism and ESG (Environmental, Social, Corporate Governance) considerations in today's interdependent world. The crisis will have created, or reinforced, a keen sense of responsibility and urgency on most issues relating to ESG (Environmental, Social, Corporate Governance) strategies. But other issues, such as consumer behavior, the future of work and mobility, and supply chain responsibility, will almost certainly move to the forefront of the investment process and become an integral part of due diligence.

4.2 *Towards a More Resilient World?*

Some behavioral changes observed during lockdowns are unlikely to be fully reversed in the post-pandemic period and some may even become permanent. Indeed, some consumption models could find long-term trend lines. Others will undoubtedly accelerate, such as online services. Some could be postponed, such as purchasing a car, while new permanent consumption patterns could emerge, such as purchases associated with greener mobility. During the lockdown period, many consumers had to learn how to do things for themselves (bake their own bread, cook from scratch, cut their hair, etc.) and felt the need to spend prudently.

To what extent are these new habits and forms of "Do it Yourself" and self-consumption taking root in the post-pandemic era? How will this impact the professional world and private life? We can already observe notable changes in consumer behavior which shows that this phenomenon is widely popular, for example in France. The IPSOS study (Ipsos 2021) carried out in February 2021 for the Leclerc group shows, for example, that 40% of consumers are adopting this new habit in order to save money.

Resilience, like any good practice, begins at home. Thus, we can reasonably assume that, in the post-pandemic era, we will all become more aware of the importance of our physical and mental resilience. The desire, driven by a greater need, to feel good physically and mentally and the need to strengthen our immune systems means that wellness and those sectors of the wellness industry able to help us achieve this will come out on top.

Additionally, the role of public health will evolve and expand. Well-being must be approached holistically; we cannot feel good individually in a world that is suffering. Therefore, taking care of yourself, an equivalence that strongly supports the promotion of the principles we talked about previously, such as stakeholder capitalism, the circular economy and ESG strategies.

At the business level, where the effects of environmental degradation on health are increasingly evident, issues such as air pollution, water management and respect for biodiversity will become paramount. Being "clean" (Schwab and Malleret 2020) will be an imperative for the industry as well as an imperative necessity imposed by the consumer.

A single change in attitude will have many different consequences, each with a particular effect on a specific industry, but ultimately impacting many industries through chain reactions.

As illustrated from the HBR-France, an increase of time spent at home can have significant and broad consequences on various sectors. The potential growth can be seen in teleworking, which drives demand for home offices, digital coordination and furnishing, while business travel and traditional office space are expected to decline. A shift in behavior can have a cascading effect on various industries and market sectors (France 2020).

4.3 French Government Resilience Plan

In order to counter the consequences linked firstly to the pandemic which is still affecting part of the world (Asia) and could continue in the years to come, but also to the war in Ukraine and the sanctions taken by France and its partners against Russia, the Government has developed an economic and social resilience plan (Gouvernement 2022).

This resilience plan has a dimension of sovereignty, independence, autonomy, ability to do it yourself (French Government 2022). Managing or managing also means knowing how to anticipate in times of great uncertainty and instability. The French Prime Minister therefore outlined the main points of this resilience plan which affects businesses as much as individuals. This includes supporting and assisting companies affected by the measures taken and inflation, facilitating the search for alternative outlets for exporting companies, securing supplies, strengthening food and energy sovereignty, and strengthening the cyber security of business administrations and infrastructure.

According to a comparative study of resilience carried out by "FM-Global Resilience Index" (FMGlobal-Resilience Index 2021), France is positioned in 19th place (cf. Fig. 4) and Slovakia in 31st place (cf. Fig. 5) of the ranking of OECD countries. Resilience is measured based on 12 criteria which are, for example, the rate of urbanization, productivity, political risks, exposure to natural risks, cyber risks, the fight against corruption, etc.

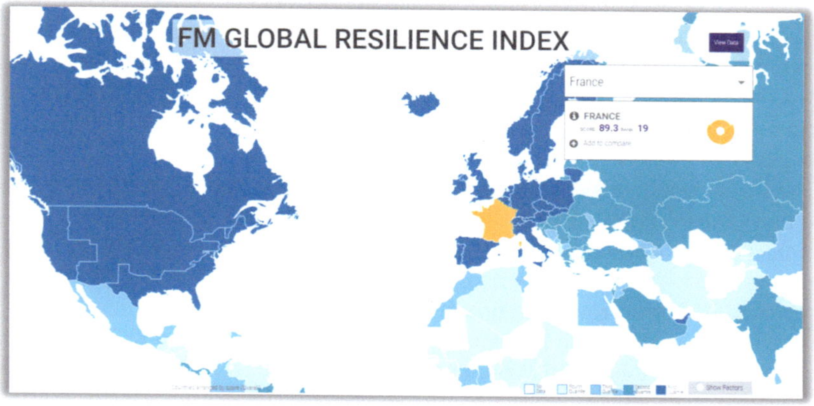

Fig. 4 France's resilience level. *Source* FM Global-Resilience Index 2021

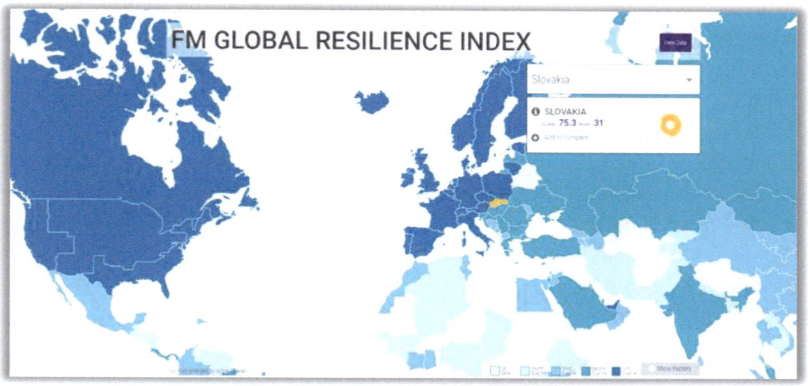

Fig. 5 Resilience level of Slovakia. *Source* FM Global-Résilience Index (2021)

5 Conclusions

Our world is weakened due to different events which can be referenced by Nassim Taleb's "black swan", but between the subprime crisis which led central banks to produce a lot of currencies (Quantitative Easing) (Schuman and Quelle 2015). This weakening does not date from today as Maurice Allais (Allais 1999) had already written and explained in 1999, just before the Internet crisis of 2001. He described the fundamental instability of our financial and monetary system, excessive debt, the factors of propagation of crises or even massive speculation which does not necessarily nourish the real market. The Italian economist and sociologist, from the end of the nineteenth century, Vilfredo Pareto (Pareto 2014) said that history never repeats itself identically but that it is certain that it always repeats itself in certain parts qualified as main.

The Covid health crisis has amplified economic and financial problems, to which have been added the shock of Brexit for the EU, the war in Ukraine which also has an impact on inflation and the shortages that we are currently experiencing, without forgetting environmental policy and global warming. The latter also has an impact on the world of tomorrow accompanied by a questioning of globalization and the entire production chain that has developed. We also have a real risk of recession or even stagflation (Monde 2022) which could soon affect our countries due to their fight against high inflation, but also the energy crisis which has, for example, hit the German economy hard (Reuters 2023b). We are also witnessing a currency war being played out between the petrodollar and the petroyuan (Pétroyuans 2023) in particular, accompanied by a new phenomenon of progressive dedollarization (Feket 2013) of the world. In addition, there is a high risk of social explosion in many countries and even famine due to the increase in food prices. But also, unfortunately, more recently the conflict in Ukraine which pits the Western world against Russia and certain of its allies, often members of the BRICS, or the Israeli-Palestinian tensions which could produce a domino effect on the rest of the world (Plessis 2002). Since Russia's military intervention in Ukraine and the numerous Western sanctions against its economy, the latter seems to have definitively turned its back on the West to give in to the anti-Western inclinations of a resurgent pan-Slavism and towards an alliance with Islam which today sees an awakening of the umma (Plessis 2004), whether Shiite or Sunni.

The gloomy picture we are witnessing portends great changes in terms of global reorganization.

We may have to move from a world that operated under a system that was able to develop with globalization and therefore often deflationary, due to relocations, robotization, computerization, or even "Robolution" (Commission européenne 2020), to a world that could be sustainably inflationary, with a new distribution of added value in global production chains, which would, for example, be more favorable to China. The risk is to perhaps witness, if tensions persist between the West and Russia but also with China, a return of a new world order breaking with the one we knew until now. We could witness a scenario of violent and brutal deglobalization (Artus 2020b) with a crushing of Western growth and the need to change the economic model, particularly for all companies that represent the values of globalization. Covid-19 has accelerated this trend of deglobalization which had been initiated in particular with the Great Recession of 2008–2009 while at the same time revealing the limits of renationalization (Krastev 2020b). The change is brutal and radical because since the 90 s, following the fall of the USSR, very efficient supply chains were then described as happy globalization (Minc 1999). Today, it is the image of a strong State that President Emmanuel Macron (Info and Macron 2020) praises but also a stronger Europe in the face of the many challenges that await it (Monde et al. 2020). However, despite the return to a national and populist logic as for example in Slovakia which was the most plausible response to a crisis of this magnitude, a more united Europe and Brussels institutions invested with additional powers in the event of an emergency could well prove, in a world lacking leadership and divided, the only realistic solution to face the challenges of tomorrow.

One thing is certain, after having experienced different crashes, such as that of 1987, the recession of 1991–1993, the Asian crisis of 1997, the Internet crash of 2000, the attacks of September 11, 2001, the Enron and WorldCom affairs, the 150 dollar barrel, the subprime crisis, the bankruptcy of Lehman Brothers or even Brexit then an epidemic followed by a crisis historic economic-social crisis and an energy crisis against a backdrop of global geopolitical tensions, we must do everything so that this global reset, described by the World Economic Forum, helps States to be more responsible, less expensive and more concerned with the good use of public money (Touati 2020c). We must do everything to ensure that this reset and this new world order are an improvement to our economic, financial, social and societal system. The European Union has its place in this great chessboard, as Jean Monnet said "Europe will be made in crises and it will be the sum of the solutions brought to these crises" (Grin 2013).

The world is on hold and what is happening now is historic.

Recognitions

This document was created as part of the project linked to the 29th conference of the ERECO-PGV network at the Faculty of Economics of Matej Bel University in Banská Bystrica in Slovakia and which consists of dealing with the different crises to which Europe is currently exposed.

References

Allais M (1999) La crise mondiale aujourd'hui : Pour de profondes réformes financières et monétaires, In: Éditions Clément Juglar, Paris. ISBN 2908735113

Artus P (2020) L'économie post-Covid: Les huit ruptures qui nous feront sortir de la crise, In: Éditions Fayard, Paris. ISBN 978–2–231–717–616

Artus P (2020) L'économie post-Covid : Les huit ruptures qui nous feront sortir de la crise, In: Éditions Fayard, Paris. ISBN 978–2–231–717–616

Bauer A (2023) Le risque de guerre civile en France, In: Le Figaro. https://www.lefigaro.fr/vox/societe/alain-bauer-le-risque-de-guerre-civile-n-est-pas-a-ecarter-d-un-revers-de-main-20230714. Accessed February 20, 2025

Bergeron H (2020) Covid-19: Une crise organisationnelle, In: Les Presses de Sciences Po, Paris. ISBN 2724626656

Boniface P (2020) Géopolitique du Covid-19: Ce que nous révèle la crise du coronavirus, In: Éditions Eyrolles, Paris. ISBN 978–2–416–00056–0

BSI Economics, La stérilisation des interventions des banques centrales: Concept, modalités et interrogations dans le cadre de l'OMT de la BCE, https://www.bsi-economics.org/38-la-sterilisation-des-interventions-des-banques-centrales-concept-modalites-et-interrogations-dans-le-cadre-de-l%EF%BF%BDomt-de-la-bce, Accessed March 27, 2025

BRICS Information Portal, News—Membership, https://infobrics.org/news/?search=membership. Accessed March 05, 2025

Camus A (1972) La Peste, In: Éditions Folio, Paris, 264 (1972). ISBN 2070360423

Center for systems Sciences and Engineering (CSSE) at the john Hopkins University, COVID-19 Dashboard by the Center for Systems Science and Engineering (CSSE) at John Hopkins University, https://gisanddata.maps.arcgis.com/apps/dashboards/bda7594740fd40299423467b48e9ecf6, Accessed February 14, 2025

Chaminade B (2022) Management VUCA : Volatilité, Incertitude, Complexité, Ambiguïté : Manager dans le nouveau normal, In: Le Mans, Éditions Gereso. ISBN 979–1039701501

Commission européenne, Bâtir l'avenir numérique de l'Europe, (2020)

Courrier International, Les Pétroyuans remplaceront-ils les Petrodollars au Moyen-Orient? (2023) https://www.courrierinternational.com/article/transactions-les-petroyuans-remplaceront-ils-les-petrodollars-au-moyen-orient. Accessed March 1, 2025

Dalio R (2021) Principles for dealing with the changing world order: why nations succeed or fail, In: Simon & Schuster Ltd, London. ISBN 1471196690

Deleersnijder H (2021) Les Grandes épidémies dans l'histoire, quand la peste, grippe espagnole, coronavirus façonnent nos sociétés. In: Mardaga, Bruxeles. ISBN 9782804709426

Duhoux J (2015) La Peste noire et ses ravages : l'Europe décimée au XIV siècle, In: Éditions Grands Evènements, Bruxelles. ISBN 2806259843

Dumontel O (2022) La littérature comme témoignage sociologique et culturel en temps de crise sanitaire, In: Romanistica Comeniana 2. ISSN 2585–8483

Du Plessis L (2002) La 3ème guerre mondiale a commencé, In: Éditions Jean-Cyrille Godefroy, Paris. ISBN 2–86553–158–9

Du Plessis L (2004) Islam Occident, la guerre totale, In: Éditions Jean-Cyrille Godefroy, Paris. ISBN 2–86553–176–7

European Central Bank, Euro Reference Exchange Rates. https://www.ecb.europa.eu/stats/policy_and_exchange_rates/euro_reference_exchange_rates/html/eurofxref-graph-usd.en.html. Accessed February 27, 2025

European Commission, Coronavirus: La Commission soutient la recherche avec 117 millions d´euros, European Comission Press Corner (2020). https://ec.europa.eu/commission/presscorner/detail/fr/ip_20_604. Accessed March 10, 2025

Edistat (2025) Statistiques de l'édition française du livre, https://www.edistat.com. Accessed January 27, 2025

Euractiv, Stockages stratégiques de métaux critiques : l'impensé européen (2023), https://www.euractiv.fr/section/energie-climat/news/stockages-strategiques-de-metaux-critiques-limpense-europeen/. Accessed March 10, 2025

Feket A (02013) Retour au standard or, In: Éditions Jardin des Livres, Paris. ISBN 2914569890

FMGlobal-Resilience Index (2021). https://www.fmglobal.com/research-and-resources/tools-and-resources/resilienceindex/explore-the-data/?&cr=FRA&sn=ex. Accessed March 10, 2025

Fondation Robert Schuman, Quelle sortie pour les politiques d'assouplissement quantitative? (2015). https://www.robert-schuman.eu/questions-d-europe/374-quelle-sortie-pour-les-politiques-d-assouplissement-quantitatif

Forum économique mondial (2021) Rapport sur les risques mondiaux, https://www.weforum.org/publications/the-global-risks-report-2020/. Accessed March 13, 2025
Forum économique mondial, The Global Risks Report 2023: 18th Edition (2024)
French Government, Summary of the Different Objectives of the Economic and Social Resilience Plan (2022). https://www.gouvernement.fr/upload/media/default/0001/01/2022_03_dossier_de_presse_-_plan_de_resilience_economique_et_sociale_-_16.03.2022.pdf. Accessed January 10, 2025
France Info, Macron et Castex font le pari d'un État fort (2020), https://www.francetvinfo.fr/politique/jean-castex/gouvernement-de-jean-castex/macron-et-castex-font-le-pari-d-un-etat-fort_4046925.html. Accessed March 27, 2025
Gouvernement, Le plan de résilience économique et sociale (2022) https://www.gouvernement.fr/info-ukraine/plan-de-resilience. Accessed January 20, 2025
Grin G (2013) Jean monnet et les crises, In: Fondation Jean Monnet pour l'Europe, Lausanne. https://jean-monnet.ch/wp /uploads/2013/11/Article_GG_crisesrelances.pdf., Accessed March 19, 2025
HBR France, Passer plus de temps à la maison: Quelles implications potentielles? (2020) https://www.hbrfrance.fr/magazine/2020/12/32263-adaptez-votre-business-aux-nouvelles-realites/. Accessed January 27, 2025
Harper K (2021)Comment l'Empire romain s'est effondré ?, In: Éditions La Découverte, Paris. ISBN 2348037149
INSEE, Tableaux de l´économie francaise—Édition 2022 (2021). https://www.insee.fr/fr/statistiques/5432509?sommaire=5435421. Accessed February 6, 2025
Ipsos, Les Français économisent en moyenne 272 euros par an grâce au DIY (2021). https://www.ipsos.com/fr-fr/les-francais-economisent-en-moyenne-272-euros-par-grace-au-do-it-yourself. Accessed March 27, 2025
Institut Montaigne, Fractures françaises 2023: Une société désenchantée. https://www.institutmontaigne.org/publications/fractures-francaises-2023-une-societe-desenchantee. Accessed March 27, 2025
Krastev I (2020) Est-ce déjà demain ?, In: Éditions Premier Parallèle, Paris ISBN 978–2–85061–054–7
Krastev I (2020) Est-ce déjà demain ?, In: Éditions Premier Parallèle, Paris. 106, ISBN 978–2–85061–054–7
Lacová Ž (2022) How is digital exclusion manifested in the labour market during the COVID-19 pandemic in Slovakia?, In: Forum Scientiae Oeconomia. Dąbrowa Górnicza : Wydawnictwo Naukowe Akademii WSB, 10(2), 129–151. ISSN 2300–5947
Lane P (2023) Le bilan de la BCE doit diminuer sans excès, In: Reuters. https://www.reuters.com/article/bce-politique-lane-idFRL6N3CA0BH Accessed March 11, 2025
La Repubblica, Libri e coronavirus: da La peste di Camus a Cecità di Saramago (2020). https://www.repubblica.it/cronaca/2020/02/27/news/libri_coronavirus_peste_camus_cecita_saramago-249694071/. Accessed March 27, 2025
Le Monde, Affaire des SMS d'Ursula von der Leyen: Mauvaise gestion administrative, selon la médiatrice de l'UE (2022). https://www.lemonde.fr/international/article/2022/07/16/affaire-des-sms-d-ursula-von-der-leyen-mauvaise-gestion-administrative-selon-la-mediatrice-de-l-ue_6135022_3210.html. Accessed March 27, 2025
Le Monde (2022) Appel à la sobriété énergétique et aides financières ciblées. https://www.lemonde.fr/planete/article/2022/09/05/crise-de-l-energie-sobriete-bouclier-energetique-solidarite-europeenne-retrouvez-l-essentiel-des-annonces-d-emmanuel-macron-lors-de-sa-conference-de-presse_6140304_3244.html. Accessed March 13, 2025
Le Monde (2022) La banque Mondiale avertit des risques de Stagflation. https://www.lemonde.fr/economie/article/2022/06/07/la-banque-mondiale-avertit-des-risques-de-stagflation_6129243_3234.html. Accessed March 1, 2025

Le Monde, Emmanuel Macron souhaite une Europe plus forte en Méditerranée (2020) https://www.lemonde.fr/international/article/2020/07/13/emmanuel-macron-souhaite-une-europe-plus-forte-en-mediterranee_6046108_3210.html. Accessed January 15, 2025

Lord Macartney (1792) Voyage à l'intérieur de la Chine et en Tartarie

Midi Libre, Guerre en Ukrain, climat: Devant l´ONU, Macron livre sa vision d´un nouvel ordre Mondial naissant (2022), https://www.midilibre.fr/2022/09/20/guerre-en-ukraine-climat-devant-lonu-macron-livre-sa-vision-dun-nouvel-ordre-mondial-naissant-10556530.php. Accessed February 10, 2025

Maffesoli M (2021) L'ère des soulèvements, In: Les Éditions du Cerf, Paris. ISBN 978–2–204–14475–8

Maffesoli M (2021) L'ère des soulèvements, In: Les Éditions du Cerf, Paris. ISBN 9782204144759

Marquet, D, Rencker É (2016) Entreprise: Muter ou périr—Face à l'ubérisation du monde, In: Éditions de l'Archipel, Paris. ISBN 28098117936

Marx K, Engels F (1999) Manifeste du communisme, In: Éditions Flammarion, Paris (1999). ISBN 2080710028

Mc Luhan M (1992) The global village: transformation in world life and media in the 21st century, In: OUP USA, New York. ISBN 0195079108

Mény Y (2000) Politique de mimétisme institutionnel : La greffe et le rejet. In: Éditions L'Harmattan, Paris. ISBN 2–7384–2091–5

Minc A (1999) La mondialisation heureuse, In: Éditions Pocket, Paris. ISBN 978–2266083331

Our World in Data, https://ourworldindata.org/, Accessed February 13, 2025

Pareto V (2014) Manuel d'économie politique, In: Éditions Hachette Livre BNF, Paris. ISBN 2013523092

Perronne Ch (2021) Y-a-t-il une erreur qu'ils n'ont pas commise ?, In: Le Livre de Poche, n.d., Paris (2021). ISBN 978–2–253–07838–8

Peya M (2020) Le désastre écologique et sanitaire de la Covid-19: L'humanité à l'épreuve des crises multidimensionnelles, In: Éditions L'Harmattan, Paris. ISBN 978–2–343–20599–1

Peyrefitte A (1980) Quand la Chine s'éveillera, le monde tremblera, In: Éditions Fayard, Paris. ISBN 2213006717

Reuters, La BCE demande aux banques de prendre en compte une nouvelle baisse des prix immobiliers, In: Les Echos (November 7, 2023). https://investir.lesechos.fr/marches-indices/devises-taux/la-bce-demande-aux-banques-de-prendre-en-compte-une-nouvelle-baisse-des-prix-immobiliers-2027290, Accessed March 27, 2025

Reuters, Allemagne, contraction de l'économie de 0.1% au troisième trimestre (2023) https://www.reuters.com/article/allemagne-economie-pib-idFRKBN31U0IW, Accessed March 17, 2025

Sapir J (2015) La Démondialisation, In: Éditions La Découverte, Paris. ISBN 978–2707186331

Schwab K (2020) Covid-19: La grande réinitialisation, In: Forum Publishing, Geneva, 240. ISBN 978–2–940631–13–1

Schwab K, Malleret T (2020) Covid-19: La grande réinitialisation, In: Forum Publishing, Geneva. ISBN 2940631131

Taleb N (2020) Le cygne noir, In: Éditions Belles Lettres, Paris. ISBN 2251451366

Touati M (2020) Reset: Quel monde pour demain ?, In: Éditions Dunod, Paris. ISBN 979–1035904869

Touati M (2020) Reset, quel nouveau monde pour demain?, In: INSEE, ACDEFI projections, 38. ISBN 9791035 904869

Touati M (2020) Reset, quel nouveau monde pour demain ?, In: Éditions Bookelis, Paris, 205. ISBN 979–10–359–0486–9

Toubiana L (2022) Covid-19: Une autre vision de l'épidémie, In: Éditions La Découverte, Paris. ISBN 9782810011117

Toute l´Europe, Qatargate, la corruption qui touche les institutions européennes. https://www.touteleurope.eu/institutions/qatargate-tout-comprendre-au-scandale-de-corruption-qui-touche-le-parlement-europeen/. Accessed February 8, 2025

Toute l'Europe, Les Gilets jaunes vont présenter une liste aux européennes (2019). https://www.touteleurope.eu/institutions/le-recap-des-europeennes-des-gilets-jaunes-vont-presenter-une-liste-aux-europeennes/. Accessed February 17, 2025

Trading economics, baltic exchange dry index—2023 Data—1985–2022 Historical—2024 Forecast – Price (2024). In TradingEconomics.com, https://tradingeconomics.com/commodity/baltic. Accessed March 10, 2025

UNICEF, Effets néfastes de la Covid-19 sur la santé mentale des enfants (2021) https://www.unicef.org/fr/communiqu%C3%A9s-de-presse/les-effets-nefastes-de-la-covid-19-sur-la-sante-mentale-des-enfants-et-des-enfants-et-des-jeunes#:~:text=Selon%20les%20estimations%20mondiales%20disponibles,pour%20cette%20tranche%20d'%C3%A2ge. Accessed March 27, 2025

Valušová A, Lacová Ž. Digital skills across Europe: can we talk about homogeneity?, In: L'Europe, ses acteurs et leurs stratégies : visions managériales croisées, Paris : L'Harmattan, 1, 1 119–135. ISBN 978–2–343–23005–4

Voxafrica, Fin de partie pour la France au Mali, chassée par le pouvoir en place (February 2022) https://voxafrica.com/fr/fin-de-partie-pour-la-france-au-mali-chassee-par-le-pouvoir-en-place/. Accessed March 14, 2025

Voy-Gillis A (2020) L'Union européenne à l'épreuve des nationalismes, In: Éditions du Rocher, Monaco, 88

Voy-Gillis A (2020) L'Union européenne à l'épreuve des nationalismes, In: Éditions du Rocher, Paris. ISBN 9782268103174

World health organization, définition: pandémie, WHO.int, https://www.who.int/fr/health-topics#P. Accessed March 14, 2025

World Economic Forum, Global Risks Report 2023 (2023), https://www.weforum.org/stories/2023/01/these-are-the-biggest-risks-facing-the-world-global-risks-2023/?utm_source=chatgpt.com. Accessed March 2, 2025

World Economic Forum. The Global Risks Report 2021 (2021), 16th Edition, https://www.weforum.org/publications/the-global-risks-report-2021. Accessed March 8, 2025

Wucker M (2016) The Gray Rhino, In: St. Martin's Press, New York. ISBN 1250180287

Functionalist and Neo-Functionalist Management of Crises and the European Integration Process: Successes, Limits and Way Forward

Francis Clavé

Abstract This article shows the influence of functionalism, then neo-functionalism, in the management of the crises that have punctuated the construction of Europe. It focuses in particular on the neo-functionalist *engrenage* (spillover) process implemented by Jacques Delors during the relaunch of European construction. The article also points out that EMU really began to be envisaged with the departure of General De Gaulle, a politician whom neofunctionalism had little sympathy for. Finally, it shows that, at the time of the Maastricht Treaty, the strong political will to create the euro led to the neglect of two fundamental problems that were to haunt the rest of European construction. First, nothing was really decided on the institutions. Moreover, the choice was made to minimally manage a strong opposition between at least two competing political economy systems: one based on Delors-style subsidiarity and the other on Tietmeyer-style subsidiarity. The result is a conflict between at least these two opposing models of Europe, which the changes brought about by the crises affecting the eurozone have failed to overcome. On precisely these issues, this article looks at ways of bringing points of view closer together while at the same time managing deep-seated divergences whose permanence can also be seen as a source of dynamism. To this end, it proposes a return to Mitrany-style functionalist economic expertise within the framework of Ostrom-style polycentric federalism.

Keywords Delors · Tietmeyer · Mitterrand · Engrenage (gear) · Functionalism · Neofunctionalism · Ordoliberalism · Subsidiarity · Spillover

---

Clavé Francis. University, Paris II Pantheon-Assas. I would like to thank the two anonymous reviewers, as well as the participants, at the ERECO 2024 conference in Banská Bystrica. Their comments, particularly those of Ghislaine Pellat, the chairwoman, have enabled me to greatly improve the present article. I would also like to thank Ghislaine Pellat for her eminent work in the translation of this contribution into English.

F. Clavé (✉)
Université Paris II, Paris, France
e-mail: Clave_F@Yahoo.fr

G. Pellat et al. (eds.), *Data-Centric Business and Applications*, Lecture Notes on Data Engineering and Communications Technologies 258,
https://doi.org/10.1007/978-3-031-95068-1_8

1 Introduction

Jean Monnet, one of the "fathers of Europe", but also one of the men whose practice of functionalism inspired neo-functionalism, maintained that the construction of Europe "will take place in crises, and will be the sum of the solutions brought to these crises (quoted in Pisany-Ferry, 2011, p. 25)". Jacques Delors, another great builder of Europe, insists in his *Memoirs* on his use, during his term as President of the European Commission, of the *engrenage* (spill-over[1]) technique, a key element of neo-functionalism. It therefore seems necessary to explain what the functionalism initiated by David Mitrany is and what distinguishes it from neo-functionalism. This trend stems from Ernst Haas's study of the practice of functionalism by the pioneers of European construction between 1950 and 1957. Similarly, to understand European integration, it is important to analyze why neofunctionalism prevailed over functionalism. This is the subject of the first part of the article, which also looks at the influence of crises (the 1968 crisis, the dollar crisis of the late '60 s and early '80 s, and François Mitterrand's 1983 turn to austerity) on European integration. In the second part, we show that the Maastricht Treaty marked a break in European construction. Indeed, from this date onwards, primacy was given to the political will for European integration, leading to a political gearing (engrenage) in which what was sought was not so much an attempt to manage profound differences in the best interests of all but the victory of one of the political governance options (neofunctionalist federalism, Mitterrandian intergovernmentalism and ordoliberal federalism) and of a system of political economy. The two oppositions often overlap. The result is a series of uncooperative games played during the crises that marked the eurozone's first twenty-four years. In the second part, after presenting the context of the treaty, we will analyze some key elements of divergence. In the final chapter, we present a number of ways of reconciling points of view and managing differences in substance.

2 From Functionalist to Neo-Functionalist Integration

This section aims to answer the following questions: What is functionalism? What unites and distinguishes neofunctionalism from functionalism? Why do neofunctionalists dislike politicians like De Gaulle? What influence did the crisis of the IMF (International Monetary System) and the events of May 1968 have on the idea of the Economic and Monetary Union? How did Delors use the economic and monetary problems of the early 1980s in the process leading up to the Single Act and the Delors Group report on the single currency?

---

[1] .Delors writes *engrenage* Hass (1968) « spill over». *Engrenage* (in English, gear or gearing) insists more that *spill over* to the mechanical aspect.

2.1 *Mitrany's Functionalism*

In this section, after studying the main features of Mitrany's functionalism, we assess its influence on Europe's founding fathers.

2.1.1 The Main Features of Functionalism

Mitrany first introduced the notion of functional theory in 1933 in *Progress of International Government,* a concept previously developed by L.T. Hobhouse and R.H. Tawney (Asworth 1999, p. 41). It is part of a perception of reason that he owes to his two mentors, L.T. Hobhouse and Graham Wallas (Asworth 1999, p. 98). For them, as for him, while reason is the best arbiter of our decisions, it is a weak tool that must be supported by institutions. A similar position is taken by Walter Lippmann, whose mentor Graham Wallas also is, in *Public Opinion.* As with Walter Lippmann, Mitrany aims to force politicians to deal realistically with the problems at hand. The aim is to prevent politicians from taking refuge in ideology, which, in his view, distorts "political judgment" (Devin 2008). At the same time, it's important to avoid experts taking refuge in abstract expertise. In this respect, David Mitrany is wary of "scientific" conceptualizations that claim to capture eternal truths (Devin 2008, p. 34). Here, too, there is a closeness to Graham Wallas (1934, p. 128) and his distrust of the "idol of the laboratory."

David Mitrany's 1943 publication of *Working Peace System* (Devin 2008, p. 3) gave new impetus to functionalism. The book was published by the *Royal Institute of International Affairs,* an English *think tank,* as part of a series devoted to thinking about the post-war world. In it, David Mitrany argues that to build a "*working peace,*" we should be wary of grand constructions such as the *League of Nations* and concentrate instead on ways of satisfying human needs. In fact, he wanted to prevent the next peace from being built like the last, based on territorial claims. On the contrary, he argues that the question of economic and social development that transcends borders should be given priority. To this end, he argues that tomorrow's world must move away from laissez-faire, as must the private/public, local and state/supranational boundaries (Mitrany 1943, p. 56). With this in mind, he argues that national and international societies must cease to be seen as two different worlds:

> But because of the legalistic structure of the state and our political outlook, which treat national and international society as two different worlds, social nature, so to speak, has not had a chance so far to take its course (Mitrany 1943, p. 42).

The desire to establish an active international society (Mitrany 1943, p. 55) led him to advocate the supranational organization of certain functions. His approach to transnational functionalism does not presuppose the forced inclusion of all nations but rather encourages the greatest possible number to participate. The management of these functions relies heavily on the functional expertise of the participants, which gives them a large part of their legitimacy. He is wary of a political approach aimed at formal equality between participants. He prefers to work based on "*working equality*" (Mitrany 1943, p. 42), considering what each participant can contribute. To this end, he advocates, for example, that in the case of an *International Investment Board*, the capital providers should play a central role, while an *International Development Commission* would see its broad lines of action drawn by the countries concerned (Mitrany 1943, p. 40). Prefiguring the neo-functionalists' theory of *engrenage* (gearing), Mitrany (1943, p. 21) believes that functions will multiply (*Every function was left to generate others gradually, like the functional subdivision of organic cells*). This raises the question of the articulation of functions (Mitrany 1943, p. 35). On this point, he distinguishes four different cases (coordination of organizations belonging to the same function, coordination of several groups of functional agencies, coordination of functional agencies with international "planning" agencies, coordination under political authority). While this last form of cooperation will be highly prized by neo-functionalists, it is not the one that most interests Mitrany, who seems wary of it. Speaking of the possibility of federalism on a functionalist basis, he writes:

> But such a federalism if it came would be the solid growth of a natural selection and evolution, tested and accepted by experience, and not a green-table creation, blown about and battered by all the wind of the politics (Mitrany 1943, p.43)

2.1.2 The Pioneers of European Construction and the Functionalist Approach

One of the main pioneers, Jean Monnet, is usually considered a functionalist (Haas 1968, p. XIX). We don't know if he ever read David Mitrany's book. On the face of it, it seems hard to imagine that he was not in contact with the thinking of Mitrany, a member of a highly influential *think tank*. On a purely factual level, between Mitrany and Monnet, who was the first Commissaire au Plan, there is a certain convergence on planning. For both men, planning must not be purely national and autarkic. In a way, it required cooperation at a higher level, of which the ECSC (European Coal and Steel Community) could be seen as an example.

Ernst Haas emphasizes that, alongside socialists such as Robert Marjolin, the founders of Europe also included Christians such as Robert Schuman. Regarding the latter, it should be noted that the *method of small steps,* based on concrete achievements, is close to functionalist thinking. Among the Christians who were not among the direct founders but who nonetheless influenced these men, we might mention François Perroux, whose 1954 work, *L'Europe sans rivages,* deals *with* themes addressed by Mitrany: mistrust of a Europe that would become a closed nation-state and the importance of limiting nationalism in particular. Another point of proximity between Mitrany and these Christians inspired by Emmanuel Mounier: a broad vision of man, which they do not limit to Homo economicus. Like Mitrany too, these men, who have often passed through the Uriage management school, insist on the role of expertise, even if they don't necessarily conceive of it in the same way. If we now look at what is happening on the German side, Adenauer is, if we are to believe Tietmeyer (1999, p. 11), if not a federalist, at least an "institutionalist," while Ludwig Erhard is a functionalist.

2.2 Neofunctionalism and the Spill-Over Effect

The concept of neo-functionalism has its roots in Ernst Haas's 1958 book *The Uniting of Europe,* a work born of observation of what was done in Europe between 1950 and 1957. In the preface to his book, the author (Haas 1968, p. XXXII) writes.

> My aim is merely the dissection of the actual "integration process" to derive propositions about its nature.

This concept was developed and clarified in Leon Lindberg's 1963 book, *The Political Dynamics of European Economic Integration*, and in Schiengold Stuart's 1965 book, *The Rule of Law in European Integration.* Compared with functionalism, which focused primarily on the notion of service to individuals in the context of post-World War II industrial society, neo-functionalism targets a much more political and institutional object, relying on "the rational action of a political and administrative elite defending its own interests (Saurugger 2020, p. 51)". By emphasizing the political and administrative elite, neo-functionalism can be seen as part of a French-style society in which the senior civil service plays a central role and sectoral or functional expertise takes second place. Moreover, as is frequently the case in France, the[2] intermediary bodies are valued, while direct contact with the people—as we shall see when we deal with General de Gaulle—is little appreciated. Compared to Mitrany, Wallas or even Lippmann, the important thing is not so much to strengthen the reason of the people, but rather that of senior civil servants and intermediary bodies. On the other hand, compared to Mitrany, there is a desire to form an opinion with a view to constructing a political object. This is perceptible when Saurugger (2020, p. 53) notes that there is in them the idea of "the formation of a transnational representation of interests which, in turn, affects the allegiances and identities of the actors and probably their political representation."

The result is a willingness to transfer loyalty from elites to this new political object (Saurugger 2020, p. 55). On this point, it should be noted that Haas is more constructivist than Mitrany, even if his constructivism is pragmatic (Rosamond 2005, p. 245). If we consider the objectives of such an object, we should note that the first supporters of Europe were to the left of the Christian Democrats or among the socialists. They included socialist or Christian trade unionists motivated by the desire to establish a large-scale, regulated economy (Haas 1968, p. 292). From this stems the problem of how to implement them in practice. Indeed, depending on how one perceives human nature, the mode of regulation is very different, as will be developed in the second part.

---

[2] We're tempted to see this notion of intermediary bodies, so prevalent in France, as a mutation of the corporatism of which some, such as François Perroux, were pre-war proponents.

The notion of spillover (a term little or not used by Mitrany) is also different. For Mitrany, *spillover* means creating new institutions when new fields arise. Haas uses the notion of spillover in Mitrany's sense to designate the creation in 1957 of EURATOM (atomic energy) or the European Commission for the common market. Geographical *spillover* (Haas 1968, p. 317), such as that which led Great Britain to seek European integration in the mid-1960s, is also close to this definition. A *spillover* that De Gaulle temporarily thwarted, as we shall see later. But neo-functionalism insists on two elements that are little present, or even absent, in Mitrany: *political spillover* and *nurtured spillover* (Saurugger 2020, p. 57). The first is based on the convictions of political-administrative elites. The second is implemented by organizations (High Authority, CJEU, Commission, etc.), which are given a great deal of tremendous (Haas 1968, p. 308) discretionary power in interpreting treaties.

2.3 Gaullism, Neo-Functionalism, the Crisis of the International Monetary System, and May 1968 in France

To understand the genesis of the concept of EMU (Economic and Monetary Union), we need to look back at the political and economic situation at the end of the 1960s. Two points are important in this respect: de Gaulle's decision to halt the process and the monetary problems linked to the events of May 1968 and the difficulties encountered by the Bretton Woods international monetary system.

2.3.1 De Gaulle and Neofunctionalism

Writing in 1968, Haas notes that De Gaulle stopped the spillover by refusing Britain's entry into the Community and practicing the policy of the empty chair. Haas doesn't seem to have much use for great men. The men they cite—Bismarck, Cavour, and Disraeli—contributed to the construction of nation-states or empires that turned out to be aggressive and nationalistic, something that both functionalism and neo-functionalism want to avoid. Generally speaking, neo-functionalists have little interest in people. Their focus is on interest groups, whose representatives they see as the dynamic elements of society. In our view, the problem lies in the fact that the resultant interests of all these groups, which technocracy will try to direct towards the common good, may diverge very sharply from what ordinary people want. This is where neofunctionalism breaks with neoliberalism à la Lippmann. A current that values great men and insists on the necessary consent of the people in elections, which constitute the "key moment in the leader's experimental approach (Milanese 2020, p. 301)". For the neo-functionalists, on the other hand, the elite is the bearer of a project that, in a way, it imposes on the people through a machinery designed to veil things to a greater or lesser extent and, if not make them desirable, at least make any opposition difficult.

From this point of view, neofunctionalism urges politicians to abandon grand politics (the experimental approach) and set themselves more modest, managerial goals (Haas 1968, p. XX). In this respect, neofunctionalism has an advantage, for, as Haas points out, great statesmen like De Gaulle are rare. In fact, the General's successors, for better or worse, depending on one's point of view, did not match his stature. Georges Pompidou, as soon as he was elected, hastened to let England join the Common Market by giving in to most of its demands, much to the dismay of Raymond Barre, then European Commissioner. It was also at this time that Delors, who was not one of De Gaulle's admirers, as he noted in his memoirs, came to the fore with Jacques Chaban-Delmas's *Nouvelle Société* project.

2.3.2 The 1968 Crisis, Pompidou and EMU Projects

The idea of EMU (Economic and Monetary Union) was born out of the crisis of the Bretton Woods system and that of 1968. As early as 1966, in response to a recession in the German economy and the monetary disorder linked to the limits of the Bretton Woods system, the German Federal Ministry realized that if it wished to pursue a policy of stability under a fixed exchange rate regime, economic policies would have to be coordinated (Tietmeyer 1999, p. 197). In the same vein, the Scientific Council of the Federal Ministry of Economics, in its report 22–24 of November 1967, recommended.

> To begin negotiations...to build a stability-oriented community, and to extend it beyond the European Economic Community to other countries that play an important role in world trade (quoted in Tietmeyer 1999, p. 198).

In 1966–1967, still in Germany, the Council of Experts (the Five Wise Men) drew up a similar project, limiting it to the EEC (European Economic Community). This document provided for "a strict monetary code of conduct imposing quasi-constitutional constraints (Tietmeyer 1999, p. 198)". Following the events of May 1968, the French franc came under attack. De Gaulle refused to devalue and continued with a strong franc policy. Things improved when, in February 1969, he decided to hold a referendum on decentralization and the Senate, which he lost on April 27, 1969. On August 8/1969, the new French president, Georges Pompidou, devalued the franc. For Raymond Barre (Djian 2001, p. 101), this devaluation of around 20% against the mark was at the root of the inflation that was to plague France from the 1970s through to the 1980s. On September 8/1969, the European Council stressed the need for economic, monetary and fiscal coordination (C2DH 2023). In December 1969, the same Council, meeting in The Hague, called for EMU (economic and monetary union) and instructed Pierre Werner to convene a group of experts to consider the issue (Djian 2001, p. 103).

During the drafting of the Werner plan, two opposing theses were put forward. Hans Tietmeyer's "economist" view, defended by the Germans, the Dutch and to some extent the Italians (Franck 2002), was that economic coordination should precede monetary coordination. This thesis is opposed by the French, Belgians and Luxembourgers, described as "monetarist" by Tietmeyer (2011, p. 4), who defines it as follows:

> It was necessary to start by linking exchange rates and building a common reserve or exchange equalization fund, and from there, as it were, to use the monetary ties to force or stimulate the economic harmonization of the fundamental process.

In the face of this opposition, Werner imposed the notion of parallelism, based on "a parallel process… of driving policy integration and monetary integration forward and linking them (Tietmeyer 2011, p. 4)". The Werner report thus included a proposal for a community body called the "decision-making center for economic policy," accountable to Parliament. This body was rejected by President Pompidou. He feared that too many constraints would prevent the country's leaders from dealing with domestic protests of the May 1968 type (Franck 2002). In fact, the federal route was also rejected. Pompidou and the French deputies, still influenced by Gaullism according to Tietmeyer (2011, p. 4), preferred to take the intergovernmental route. It will be seen later that, on this point, François Mitterrand adopted the same political line at Maastricht.

The Werner report only led to the monetary snake, which was rendered inoperative by the end of the Bretton Woods system and the 1973 oil crisis. When Giscard d'Estaing and Helmut Schmidt created the EMS in 1979, Helmut Schmidt had to contend with the Bundesbank, which defended the "economist thesis" and more or less imposed what Tietmeyer calls a "strongly monetarist stance," including measures to ensure that the currency peg guided and constrained national policy (Tietmeyer 2011, p. 5).

2.4 *Engrenage (Gear) Theory and the Relaunch of European Integration in 1985*

How did Delors situate himself in relation to neo-functionalism, of which *engrenage* (gearing) was a key instrument (Rosamond 2005, p. 246–251)? How did he use this technique in the process leading up to the Single Act and then in that leading up to the Delors Committee's report on the single currency? These are the questions addressed below.

2.4.1 Delors, Neo-Functionalism and the *Engrenage* (Gear) Theory

Delors, like Jean Monnet and Robert Marjolin (another great European), was at the Commissariat général au plan from May 1962 (Delors 2003, p. 46) to 1969 (Delors 2003, p. 86), i.e., at the end of the first period (1945–1965), devoted to the fight against malthusianism, and at the start of the second period (1966–1975), when the aim was to integrate the social dimension into a plan that had hitherto been mainly economic. In short, at a time when the plan was moving away from functionalism and tending to blend economic and social functions. For Delors, as in neo-functionalism, leadership is not directed towards the people but towards the management of interest groups; at least, that's where his experience takes him:

> From a methodological point of view, there's a parallel with the two other positions I've held in the past. Firstly, at the Commissariat Général du Plan, where maieutics are essential, and secondly, in union negotiation, which also involves the appropriation of ideas by the players so that they can claim them (Delors 2003, p. 109).

While, like the neofunctionalists, Delors (2003, p. 251) emphasizes the importance of interest groups, he also gives pride of place to the regions. This attitude is partly due to his belief that France remains too centralized. However, the question of decentralization remains more ambiguous than it might seem. Be that as it may, for Delors, the European Commission, like the Commissariat au plan, had to keep special interests in check so that decisions were taken "in the sole European interest" (Delors 2003, p. 253).

In his 1958 book, Haas (1968, p. xxxiii) notes that unless the Commission asserts itself and creates incentives, its role in the integration dynamic will be weak. Indeed, it plays a key role in the institutional machinery. With Delors, a man whom Margaret Thatcher considers in her memoirs "as a new type of president, a major player in the game (Delors 2003, p. 236)", the Commission's role will be strong and will incite action, or more precisely, the spiral to which functionalism tends to reduce it. Functionalist theory," he notes, "wants to prove movement in motion, with achievements already made automatically calling for others, as in a gear (*engrenage*) (Delors 2003, p. 175)". For Rosamond (2005, p. 244), Haas's *spill-over* approach seeks to "capture the process by which the expectations of social actors lead them to ask for further integration". Delors, more proactive than Haas, insisted on the "trigger" element: for him, it was a matter of creating expectations that would set the movement in motion. Thus, as soon as he took over the presidency of the Commission, during talks with the heads of state, he emphasized the benefits that would flow from "a large common area without borders, with free movement of goods, services, capital and, one day, people" (Delors 2003, p. 155). He also emphasized that these advances would solve the economic problems of the time (stagnation, rising unemployment) (Delors 2003, p. 205). Once this preparatory phase had been completed, he was able to have the 1992 objective adopted by the European Council on March 29 and 30, which he saw as the start of the first phase of the process leading to the Single European Act. It should be noted here that in his memoirs, Jacques Delors considers that during his presidency of the European Commission, the *engrenage* (gearing-up) phase mainly extended from 1985–1989 (Delors 2003, p. 405) and comprised two sub-phases: the Single Act and its annexes (notably the Delors I package), and the creation of EMU. Before turning to these points, it should be noted that the "Delors's package," which in effect consists of aids paid to regions and states within the framework of various policies, constitutes both the lubricant of the machinery and an incentive to enter it. If this "generous" policy is possible, it is also because, on the resources side, Helmut

Kohl has accepted that his country (and he will not be the only one) make an effort (Delors, 2004, p. 239).

2.4.2 The Second Phase of the *Engrenage* (Gear) the Delors Committee Report

Before turning to the content of the report, it would be useful to set the context for a better understanding of the issues at stake.

Background

In monetary terms, the dollar was highly unstable in the 1980s. With the arrival of Ronald Reagan in 1981, the dollar appreciated sharply. In the Plaza Agreement of 1985, countries agreed to depreciate the dollar against the French franc, the pound sterling, the yen and the mark. At the Louvre agreements in February 1987, countries tried, on the contrary, to stabilize the dollar, which was falling too much. All of these factors prompted Europe to seek a degree of monetary autonomy. In France, François Mitterrand's 1983 austerity policy, strongly supported by Jacques Delors, was a key factor in his commitment to European integration. France decided to remain in the EMS (European Monetary System) and adopt an anti-inflationary policy close to the Bundesbank model (Tietmeyer 2011, p. 7). But remaining in the EMS posed a number of problems for François Mitterrand, French politicians and senior civil servants. The devaluation of the franc against the Deutsche Mark, which from an economic point of view was a relatively neutral technical adjustment, was seen by public opinion and the politico-administrative class as, if not a "defeat" against Germany, at least a humiliation for the national pride. Moreover, the arrival in government in 1986 of Jacques Chirac and Édouard Balladur facilitated acceptance of the liberalization of capital movements advocated by the British, Germans and central bankers (Palayret 2017). Once capital liberalization had been complete, the Single Act acted "as a lever in favor of EMU (Palayret 2017, p. 132)". In December 1987, Delors proposed to the European Council the creation of a group of wise men to study the question of a single currency (Delors 2003, p. 331). For him, this date marked "the start of a new *engrenage* (gear) after the three stages of the 1992 objective, the Single European Act and the Delors package" (Delors 2003, p. 331). On January 8, 1988, Édouard Balladur, the French Finance Minister, sent his European partners a memorandum in which he argued that the logic of the Single Act "would be to create a single-currency zone [...] in which there would be a common central institution and federal banks in each country" (Palayret 2017, p.132). In early 1988, the German Foreign Minister, Hans-Dietrich Genscher, published a personal document entitled "A European Currency and a European Central Bank." In it, he advocated an independent central bank similar to the Bundesbank and proposed that the Hanover Summit appoint a group of experts to deal with the issue (Endo 1999, p. 158). At the beginning of 1988, Kohl invited Delors to his home in Ludwigshafen to tell him that he should chair the group (Delors 2003, p. 332). The Hanover European Council of June 27–28, 1988, approved the project (Delors 2003, p. 334) and entrusted Delors with the report.

The Delors Committee Report

EMU is a project in which Jacques Delors was deeply involved. As soon as he was appointed, he detached the monetary sector from the economic one and took it over himself. Alois Pfeiffer, the Commissioner in charge of the sector, saw his field reduced to the economy alone (Delors 2003, p. 193). This detachment of monetary affairs from the economy may seem benign, but it can also be seen as the beginning of a process that led to money being approached primarily from a political-monetary angle without really taking economic realities into account. The Delors Committee tasked with drawing up a single currency project was made up of another commissioner, Frans Andriessen (external relations), three independent personalities and the governors of the central banks of the countries concerned. Among them, Karl Otto Pölh of the Bundesbank and Jacques de Larosière of the Banque de France will play an important role. Pölh, because his agreement was seen as a necessity by the French, and de Larosière because he played a key role in France's acceptance of the creation of an independent European Central Bank (Tietmeyer 2011, p. 9). While the Committee was inspired by the Werner report (Delors 2003, p. 336), it differed in two respects. Firstly, the number of member countries of the European Union is higher, and not all countries can or want (the British) to participate in the single currency. So, at least for the Germans, one can think that convergence criteria were also selection criteria (Tietmeyer 2011, p. 9). Moreover, remembering the opposition to the Werner plan, the Delors Group report makes little mention of the "control of economic and financial (budgetary) policy" to be implemented in parallel with monetary policy (Tietmeyer 2011, p. 10). Thus, Tietmeyer argues that if the Delors report raised a problem, it was not well resolved in the Maastricht Treaty, as France does not accept the "supra nationalization" of national fiscal policy (Tietmeyer 2011, p. 9).

Karl Otto Pölh introduced the maximum convergence criteria, which Delors accepted (Endo 1999, p. 164). It should be noted, however, that the 3% budget deficit threshold comes from the French (Abbati 2017, p. 188). However, Pölh signs the report without being really convinced. It's likely that he follows a line close to that of Hans Tietmeyer, which has just been outlined. It is also likely that he is in line with the parallelism of the Werner report, which states that political and monetary integration go hand in hand. Be that as it may, his attitude worried Paris to such an extent that, although the committee completed its work in June 1989, at the Madrid summit in September 1989, Mitterrand and Dumas still feared that Pölh would dissociate himself from the Delors report (Palayret 2017, p. 142).

3 From the Neo-Functionalist Political *Engrenage* (Gear) to a Return to More Economic Functionalism

Three questions punctuate this section: why didn't Delors recognize himself in the Maastricht Treaty? What are the main principles of political economy on which ordoliberals, and mainstream French politicians and economists diverge? How can we reconcile these points of view?

3.1 The Maastricht Treaty and Opposing Views on Political Institutions

Jacques Delors considers that the Maastricht Treaty, which he does not claim, marks the end of his policy of *engrenage* (gearing up). If we ask ourselves why he thinks this, we have to admit that it has more to do with differences over political institutions than economics.

3.1.1 From Economic to Political Engrenage (Gear)

At Maastricht, Jacques Delors and François Mitterrand were at odds over the form Europe should take: neofunctionalist federalism or intergovernmentalism. It will also be seen that this opposition led the Germans to think that their vision of European federalism was a little *naïve* and to wonder how this could be remedied.

Delors' Politicization

The creation of the euro could be seen by Delors as a triumph, given that the single currency had been one of the major objectives of many Europeans, in France and Germany, since the second half of the 1960s. However, he does not recognize himself in this treaty. There are several reasons for this. Firstly, Delors considers that, on this occasion, "the people are bursting into the often-abstract theater of European construction" and welcomes the fact that he has been "calling for years for a public debate on Europe (Delors 2003, p. 405)". However, beyond these words, things are not as simple as one might think. First, it's worth noting that he uses the word "debate" rather than "participation in decision-making," in line with the neo-functionalists for whom the people are not really the key participants in decision-making. Delors also advocated ratification of the treaty under Article 235, where the agreement of the Council of Ministers would suffice. It was Hans Tietmeyer who convinced Chancellor Kohl to ask for ratification under Article 236, which led to votes in which the people were able to have their say (Tietmeyer 2011, p. 7). In these votes, the people were less than enthusiastic. In Denmark, the changeover to the euro was rejected. In France, the referendum ended in a victory for the yes side, without the vote being an outright triumph.

If we now turn to institutional matters, it's clear that Delors increasingly moved away from the functionalist method, which emphasized the importance of economic successes, and, in a very neo-functionalist way, sought above all to strengthen the powers of the Commission, as we saw with the Single European Act. Eric Bussières, in a 2021 article entitled *Jacques Delors et l'Europe: vers la politisation des enjeux économiques (1985–1995)*, notes that during his terms in office, Jacques Delors abandoned Jean Monnet's vertical pillar structure in favor of horizontal lines of force based on common policies; in short, he strengthened the central power of the Commission, harming the decentralized powers of the individual countries. In so doing, he irritated François Mitterrand, who imposed a more intergovernmental vision of European institutions at the 1991 Intergovernmental Conference. To this end, the French president and other heads of state commissioned the secretariat of the European Council, assisted by Pierre de Boissieu of the Quai d'Orsay and the Luxembourg presidency, to draft the final document. Delors and the Commission were kept out of the loop. This situation was made possible by Delors' declining influence over the Commission. Delors was challenged by his administration, which felt sidelined by the highly centralist (Jacobinist for Endo (1999, p. 175)) attitude of the duo he formed with his cabinet director Pascal Lamy. Under these conditions, he was all the less able to oppose the intergovernmental path supported by François Mitterrand, since the project put together by the European Council secretariat presented a pillar structure (economic pillar, justice pillar and foreign policy pillar) designed to counter the Delors-style spillover (engrenage) aimed at strengthening the powers and increasing the functions of the Commission (Endo 1999, p. 179). It's hardly surprising, then, that Delors told the newspaper *Le Nouvel Economiste* on March 25, 1994, that, while he claimed responsibility for "the report on Economic and Monetary Union," he made it clear that Maastricht "is not my work; it is the creation of governments" (Bussières 2021, p. 19).

Maastricht and Mitterrand's Vision of the Engrenage (Gear)

When the Delors report was presented at the European Summit in Madrid (June 26–27, 1989), the French delegation endorsed the rapid convening of the IGC (Intergovernmental Conference), but its position on the creation of the euro was far from settled. The French delegation rejected the creation of a European reserve fund proposed by Jacques de Larosière and Pierre Bérégovoy. Similarly, the convergence criteria and mutual surveillance adopted hardly convinced Treasury officials (Palayret 2017, p. 141). However, the Finance Ministry's proposal for a strengthened Ecofin "which would govern the relationship between structural, fiscal and social policies (Palayret 2017, p. 140)" and the idea of an economic government were rejected.

Mitterrand's position evolved and strengthened with the fall of the Berlin Wall in November 1989 and the subsequent start of German reunification. The French president then decided squarely in favor of the Euro, abandoning the alternative avenues explored by the Treasury. Germany's unilateral recognition of Slovenia and Croatia in late December 1991 further strengthened his resolve (Abbati 2017, p. 187). From that date on, François Mitterrand irrevocably converted to the single currency and accepted the idea of an independent central bank. He aimed to ensure that the single currency became a reality. The result was France's insistence on a three-stage process for creating the euro and a timetable for making the process irreversible. In this article, that approach is referred to as the "Mitterrand political *engrenage* (gear)."For the President, it is also a question of influencing the management of the euro and, to this end, enlarging it sufficiently to ensure that it does not become a mark zone.

However, for a German like Tietmeyer, the Mitterrand option created "a gap between the extent of monetary integration and that of political integration (Tietmeyer 1999, p. 234)". As a result, the German position evolved from a certain *naïve* federalism aimed at achieving a United States of Europe to an approach closer to "French orientations (Tietmeyer 1999, p. 234)". The most striking consequence of this decision is that, for Hans Tietmeyer (1999, p. 246) and perhaps more generally for many Germans, "supranational monetary policy can only be apolitical, i.e., removed from the political influence of nations." This quote can also be seen as a justification for the *no bailout* clause.

3.1.2 The Continuing Dominance of Politics Over Economics in the March Towards the Euro

From the mid-1990s onwards, the Delors-inspired ordoliberal policy of a strong franc or competitive disinflation began to be challenged. It was challenged by Jean-Paul Fitoussi in a 1995 book entitled Le *débat interdit* (The *Forbidden Debate*). It is also being challenged by politicians. In the 1995 presidential election, candidate Chirac campaigned on the topic of *la fracture sociale* (social divide), distancing himself from the economic policy followed until then. However, once elected, Chirac was determined to respect the convergence criteria and promote the Juppé plan, although economic policy underwent a shift away from the strong franc policy followed until then. In a November 29, 1996, newspaper article entitled "Le rêve débusqué du franc fort dans l'Euro faible" Paul Fabra denounced "the Giscard proposal to bury the strong franc policy" as well as the Juppé government's "strong franc in a weak euro."

This development is worrying the Germans. As early as November 7, 1995 (Majocchi 2017, p. 199), the German Finance Minister proposed to the Bundestag the creation of a stability pact that would limit the deficit to 3% in the case of a negative cycle and 1% in the normal case. The European Commission considers that 1% is too low for certain countries, as is 3% in the event of a negative cycle. It proposed, and this was accepted, the 3% rate as the upper limit in "normal" times (Majocchi 2017, p. 200). On June 17/1997, the final version of the Pact was adopted at the Amsterdam European Council. It also provided for a multilateral surveillance framework and sanctions in the event of non-compliance (Majocchi, 2017, p. 201–202). Yet, Tietmeyer (2011, p. 9) is concerned about how the pact will be implemented. He fears, and the sequel will prove him right, that the pact will not be applied strictly, but in a political manner.

Similarly, entry into the Eurozone is essentially based on a political assessment of the criteria. Caught up in Mitterrand's three-phase *engrenage* (gear), the Germans, who are nearing completion of reunification, are all the less able to refuse or negotiate vigorously, given that reunification has led to a rise in interest rates, from which their partners have suffered. But they certainly understood what Mitterrand and French politicians wanted to do and the probable long-term struggle between the two political economy systems discussed above. Under these conditions, they felt it prudent to take precautions. Hence, the extreme care taken when creating institutions. For Tietmeyer, however, the decisive factor in German acceptance seems to lie in the fact that senior civil servants and politicians are politically in favor of Europe and of the euro. They also know, like François Mitterrand and Felipe González, that they are playing their part in history (Pisany-Ferry 2011, p. 35).

For political leaders, with the possible exception of the Germans, joining the euro is also a question of prestige, linked to their participation in a strong, world-class currency. Under these conditions, economic aspects carry little weight, even if political leaders "have had to make major political efforts" (Pisany-Ferry 2011, p. 59) to meet the convergence criteria in part. As for the economic future, except for the UK, which will not be joining the euro, Pisany-Ferry (2011, p. 62) notes that no "country is carrying out a stress test to assess the capacity of its economy to function in a monetary union regime." Virtually no country is "seriously considering what this might mean for its own economic policy," nor is it "launching a program of reforms specifically designed to prepare the economy for it (Pisany-Ferry 2011, p. 62)". France, on the other hand, embarked on a program of shorter working hours at the end of the 1990s, when in early 2000 the Schröder government launched reforms aimed at boosting its competitiveness. These reforms, similar to those advocated by Tietmeyer, ensured Germany's economic preeminence until 2022, when war broke out in Ukraine.

3.2 Issues Blocking the Establishment of a Relatively Coherent Political Economy System

The creation of the euro was both the result of circumstances and the shared political will of the leaders of the participating countries. With this in mind, it was decided to overcome differences of opinion. Nowadays, if we want to move things forward, it seems useful, even necessary, to analyze, if not all divergences, at least those which, directly or indirectly, condition the others. This is the aim of this section, which focuses on the divergence on subsidiarity between Delors and French leaders in general, and Tietmeyer and ordoliberalism. A divergence that underpins political and economic oppositions, notably on how to optimize the eurozone.

3.2.1 Subsidiarity in Delors and Tietmeyer: Oppositions of an *Anthropological* Nature

Endo (1999, p. 88) argues that Delors used the word subsidiarity, especially from 1992 onwards, when he began to perceive that his insistence on institutional reforms to strengthen the Commission were not being well received, and that it was time for a tactical retreat (Endo 1999, p. 88, p. 178, and p. 251). For him, and for the French in general, the word subsidiarity is used when centralization takes place, to specify that this is not "à la française" or "jacobine," but "good centralization."[3] Yet when you read Bussières's article, you can't help thinking that Delors is reasoning as if Europe should be built on the model of France alone. In this respect, it is very interesting to consider an important innovation introduced by Delors, according to Bussière (2021, p. 15), namely the shift from a function-based architecture to a network architecture. This highly verticalized network requires that links between states pass through Brussels, as in the centralized architecture of the French state. On this point, one can note that the Delors's opposition to the Eureka program is almost entirely due to its intergovernmental origin, and thus to the fact that it does not emanate from the center. A similar concern can be observed when Delors and the Commission want to create an industrial base. To this end, they intend to promote intra-European mergers, in short, to follow the path taken by France from 1945 onwards. The French approach was justified by that country's malthusianism, a problem that has not affected Germany or the Netherlands, to name but two. The same applies to social relations. Delors remains deeply influenced by his experience with the French government and Jacques Chaban-Delmas. This is how he came to support the idea of a plan with a "social dimension based on capital-labor relations arbitrated by the public authorities under the aegis of the European Commission (Bussière 2021, p.) without really considering the particularity of social relations in France compared to their German or Dutch counterparts, to name but two.

For Tietmeyer, unlike Delors, subsidiarity is part of an already profoundly decentralized world. He notes, for example, that "in the German federal system, important areas are the responsibility of the Länders, particularly those of internal security, culture and education (Tietmeyer 1999, p. 18)". But, most importantly, his perception of human nature is in line with his subsidiarity. Indeed, he (Tietmeyer 1999, p. 7) argues that, as far as possible, it should be men who make the decisions, not a "better-informed patriarchal state based on the principle of meritocracy." To understand this point, it's worth noting that Tietmeyer, like Hayek, is opposed to "constructivist rationalism." In the latter, the elite and experts have the means to know everything and therefore have no need to call on the intelligence of ordinary people. In Hayek's view, on the other hand (Tietmeyer 1999, p. 7–8), grassroots individuals have information that only they possess, by the very fact of their position, however modest it may be. It follows that the elite do not have all the information they need to make the best decisions. This has two consequences. Firstly, all men are entitled to participate

[3] On this concept of centralization or of decentralization seen as good centralization, see Rosanvallon (1985, p. 58).

in public decisions, and thus in what Hannah Arendt (1967, p. 171) calls "public happiness." Secondly, the notion of social justice is different. Indeed, if, outside the elite, participation in economic action does not require the active participation of individuals, it is appropriate to neglect commutative justice to focus on distributive justice, which tends to encourage inactivity, which in turn reinforces the elite's vision of the common man or woman. On this point, on the contrary, Tietmeyer (1999, p. 38) stresses the need to limit redistributive justice in favor of what he calls productive justice, while France is more in line with the former. A situation that suits French liberalism very well for it is "liberalism by the State" (Jaume 1997, p. 11), where the "self-effacing individual (individu effacé)" needs the protection of the elite. The reader will also find a similar way of thinking in François Perroux. It follows that, while the French insist on the "protective euro,"Hans Tietmeyer stresses the need for competition and emulation between member countries. Europe is more a place for competition than protection. A point he emphasizes both in his book on ordoliberalism and in a 2011 interview. In this interview, he argues that if EMU as a whole were to adopt "an egalitarian economic and financial policy, geared, as it were, not to the best performer, but to the average (Tietmeyer 2011, p. 12)", "that would mean that, in Europe, we would certainly lose global competitiveness (Tietmeyer 2011, p. 12)". In a word, Hans Tietmeyer's idea of a more decentralized Europe, with more trust in individuals than in neo-functionalist federalism. While, as we shall see, the latter leads to a vision of federalism close to the current American model, to which French economists refer fairly systematically when they talk about managing the euro, Hans Tietmeyer's federalist model is more "Ostrom-like," as we shall see later.

3.2.2 Problems Linked to the Non-Optimality of the Eurozone

What are the economic reasons for several countries to have the same currency? Historically, the first argument is that of economies of scale linked to a larger market. However, for Stiglitz (2016, p. 65), these 18th- and 19th-century analyses are hardly operative in an economic area where the free movement of goods, people and capital already exists. Since 1961, i.e., practically since the idea of monetary union in Europe was first mooted, the most widely used economic analysis has been based on the theory of the optimal currency zone developed by Mundell (1961) and Mackinnon (1963) (Bourguinat 1992, p. 4). For a currency zone to be optimal, it must simultaneously experience high mobility of labor and capital factors between members and high external openness (Bourguinat 1992, p. 5–6). In the eurozone, labor mobility remains low. Stiglitz (2016, p. 118) argues that it is simultaneously lower than between American states, and that different historical and linguistic factors modify its terms. For a currency zone to be optimal, Bourguinat (1973 and 1992, p. 10) also requires "identical (or close) preferences" in terms of collective goods and key objectives. On this point, it will be seen that the convergence hoped for by Bourguinat between France and Germany did not materialize, for reasons explained below.

Optimal currency zone theory distinguishes between microeconomic and macroeconomic effects. Microeconomic effects are generally favorable, due to lower transaction costs. From this point of view, stable exchange rates are an economic plus. This is why, back in 1966, when the Bretton Woods system became very unstable, the Germans wanted to create a monetary zone, and why Giscard d'Estaing and Helmut Schmidt later created the European Monetary System. At the macroeconomic level, the advantage lies in the fact that, in the event of an adverse shock, the countries of a common currency are better protected, as the risks are better spread (McKinnon 2002, p. 5). Finally, countries participating in a single currency have greater international influence, helping them to defend their interests more effectively. One of the major problems of a monetary zone with a single currency is asymmetric shocks. On this point, studies have indicated that European countries are more subject to asymmetric shocks than American states (Sénégas 2010, p. 385), which makes adjustment more difficult insofar as the exchange rate weapon has been abandoned. On this point, Stiglitz (2016, p. 74) points out that the withdrawal of the exchange rate adjustment mechanism "makes the economic system less efficient." This is all the truer when countries' political economy systems diverge on crucial points, as is the case between France and Germany. The negotiators, notably the senior officials of the French Ministry of Economy and Finance, knew this, which is why they worked on alternative solutions. This question forms the backdrop to all discussions on the "economic vision"/"monetary vision" opposition, and on the convergence criteria. But politicians, especially the French, do not want this monetary adjustment mechanism because changes in parity are perceptible to their fellow citizens and put at their expense. On the contrary, lower growth, especially when it affects all members of the eurozone, is less directly visible to the public.

Partly as a result of the desire to create the eurozone, developments in optimal currency zone theory from the 1990s onwards have sought to determine whether the forces internal to a currency zone tend to drive it towards greater optimality (Sénégas 2010, p. 381). The results of these studies tend to show the opposite, particularly in the industrial sector. In fact, a single currency, without an exchange rate, prevents divergent economies from adjusting, and if they don't, industry will concentrate in a few areas, as will happen in the 2000s. This was already Krugman's thesis at the creation of the Euro (Pisany-Ferry 2011, p. 31). If we can't rely on the exchange rate and labor mobility to adjust, we can rely on flexibility, which is the position of Tietmeyer and ordoliberalism. The other solution (the two are not mutually exclusive) is to use a large federal budget and make transfers. This is France's more or less official position. As far as transfers are concerned, France and the southern countries are philosophically in favor, while the northern countries are reluctant (Brunnermeier 2016, p. 5). Moreover, for the northern countries, transfers can only take place at the federal level, whereas France favors the intergovernmental approach. In short, countries differ profoundly on the measures to be adopted.

3.2.3 Political Currency/non-Political Currency Versus Rule/choice

The divergence we have just analyzed between the French way of optimizing the eurozone and the ordoliberal version is also perceptible in the approach to what is called in France the rule/choice opposition, which Tietmeyer addresses under the theme of political currency/apolitical currency. Firstly, it should be remembered that, for Tietmeyer, the euro must be apolitical because it is not part of "a vast political union conceived in the manner of a state." Under these conditions, he believes it should be removed "from the political influence of the nations" and entrusted to an institution responsible for guaranteeing the currency's value (Tietmeyer 1999, p. 245). But this is only one aspect of the problem. On the substance, his position is in line with the way ordoliberal perceive the political game. For them, the danger lies in the fact that the political arena is dominated by interest groups who have the means to seize control of money creation. On this point, it's worth emphasizing the sharp divergence from the neo-functionalists in general, and Jacques Delors in particular. The latter rely on activist groups (political parties, trade unions, the state apparatus, NGOs, etc.), counting on the "enlightened despotism" of senior civil servants or virtuous politicians to harmonize divergent interests. As we've seen, subsidiarity among neofunctionalists essentially concerns activists who are members of pressure groups. Ordoliberals, on the other hand, are concerned with ordinary people. They are the ones who need to be protected from the rentier actions of activists. This is made abundantly clear when Tietmeyer discusses Nicolas Oresme. He argues that, while for the latter, money belongs to the community, this does not mean that it should be entrusted to Parliament, as it would then be the responsibility of an institution capable of abuse through the possible subversion of this institution by interest groups. Hence, in ordoliberalism, the importance of entrusting money to an institution whose purpose is to ensure that groups do not seize control of money creation. The emphasis on the stability of money stems from the fact that, for ordoliberals, it facilitates the actions of those who do not belong to the elite. Behind this lies the desire to avoid central bank capture, from which we shall see that the ECB is not really immune. Another point should be noted. For Tietmeyer and the ordoliberals, price stability takes precedence over exchange rate stability. Indeed, he points out that it was to guarantee price stability that Germany adopted a strong mark policy in the 1970s to protect itself from imported inflation (Tietmeyer 1999, p. 151).

On the contrary, in France (and, it seems, in English-speaking countries), the emphasis is on the opposition between rules and discretion. This is due to the importance attached to Keynesian-inspired cyclical policies, which hardly inspired Tietmeyer, a man marked by Karl Schiller, German Finance Minister from 1966 to 1972, who, after having favored "enlightened" regulation, soon became disillusioned. Tietmeyer also concluded from Chancellor Schmidt's experience that cyclical policies were not conducive to economic adjustment. In his experience, they entail substantial budgetary costs, which disrupt adjustment mechanisms by reducing the amount of capital available for investment. This brings us back to the question of subsidiarity, insofar as adaptation is everybody's business, whereas economic policies depend on the ruling class. It's worth noting that, for Tietmeyer, monetary stability is necessary to enable individuals to adapt to a fundamentally unstable world. It should be noted here that ordoliberals, like Hayek on this point, do not place the notion of equilibrium at the heart of their thinking, as Walras and the neoclassicals did. For them, the world is first and foremost one of permanent adaptation.

3.3 Create Convergence and Organize Competition Despite Deep Differences

The Maastricht Treaty has led to a Europe in which at least two major visions of the European economy and social model clash, against a backdrop of diverging views on subsidiarity. These models come into conflict with each crisis in the eurozone. Behind their opposition lies another issue: whether Europe, like France, should be built on a single model, or whether greater diversity is not as desirable as it is necessary. These points are addressed in the first phase, before we analyze how to remedy them in the second.

3.3.1 Euro Zone and Non-Cooperative Games

In this section, we begin by analyzing the confrontation between two camps during the crises, and the resulting institutional and policy implications. It then examines, through two cases—the Banking Union and the electricity market—whether we really need to move towards a single model or not.

The Debt Crisis

The public debt crisis of the late 2000s mainly affected Greece, Ireland, Spain and Portugal. On this occasion, the Stability Pact was strengthened by the six-pack and then the two-pack, which introduced new tools for economic coordination between members (Majocchi 2017, p. 203). However, the rules of the pact were above all strengthened by the Treaty on Stability, Coordination and Economic Governance (TSCG) of March 2012 (Majocchi 2017, p. 203), a text that marks a certain return to the German position of the late 1990s. Thus, the structural deficit net of the influence of the cycle must not exceed 0.5% of GDP, and countries whose debt exceeds 60% of GDP must reduce the gap by 5% per year.

On the other hand, in contradiction with the *no bailout* clause, in 2010, a European Financial Stability Facility was created for countries in economic and financial difficulty. It was succeeded in 2012 by the ESM (European Stability Mechanism), an intergovernmental institution based in Luxembourg. This organization, which can draw on a strike force of up to 700 billion euros (80 in equity and 620 that can be mobilized from countries), is reserved for members of the eurozone. To benefit from it, a request must be made to the European Commission, which examines the matter with the ECB. The decision rests with the Governing Council, made up of eurozone finance ministers. The decision is taken either unanimously, or by a qualified majority of 85% of votes cast. Votes and contributions are proportional to the economic power of each country. Aid is subject to conditions, including the need to subscribe to the TSCG. Beneficiary countries include Ireland, Portugal, Spain, Greece and Cyprus.

While on these points the ordoliberal position is sometimes preserved, even strengthened, and sometimes partially weakened, the same cannot be said of the ECB. This is because, as Mitterrand and French politicians intended, in an enlarged eurozone where each country has one vote, the ordoliberal voice is in the minority at the ECB. Thus, in 2009, the ECB Governing Council decided to finance banks directly and governments indirectly through the purchase of sovereign debt (Abbati 2017, p. 191). This agreement was challenged before the German Federal Court, which ruled that "no further integration process, nor any delegation of sovereignty to the EU, can take place in the future within the constitutional framework currently in force (Abbati 2017, p. 191)". This decision does, however, allow J.-C. Trichet, President of the ECB, to launch the SMP (Securities Market Program) in 2010. Under this mechanism, central banks outside the ECB are authorized to buy sovereign debt on secondary markets. The Bundesbank, which voted against, was outvoted. The same will be true when the OMT (Opérations Monétaires sur Titres) replaces this mechanism. However, it was not until the arrival of Mario Draghi that the situation changed radically. The ECB's primary objective is no longer stability, but the preservation of the eurozone at all costs, "whatever it takes to preserve the Euro" (Abbati 2017, p. 193). While it promoted LTRO (Long Term Refinancing Operations) to help banks in 2011, it was with the *Quantitative Easing* or *Asset Purchase Program* launched during the third phase of the Greek crisis that the situation changed radically. Under this program, the ECB itself buys secondary government debt, albeit to a much lesser extent (20%) than national central banks. In July 2020, the German Federal Court (Zarka 2020) ruled that the ECB had not explained why its policy was proportionate to the dangers and had therefore failed to justify its decision. In passing, it criticized the CJEU, which had validated the process. The result is a conflict between the two jurisdictions that remains latent.

The crises have also shown that the ECB can be captured by special interests, first and foremost by bankers and by financiers. This is an age-old question, to which both ordoliberals and Milton Friedman tried to find solutions. While the ordoliberal process seemed to work within the framework of the Bundesbank, the compromises from which the ECB emerged seem not to have protected it in this area. Indeed, Stiglitz (2016, pp. 193–196) argues that during the debt crisis, the big banks sometimes captured the ECB, allowing them to be relatively spared.

The Covid Pandemic

The covid pandemic was declared a health emergency by the WHO from January 30, 2020, to 2023. In Europe, however, the worst of the crisis can be considered to have passed by the end of 2021. Initially, France and other southern European countries insisted on a European recovery plan. Following their request, in May 2020 the Commission proposed a plan entitled NextGenerationEU, financed by a loan of up to 800 billion euros. At the same time, the Stability and Growth Pact is suspended from 22/03/2020 to 31/12/2023. On this occasion, France and other countries are calling for the Pact to be both reviewed and lightened. In line with France's traditional position, Paris, supported by other countries, wanted negotiations on deficits to take place solely on a bilateral State-Commission basis, while Germany and some other countries holding a stricter vision, less subject to the Commission's discretion, wanted everyone to remain involved in the process. And they got it. Similarly, France was not in favor of maintaining the 3% deficit and 60% debt thresholds, which will ultimately be retained. The only change is that the debt reduction requirement will be relaxed (Columbus 2023). These new measures will only apply in 2025. In 2024, the previous version of the pact will apply. There is every reason to believe that, until a serious crisis hits, France will continue its "strategic application of the laws" of the Stability Pact (Clavé 2021b, p. 18).

The Question of the Uniqueness of Economic Rules and Processes

The diversity of economic realities and options serves as a backdrop to two issues that have recently come up, namely electricity prices and the banking union. In the first case, there is opposition between France and Germany, which fears that the low production cost of French nuclear power will distort competition. This is a one-sided view. In fact, it could be argued that France's 35-h working week and social security system weigh down production costs to such an extent that the relevant assessment of competition should be made at a more global level, to take account of the structural choices made by countries.

The same question is currently being asked about the decisions to be taken regarding the progress to be made on the Banking Union. Let's recall that this union, created by the European Council of June 28 and 29, 2012, includes a Single Supervisory Mechanism (SSM) operating within the ECB. The MSU monitors compliance with prudential standards. Its Chairman and Vice Chairman are appointed by the EU Council on a proposal from the ECB and Parliament. It acts in liaison with the national supervisory authorities. Today, the debate focuses on deposit insurance, with two opposing camps. On the one hand, northern countries want to make bankers more accountable, while southern countries are less interested in limiting bank risks than in pooling the damage caused (Perrotte 2021). In a word, on the one hand, we want to make players more accountable, while on the other, in line with the notion of the "self-effacing individual (individu effacé)", we want to protect the weak. But, in doing so, we avoid bankers, who are not particularly poor, bearing the responsibility for their actions. In such a case, we might well ask whether it would not be better to create a dual system based on the preferences of individual countries. Indeed, a single model risks pushing all banks towards a strong aversion to entrepreneurial risk, which would weigh on growth (Clavé 2020, p. 51). In reality, the banking issue is very broad and deserves to be examined in greater detail. Indeed, while the French authorities advocate large banks, economists are more divided on the question of optimal bank size. In contrast to the French, the Germans are more concerned about regional banks, which are smaller but more active in the real economy (Grésillons 2018). When European leaders discuss banking union, they often think about the circulation of European savings. Some would like it to circulate more efficiently, and in particular to disregard differences in structures and preferences between countries, a position that may seem contrary to the proper allocation of savings and therefore contrary to good economic dynamics.

3.3.2 Creating Convergence and/or Managing Divergence: Mitranian Expertise and Ostromian Federalism

In the preceding sections, we have analyzed the central oppositions weighing on the Eurozone. The question now is how to deal with them. Firstly, it's worth noting that this seems to be a good time to reflect on the subject. Indeed, Europe is undergoing a new enlargement process, which is leading to reflections on institutions (Balladur 2023), while the issue of ecological transition is still to be addressed, and the Ukrainian crisis is provoking, particularly in Germany, a change in the options taken at the very beginning of the 2000s, and therefore reflections on the future. Returning to the question of what to do? The path explored here is that of a return to Mitranian expertise within the framework of Ostromian federalism.

On the Desirability of a Return to Functionalist-Inspired Expertise

Following the Single European Act, politics has begun the main focus stage among the concerns of Europe's leaders. The problem is that both the economy and money have their own logic, which must be considered if they are to be tamed. So, a return to Mitranian functionalism, with greater emphasis on economic expertise, seems a necessity. However, this return must not be made with the sole aim of consuming goods (Clavé 2017, p. 191). The period is different from that of the post-World War II era. Today, the central problem is that of environmental constraints in a world where the western way of life is not really transposable to the eight billion men and women who inhabit the earth. This return to economic expertise is all the more necessary given that, these days, there is nobody in the eurozone concerned with the economy as a whole. In particular, the question of the goals pursued is largely neglected. Yet, this is a point that deserves to be studied, given the need to analyze the sharp divergence between France's consumer-oriented economy and Germany's more production-oriented one.

As far as expertise is concerned, it is possible to object that the ECB is well-endowed in this field. However, this is a very fragmented expertise, focused on money and finance. On this point, it should be noted that in the ordoliberal tradition, centers of expertise have always existed within both the Ministry of Finance and the government (the five experts). In Europe, nothing of the sort. The Ecofin Council, which brings together the finance ministers, focuses mainly on short-term budget monitoring. On this point, it is possible that changes to the Stability Pact, by making the process more technical, will further obscure matters and prevent people from perceiving the logic that would make it possible for them to adhere to it. There is a risk that, at European level, we will be faced with the same issues as the French Ministry of Finance, whose main concern is to finance the decisions taken by politicians, often in the context of very short-term management of the country. It's important to realize that politicians are not only subject to short-term constraints—the next elections—but are also encouraged in their short-termist vision by their communications advisors. It therefore seems necessary to introduce economic expertise into the eurozone, with a pragmatic, Mitrany-style remit to make politicians aware of the medium/long-term perspective, and to do so in a way that is comprehensible to the public. In truth, given the differences between Europe's political economy systems, we shouldn't necessarily be looking for a single model, but rather for solutions combining convergence and the organization of competition when divergences are too great. From the point of view of convergence, France would have much to gain, both politically and in terms of social spending, by drawing inspiration from Tietmeyer-style subsidiarity. As this is likely to be a lengthy process, countries should take greater responsibility for the resulting financial imbalances in their particular options. To this end, returns to the markets should be envisaged to alleviate the political pressure placed on decision-makers, or rather on the facilitators of agreement at the European level.

How can this expertise be introduced? This can be done with the professional's expertise of the Commission or in the form of pooled expertise, i.e., involving experts and researchers from different countries appointed for given periods of time by European institutions. The first approach runs the risk of creating a bureaucracy in charge of the economy, which would be tempted to govern the economy in place of elected leaders and standardize everything. So, it doesn't seem like a good idea. In Germany and the USA, presidents and chancellors are assisted by a council of economists. The idea seems worth transposing. The President of the European Council and the President of the European Commission could be given a council of economists to raise their awareness of these issues and to fuel debate.

Overcoming the Difficulties of Ostrom-Style Federalism

If we want to build a Europe as broad as the one, we seem to be heading towards, it will be difficult to continue working with highly centralizing institutions seeking to impose a single political economy framework. It seems futile to hope that France will soon align itself with the ordoliberal vision. The opposite seems more possible, but not really desirable. Indeed, Germany, under Prussian rule, experienced a high degree of centralism, and could return to it, but would this be a good thing? Is there not a risk of clashes between centralists to secure the leadership of Europe? It's also important to understand that, with a French-style model, the euro would become the franc of the twentieth century. It's not certain that countries entered the eurozone to share a weak currency. Finally, it has to be said that, throughout history, what has always been a source of strength (and sometimes also weakness, when divergences turned into global conflict) has been the diversity of experiences, and the possibility of aligning with those who had the best results. China, a centralized state, did not show the same dynamism as Europe in the nineteenth century. It went through a period of decline, which it only recovered from at the end of the twentieth century. Nevertheless, when economists in Europe refer to federalism, they have in mind the centralizing federalism that developed in the United States after President Wilson, not the polycentric federalism of the Ostroms.

What is federalism, à la Ostrom? Why might it be the answer to the problems posed? To answer these questions, let us first note that this type of federalism is based on the triptych of competition, cooperation and conflict resolution (Clavé 2021a, p. 235). A model well suited to the management of the situation in Europe as analyzed above. From this point of view, Ostrom's interest in the *ideas-to-deeds* link (Clavé (2020, p. 53) seems compatible with a Mitrany-style expertise that distrusts the abstract and gives precedence to the study of what is. In a way, both Mitrany and Ostroms are in line with a Dewey-style perspective on inquiry. Finally, as we have seen in two other studies (Clavé 2020 and 2021a), such federalism is conducive to entrepreneurial dynamism. This is because it requires the dynamism of human beings and political artisans (a concept developed by Vincent Ostrom) to resolve differences between centers of power. Federalism à la Ostrom is, in fact, a polycentric federalism, based on autonomous decision-making centers that are as much in competition as in cooperation. In this sense, it is also harmonizing with Mitranian functionalism, based on expertise by function, where relations between functions are left to the initiative of citizens or their representatives. It is, however, a demanding federalism, based on a more Tietmeyer-style subsidiarity than in the current model. It requires a shift from "power over" to "power with." On the other hand, it has the advantage of avoiding the costs incurred by "power over" and centralization (Clavé, 2023, p. 274). Costs that are all too often poorly assessed. On this point, Herbert Simon, in his 1947 book *Administrative Behaviour*, argues that the centralizing administrative principles developed in the United States since Wilson lack logical coherence, and that there is no empirical evidence that they are the most efficient (Ostrom E&V 2015, p. 615). The Ostroms, for their part, argue that polycentric institutional arrangements can generate superior outcomes to monocentric ones (Ostrom 2015, p. 9). Federalism à la Ostrom thus has several advantages. It starts from a vision of human beings close to that of ordoliberalism, of which it is one of the strong points. Through its mechanism for resolving divergences, it allows us to resolve the need to take account of French realities, which led Presidents Pompidou and Mitterrand towards intergovernmentalism, in a way other than that of the domination (the French economic model extended to Europe).

4 Conclusion

The creation of the European Union was initially based on a functionalist approach, founded on concrete economic expertise and a function-based approach to problems. Functionalism was soon succeeded by neo-functionalism, based on the management of interest groups by an elite of leaders in favor of European construction. Neo-functionalism also insists on *engrenage* (gearing up), an instrument Delors used during the relaunch of Europe from 1985 to the Maastricht Treaty. A treaty that led to a political *engrenage* (spillover), in which the willingness of the leaders of the various countries to participate in the management of a strong and influential currency, as well as the feeling of being part of a page in history, took precedence over a double divergence. Political divergence between a federalism inspired by neo-functionalism, a federalism more marked by ordoliberal thinking based on the valorization of individual action, and Mitterrandian intergovernmentalism. Economic opposition between a "French" model, based on a "self-effacing individual (individual effacé)," and an ordoliberal system with much more confidence in human beings, whose action is to be encouraged. Throughout the first two decades, these oppositions, deliberately kept in the shadows, constantly disrupted the functioning of the eurozone. The German and French models continued their struggle to secure preeminence over the zone. With each crisis, each protagonist tries to impose his or her vision, without the *status quo* really changing. In these conditions, we suggest a way of dealing positively with these divergences through Mitrany-style expertise capable of identifying a few major objectives and studying the means of achieving them, and Ostrom-style federalism responsible for positively managing intrinsic divergences. It's an approach that could revitalize a positive dynamic, provided everyone accepts that the European Union does not imply a single economic and political model, but rather a certain coexistence marked by both cooperation and competition.

References

Abbati CD (2017) The Euro between birth problems and identity uncertainties (1991–2016). In Preda D (ed), The History of European Monetary Union, Euroclio PIE Peter Lang Brussels

Arendt H (1967) [1963] Essai sur la révolution. Paris : nrf Gallimard

Asworth LM (1999) Creating international studies (Angel, Mitrany and the Liberal tradition). Ashgate Aldershot

Balladur E (2023) L'Europe et notre souveraineté : L'Europe est nécessaire, la France aussi. Paris: Fondapol. https://www.fondapol.org/dans-les-medias/edouard-balladur-la-france-t-demeurer-souveraine-et-le-peuple-francais-maitre-de-son-destin/

Bourguinat H (1973) Des moyens de l'intégration monétaire européenne. *Actualité économique*, Montreal, October-December

Bourguinat H (1992) Maastricht et la théorie des zones monétaires optimales. Rev Française D'économie 7(3):3–27

Brunnermeier MK, James H, Landau, J.-. P. (2016) The Euro and the Battle of Ideas. Princeton University Press, Princeton

Bussière É (2021) Jacques Delors and Europe : towards the politicization of economic issues (1985–1995)? *Studici Storici. Rivista trimestriale dell'Instituto* Gramsci, 1/2021, 159–187. https://www.rivisteweb.it/doi/https://doi.org/10.7375/100248.
C2DH (2023) Le plan Barre. Luxembourg : University of Luxembourg. https://www.cvce.eu/education/unit-content/-/unit/7124614a-42f3-4ced-add8-a5fb3428f21c/96ed3cc1-215e-470e-9abf-316277829d74
Clavé F (2017) La croissance en zone euro et les limites de l'articulation de l'économique et du politique dans l'approche fonctionnaliste de l'intégration européenne. Revue internationale des économistes de langue française, 2(2), 177–194. https://aielf.org/wp-content/uploads/2016/03/RIELF-17-02-textes.pdf
Clavé F (2020) Une Europe polycentrique d'inspiration ostromienne serait-elle compatible avec une économie entrepreneuriale européenne capable de répondre aux défis de demain ? Proceedings of the Grenoble 2020 ERECO-PGV conférence
Clavé F (2021a) Le recours à deux concepts clés des Ostroms (méthode IAD et fédéralisme polycentrique) pourrait-il contribuer au dynamisme entrepreneurial en Europe. In Pellat C., Richet X., L'Europe, ses acteurs et leurs stratégies, visions managériales croisées, Paris: L'Harmattan.
Clavé F (2021b) Étude des incitatifs et des freins à la découverte des vaccins contre la Covid en France avec le cadre ADI (Analyse et développement institutionnel). Study of incentives and disincentives to Covid vaccine discovery using the IAD framework. Management et gouvernance, 26(2), 15–28
Clavé F (2023) L'approche structuraliste française du développement des années 1950–1960 face à l'approche de l'école d'économie politique des Ostroms : proximités et divergences" in Redslob A, Gemayel J, Malaga K (eds), Monnaie, Finances, Développements. Beirut: Éditions de l'USJ
Collomp F (2023) Comment Berlin a fait plier Paris sur des règles budgétaires strictes pour la zone Euro. Le Figaro du 21/12/2023
Delors J (2003) Mémoires (interview with Jean-Louis Arnaud). Plon, Paris
Devin G (2008) What remains of international functionalism? Rereading David Mitrany (1888–1975). Crit Int 1(38):137–152
Djian J-M (2001) Raymond Barre. Flammarion, Paris
Endo K (1999) The Presidency of the European Commission under Jacques Delors. ST-Martin Press NY, St-Anthony Series
Eurostat (2023) Euroindicators, 2/2023, January 6. https://ec.europa.eu/eurostat/fr/web/main/publications
Fitoussi J-P (1995) Le débat interdit. Arléa, Paris
Franck R (2002) Hans Tietmeyer et le plan Werner : le débat politique franco-allemand in le rôle des ministres des finances et de l'économie dans la construction européenne (1957–1978), Paris, Institut de gestion publique et de développement économique/Comité pour l'histoire économique et financière. https://books.openedition.org/igpde/7898?lang=fr
Grésillons G (2018) Paris and Berlin find a first compromise for the Banking Union. Le Monde du 28/05/2018
Haas E (1958, 1968) The Uniting of Europe : Political, Social and Economic Forces, second edition, Stanford: Stanford University Press
Jaume L (1997) L'individu effacé. Fayard, Paris
Larosière J (2021) 40 ans d'égarements économiques : quelques idées pour s'en sortir. Odile Jacob, Paris
Liboreiro J (2023) L'économie de l'UE perd de son dynamisme en raison de la guerre en Ukraine, de l'inflation, des catastrophes naturelles et de la hausse du taux d'intérêt. Euronews du 25/10/2023. https://fr.euronews.com/business/2023/09/11/leconomie-de-lue-perd-de-son-dynamisme-en-raison-de-la-guerre-en-ukraine-de-linflation-des
Lippmann W (1922, 1998). Public opinion. New Brunswick: Transaction Publishers
Maes I (2017) Robert Triffin and the process of European monetary integration. In Preata D (ed), The History of European Monetary Union, PIE Peter Lang Brussels

Majocchi A (2017) Fiscal Discipline and Growth within EMU in Preda D (ed), The History of European Monetary Union, Euroclio PIE Peter Lang Brussels
Malingre V, Ricard P, Wieder T (2023) La brouille entre Paris et Berlin pèse sur l'UE. Le Monde, 8 and 9/10/2023
McKinnon R (1961, 1963) Optimum currency areas, the American economic review, 53(4):717–725
McKinnon RI (2002) Optimum currency areas and the European Experience) in Economics of Transition, Wiley online library, https://doi.org/10.111/1468-0351.00115
Milanese A (2020) Walter Lippmann d'un néolibéralisme à l'autre : changement social et leadership libéral. Classiques Garnier, Paris
Mitrany D (1943) A working peace system. The Royal Institute of International Affairs and Oxford University Press, London
Mundell RA (1961) A theory of optimum currency areas, the American economic review, 51(4), 657–665
Ostrom E (2015) A Long Polycentric Journey in Cole Daniel H, McGinnis Michael D (eds) (2015), Elinor Ostrom and the Bloomington School of Political Economy, volume I, Polycentricity in Public Administration and Political Science. London: Lexington Books
Ostron V, Elinor O (2015) Public choice: a different approach to the study of public administration in cole Daniel H, McGinnis Michael D (eds) (2015). Elinor Ostrom and the Bloomington School of Political Economy, I. Polycentricity in Public Administration and Political Science. London, Lexington Books, p 362
Palayret J-M (2017) La présidence Mitterrand, fer de lance de l'Union monétaire européenne (1988–1992). In Preda D (ed), The History of European Monetary Union, Euroclio PIE Peter Lang Brussels
Perrotte D (2021) Les Européens restent divisés sur l'Union bancaire. Le Échos, 21/06/2021
Pisani-Ferry J (2011) Le réveil des démons. Fayard, Paris
Rosamond B (2005) The uniting of Europe and the foundations of EU studies: Revisiting the neofunctionalism of Ernst B. Haas. J Eur Public Policy 12(2):237–254. https://doi.org/10.1080/13501760500043928
Rosanvallon P (1985) Le moment Guizot. Nrf Gallimard, Paris
Saurugger S (2020) Théories et concepts de l'intégration européenne. Presse de la Fondation nationale de SciencesPo, Paris
Sénégas M-A (2010) La théorie des zones monétaires optimales au regard de l'Euro: Quels enseinements après dix années d'union économique et monétaire en Europe? Revue D'économie Politique 120(2):379–419
Tietmeyer H (2011) Transcript of the February 10, 2011 interview with Hans Tietmeyer. CVCE www.cvce.eu
Tietmeyer H (1999) Économie sociale de marché et stabilité monétaire. Economica, Paris
Toute l'Europe (2023) Growth in the European Union. 27/04/2023. https:// toute l'Europe.eu
Wallas G (1934) Social Judgment. George Allen & Unwin
Zarka J-C (2020) L'arrêt du 5 mai 2020 de la Cour constitutionnelle fédérale allemande concernant le programme PSPP de la BCE. Lextenso of 03/07/2020 https://www.actu-juridique.fr/international/international-etrangers/droits-europeen-ue/larret-du-5-mai-2020-de-la-cour-constitutionnelle-federale-dallemagne-concernant-le-programme-pspp-de-la-banque-centrale-europeenne/

Cross Border Project Implementation Crisis During and After Pandemic

René Pawera, Marian Šuplata, Lívia Bott Domonkos, and Sabina Rýzková

Abstract The article examines the cross-border projects in crisis time of the EU, situated in the transboundary areas of Hungary and Slovakia. Behind the EU project related statistical data comprises additional information gathered through field research among applicants representing project distribution influenced by Joint Risk Management. It centers on the socio-economic impacts of suburbanisation, examining its influence on the community of suburban residents in Slovakia. Comparing cross- border project related priorities and implementation information in cross border project management prior to and following the year 2020 can help to forecast the prevailing patterns shaped by the ongoing pandemic, war and crisis. Added value is the prognosis of joint risk management behaviour in cross border project regions from the perspective of past and current applicants primarily amidst Hungary and Slovakia following the aftermath of the COVID-19 pandemic. Applicants profile is compared during the new program period of the EU from 2021 to 2027,with hypothesis, that there could exists considerable potential for the advancement of cross-border projects under INTERREG HU SK in terms of preventing risk and crisis. Interreg originated as a Community Initiative with a modest budget of 1 billion EUR. Subsequently, it expanded its scope to include transnational and interregional collaboration. However, during this period, the European Union did not experience the level of crisis seen since 2019. In the 2014–2020 period, European Territorial Cooperation emerged as one of the primary objectives of Cohesion Policy, alongside

R. Pawera · M. Šuplata (✉) · L. B. Domonkos · S. Rýzková
Faculty of Management, Department of International Management, Comenius University, Bratislava, Slovakia
e-mail: marian.suplata@fm.uniba.sk

R. Pawera
e-mail: rene.pawera@fm.uniba.sk

L. B. Domonkos
e-mail: livia.bott.domonkos@fm.uniba.sk

S. Rýzková
e-mail: ryzkova5@uniba.sk

G. Pellat et al. (eds.), *Data-Centric Business and Applications*, Lecture Notes on Data Engineering and Communications Technologies 258,
https://doi.org/10.1007/978-3-031-95068-1_9

investment for Growth and Jobs. Over time, Interreg has evolved into the European Union's principal tool for financially supporting cooperation between partner countries across borders, with the overarching objective of collectively addressing common obstacles and exploring collective remedies is essential, whether in the fields of healthcare, research and education, or transportation or renewable energy sources, however the priorities should adjust to changing EU and crisis prevention. The article farther contributes to the explanation and comparison of territorial strategy on INTERREG, from background materials of Cross-Border Review.

Keywords Regional development · Innovative strategy · Project impact · Territorial strategy · New programme period · Cohesion · Cross-border policy

1 Introduction

The article provides an insight into the new NextGeneration EU Strategy transiting from old 2014-2020 EU programme period facing risks, crisis, obstacles in EU region, following territorial strategy. Comparative analysis beyond past and future "Plans of New Structurisation" are describing major crisis comparison and risks between EU sustainability and EU project related issues. Analysis focus on challenges about the sustainable transition between two programming period's standards and evoke collaboration prior to and following the year 2020 to forecast prevailing patterns impacted by the pandemic crisis and potential EU sustainability risks in terms of EU project implementation, realization, monitoring, and strategies. Added comparison is resulting possible collaboration possibilities by measuring joint risk management strategy influenced following the COVID-19 pandemic during the new EU program period from 2021 to 2027. As for specific new generation example research provides wide potential of Calls for proposals based on needs, addresses specific needs of Sectors, Ministries, Geopolitics, Regional Sustainability for example in. INTERREG HU SK. The goal of the article is to present potential project implementation crisis and to address mutual difficulties collectively and discover collaborative resolutions in other EU member states based NextGenerationEU plan implemented in Slovakia. Moreover, in the field of healthcare, transport, research and education or sustainable energy, economy provide research-based data analysing Risk management categories as part of specific Eu development strategies. The upcoming Interreg initiatives, backed by financial assistance from the European Regional Development Fund, provide support development of the European Union's territory at different levels. Nowadays, started their new programme period just currently with new Calls of proposals, strongly reacting on Priority Axis of Joint Risk Management and Social Cooperation, reacting properly on the after-pandemic, war influenced and long-term migration of EU's challenges. NextGenerationEU according to their description than a recovery plan, chance to come over emerging resilient post-pandemic, reshaping the economies of the EU, fostering prospects and employment for the European region where the vision to value together €806.9 billion can contribute to a smarter, greener, more digital and more resilient Europe.

The selection of the research subject is deemed appropriate due to its relevance in the context of cross-border crisis was detected in the years of 2020–2023, which is a common phenomenon of current post-covid project implementation period, even though the fact, that two programming periods were active. The ending programme period of 2014–2020 + 3 (n+3 rule) expanding the period until 2023 and the new programme period with the shifting level of new era influenced by NextGenerationEU.

Project globalization is currently influential and significant, taking on new dimensions within the INTERREG programs. This research focuses on INTERREG HU SK programme projects (Hungarian-Slovak interregional cooperation), specifically examining it as the longest border collaboration Slovakia among bordering countries, as Hungary, which the Slovak Republic disposes. Furthermore, if joint risk management is applied on one side of the border, it logically can have serious impact on the Hungarian border region, which means complementarity, subsidiarity and cohesion, as it comes from the original idea of the EU projects and call for proposals (Anderson 2015).

Interreg originated as a Community Initiative in 1990 with a modest budget of 1 billion EUR, dedicated solely to cross-border cooperation. Subsequently, it expanded to encompass transnational and interregional collaboration. In the 2014–2020 period, European Territorial Cooperation emerged as one of Cohesion Policy's dual objectives, alongside investment in Growth and Jobs. Over time, Interreg has evolved into the primary tool of the European Union for fostering collaboration among partners spanning borders. Its objective remains addressing shared challenges collectively and discovering mutual solutions, whether in areas such as health, research, education, transportation, or sustainable energy. The funding for Interreg programs is provided by the European Regional Development Fund, aiming to support the balanced development of the European Union's territory across various levels.[1]

The selection of the research subject is warranted, given that cross-border agglomeration is a prevalent occurrence in today's era of globalization. The concept is complex and of great significance in the present day, and it has been given renewed emphasis through INTERREG programs. Our focus will be on INTERREG HU SK,[2] as the extensive border area surrounding Bratislava significantly influences the border area of Hungary.

In 2007, following the effective execution of the INTERREG Community initiative from 1990 to 2006, European Territorial Cooperation, commonly known as INTERREG, emerged as a key focus of EU Cohesion Policy. As a result, the European Commission evaluates studies and analyses that assess the execution of INTERREG programs, covering cross-border, transnational and interregional strands. Cross-border development strategies have gained popularity as effective mechanisms for driving social and economic progress in borders within Europe and regions that span

[1] Inforegio–Regional Development and Cohesion Policy 2021–2027 (europa.eu).

[2] Interregional cooperation between Hungary and Slovakia.

across borders. This strategic structure offers a more cohesive vision with future expectation, outlining objectives and advancement priorities for the region. Additionally, such a framework facilitates planned and integrated actions, as opposed to sporadic and uncoordinated interventions. Overall, cross-border strategies represent a innate progression beyond addressing constantly challenges and provide a valuable plan for the formulation of multi-annual cross-border policies and financial support programs from various financial sources (Medeiros 2018).

During the period spanning from 2014 to 2020, the INTERREG programs financially supported a diverse array of initiatives with the goal of advancing collaboration across territories and the advancement of regional development within the European Union. These initiatives covered a broad spectrum of thematic areas, addressing issues ranging from economic growth, innovation, and environmental sustainability to social inclusion and cross-border collaboration. The funded projects facilitated collaboration among diverse regions, countries, and stakeholders, promoting the exchange of knowledge, best practices, and experiences. Their overarching goal was to overcome territorial barriers and promote the harmonious advancement of border regions. The funded projects took various forms and scopes. Some focused on promoting entrepreneurship and supporting small and medium-sized enterprises by establishing cross-border networks and initiating business development efforts. Others concentrated on environmental protection and sustainable resource management, endorsing initiatives such as renewable energy, waste management, and ecosystem conservation. Furthermore, the Interreg programs allocated funds to projects in areas such as research and innovation, education and training, cultural heritage preservation, social inclusion, and healthcare cooperation. These projects aimed to enhance collaboration and the exchange of expertise across borders, driving regional development and fostering a sense of shared identity and cooperation among the participating regions. In conclusion, the projects supported by the Interreg 2014–2020 programs played a crucial role in strengthening European territorial cooperation. They addressed common challenges and made significant contributions to the sustainable and balanced development of the regions involved (Interreg Europe 2021).

During the upcoming programming period from 2021 to 2027, there are plans to sustain INTERREG initiatives aimed at fostering collaboration across European territories. These programs will concentrate on various aspects, including cross-border interactions, interregional cooperation, and the reinforcement of convergence and competitiveness. The specific priorities and actions for this new INTERREG phase are still under development, typically led by the European Commission and regional or national administrations. Looking ahead, the emphasis is expected to persist on addressing evolving socio-economic needs, overcoming legal hurdles, and advocating for a positive attitude towards cross-border living. Building upon the accomplishments of the preceding period, the INTERREG programs for 2021–2027 aim to amplify collaboration and integration among border regions. The stated objectives for this period include prioritizing competitiveness, supporting environmental sustainability, promoting the use of digital technology, encouraging inclusivity within society, and improving the overall well-being in the involved regions (Interreg Europe 2022).

The aim of several specific thematic projects of the call HU-SK aims to offer an extensive range of services to Slovak entrepreneurs, facilitating the establishment of new business collaborations with partners on the Hungarian front (BIC Bratislava).

Interreg programs exert an indirect influence on the process of urbanization and associated land-use practices. They tackle significant challenges within the realm of sustainable development, encompassing issues like environmental protection and the promotion of sustainable development. URBACT III[3] has explicitly endorsed the concept of polycentric urban structures, emphasizing the importance of small- and medium-sized cities as well as the connections between urban and rural areas. This endorsement specifically calls for the implementation of cohesive policies to address rejuvenating urban areas and managing the expansion of urban development. Moreover, the overarching objective of the program is to facilitate integrated sustainable urban development. Consequently, the various projects funded under this initiative are anticipated to make diverse contributions to sustainable urbanization. These contributions may involve the integration of transport planning and land-use planning, the encouragement of brownfield redevelopment, the promotion of green infrastructure, and the management of urban soil, among other aspects (Reitel et al. 2018).

The Benefits Derived from European Territorial Cooperation Programs are most visible in the new programme period, which can be influenced by the following characteristics:

- European added value emerges through Cross-Border Cooperation, contributing to the advancement of security, freedom, peace, and the protection of human rights;
- Political benefits include fostering European partnership, integration, cohesion and subsidiarity;
- Institutional advantages involve citizens and authorities, ensuring understanding of neighbouring regions and fostering long-term cooperation;
- Economic gains are realized by tapping into the inherent potential of cross-border areas, engaging all economic stakeholders, facilitating labor market access, and aligning qualifications;
- Sociocultural benefits manifest through the sharing of knowledge about neighbouring regions, their cultures, and the comprehensive understanding of their languages. The establishment of a functional international setting for trade, business and services relies on effective sociocultural cooperation (Guillermo-Ramirez 2018).

The Slovak Plan of Recovery delineates the EU's latest programming period (Plan of Recovery n.d.). The EU Commission introduced a new regional approach

[3] URBACT helps cities to develop an integrated set of actions for sustainable change.

for cross-border areas, incorporating local and microregional policies, with a foundation in cohesion policy.[4] The strategy of cohesion and subsidiarity includes more steps into the development of common territorial strategies in European Territorial Programmes.[5] The strategy across borders is shaped by innovation, programs, and topics, as indicated in the extensive description of the Interreg program.

The principle of subsidiarity is defined in Article 5(3) of the Treaty on European Union. It aims to ensure that decisions are taken at the level closest to citizens and that there is constant scrutiny to verify that action at European Union (EU) level is justified in the light of the options available at national, regional or local level.

Specifically, this is the principle that the EU does not act (except in areas falling within its exclusive competence) unless its interventions are more effective than those at national, regional or local level.

This principle is closely linked to the principle of proportionality, which requires that any EU action should not go beyond what is necessary to achieve the objectives of the Treaties. Another related principle, the principle of conferral of powers, states that any policy areas not explicitly agreed by all EU Member States in the Treaties remain within their competence (Pawera et al. 2015).

The social, political and economic landscapes of our cross-border region will undergo significant transformations due to the repercussions of the COVID-19 pandemic. These effects are poised to shape the trajectory of regional development and innovation in various projects. The pandemic has brought to light both specific and overarching challenges within the European Union, Slovakia, the cross-border region, and the global community at large.

The effects and the rise of innovation pose a longstanding challenge, placing a strain on the complete social structure by introducing a range of hardship, spanning from issues related to physical health to those linked with social and economic progress. The outcomes are uncertain and unforeseeable, yet the evident social, political, and economic repercussions of the ongoing pandemic are already manifesting their nature.

An increasing number of scientists and specialists are suggesting that humanity has reached a new phase in shaping the global order. What will the post-pandemic era bring for a society that transcends borders in the upcoming program period? How will socio-economic disparities evolve between 2021 and 2027? In the aftermath of the crisis, How will both local and international communities react and develop in relation to project awareness and enthusiasm for participation?

[4] INTERREG Programme Document 2021–2027 https://www.skhu.eu/programme/interreg-husk-2021-2027.

[5] European Commission, DG Region Urban Policy, 2023.

2 Literature Review

NextGenerationEU (NGEU) represents the European Union's temporary recovery initiative amounting to €800 billion, designed to facilitate economic recovery in the aftermath of the COVID-19 pandemic. Its primary focus is on fostering a more environmentally sustainable, digitally advanced, and resilient future. At the core of NGEU lies the Recovery and Resilience Facility, which extends grants and loans totalling €723.8 billion at current prices to support reforms and investments in EU Member States. A significant portion, up to €338 billion, is allocated as grants, while the remainder is designated as loans, with the understanding that these will be repaid by the respective Member States. If Member States do not utilize the full allocation of loans, the remaining funds will contribute to financing REPowerEU. These financial resources, disbursed through the Recovery and Resilience Facility, align with the national Recovery and Resilience plans of Member States, outlining pathways for reforms and investments aimed at enhancing the ecological, digital, and overall resilience of EU economies (European Commision (s.a.) 2023). The primary element of the NGEU initiative is the Recovery and Resilience Facility (RRF), which allocates €672.5 billion, comprising up to €312.5 billion in grants and €360 billion in loans (NGEU tracker (s.a.) 2023).

The scale of the Covid-19 shocks to the European and global economies far exceeds that of the financial crisis a decade ago, which was already historically significant. The pandemic has affected the economy through various channels, including heightened uncertainty, declines in specific areas of household spending, supply chain disruptions, and the profound consequences of measures to curb the virus, such as lockdowns, social distancing, and travel restrictions. Unlike the recession following the financial crisis, caused by demand shocks from widespread deleveraging efforts, the current economic downturn is characterized by both demand and supply shocks occurring simultaneously. The European Council approved a €750 billion package to provide financial assistance to governments grappling with pandemic-related challenges. The uniqueness of this package lies in its funding mechanism, relying on the issuance of EU bonds backed by the EU budget, including the repayment of debts covered by a slightly increased EU budget. The initiative, known as NGEU, is a commendable undertaking with a three-fold objective: stimulating aggregate demand, aiding the most severely affected countries to foster cohesion, and enhancing the Union's economic growth potential (Codogno and Noord 2020).

The EU has adopted a diverse borrowing approach for the NGEU, incorporating various types of securities and market access methods. NGEU represents a significant boost to the Union's fiscal capability, achieved through the issuance of public debt for which the EU assumes responsibility, without involving any fundamental capital transfers among Member States. This fiscal capacity, though fragile, extraordinary, and temporary, is intended to address the socio-economic consequences of the pandemic. Despite its primary role in pandemic recovery, NGEU not only offers relief but suggests a restructuring of the economic framework of Member States, too.

Consequently, it serves as a tool supporting reforms recommended and investments within the European Semester (Rayo 2021).

The funds allocated for the NGEU initiative will be distributed across three main pillars. Firstly, it will help Member States through financial support and reforms. Secondly, it aims to jumpstart the EU economy by supporting private investments and addressing crisis-related lessons.

A Recovery and Resilience Facility, totalling €560 billion, is established to financially support investments and reforms, particularly focusing on digital and green transitions, along with the resilience of national economies. This facility, integrated into the European Semester, will have a grant facility of up to €310 billion and the ability to offer €250 billion in loans. While accessible to all Member States, priority will be given to those most affected and in need of resilience. A €55 billion augmentation of existing cohesion policy programs in accordance with the latest REACT-EU program, distribution is determined by the degree of socio-economic severity impacts, is proposed. Additionally, there's a plan to enhance the Just Transition Fund to €40 billion, aiding Member States in accelerating their transition towards climate neutrality. The European Agricultural Fund for Rural Development will receive a reinforcement of €15 billion to assist rural regions in implementing essential structural adjustments in harmony with the European Green Deal. The Solvency Support Instrument, with a budget of €31 billion, we will leverage private resources to promptly aid sustainable European enterprises, aiming to unlock €300 billion in solvency support. InvestEU, Europe's primary investment program and will be upgraded to €15.3 billion. A new Strategic Investment Facility, integrated into InvestEU, is introduced to generate up to €150 billion in investments, particularly in sectors related to the digital and green transition. EU4Health,[6] with a budget of €9.4 billion, is introduced to enhance health security and get ready for upcoming health emergencies. There's also a €2 billion reinforcement for rescEU, the Union's Civil Protection Mechanism, to prepare for and respond to future crises. A substantial amount of €94.4 billion is earmarked for Horizon Europe, reinforcing it to fund crucial research in health, resilience, and the green and digital transitions. Furthermore, an additional €16.5 billion is allocated for external action, including humanitarian aid, to support Europe's global partners. Lastly, other programs of EU and instruments will be enhanced to be in accordance with recovery needs and strategic priorities, making the EU budget more adaptable and agile (Danieles 2020).

The Recovery and Resilience Facility constitutes the majority of the NGEU initiative, accounting for approximately 90% of the overall funding. The Council establishes facility, which allocates €672.5 billion in the form of both loans and grants. These funds are intended to assist member states in addressing the social and economic repercussions of the COVID-19 pandemic while facilitating their transition to greener and more digitally advanced economies, promoting sustainability and resilience (European Council (s.a.) 2023).

[6] A new Health Programme.

3 Data and Methodology

The article provides an insight into the 2014–2020 programme periods obstacles in border region, following territorial strategy and pandemic limits. Survey beyond past and future applicants describes a comparison between cross-border related issues, topics, territorial challenges about the cross -border life standards to predict current trends influenced by the pandemic and the potential crisis in the EU involves examining developments both prior to and following the year 2020. The added value stems from assessing the disparity in demographic and economic progress in cross-border regions, primarily between Slovakia and Hungary, post the COVID-19 pandemic during the new EU program period of 2021–2027. This is facilitated by the expansive possible of INTERREG HU SK cross-border project development.

The article aims to:

(1) highlight potential challenges in project implementation during crises;
(2) address common challenges collaboratively, seeking shared solutions among EU member states through the INTERREG HU SK territorial strategy, and
(3) evaluate and compare aspects such as education, research, health, transport, and sustainable energy.

The Interreg programmes receive funding from the European Regional Development Fund to promote the balanced development of the European Union's territory across various dimensions and started their new programme period just currently in 2023 with new Calls of proposals, also reacting on Priority Axis as Social Cooperation and Joint Risk Management, reacting properly on the after-pandemic status of EU crisis.

This study utilizes yearly panel data for the two neighbouring countries from NextGeneration EU plan in two implementation periods from 2014 to 2020 (n = 188 projects) and later between 2021 and 2023 in the beginning of the implementation period. The rationale behind the chosen period was that this period is consistent with EUs programming period and political upheavals, civil war, and unrest within the neighbouring (Slovakia and Hungary) countries, which led to the increase in defence expenditure of these countries. Secondly, this period also witnessed as crisis in developing and implementing regional development programmes The third reason for the chosen period was data availability in these countries and from Eurostat. The model of assessing the willingness to participate in projects is based on Human Development Index, where in development socially oriented projects are measured more as main economic indicators like GDP, inflation, unemployment rate etc. In line with this, the study established its goals with the following hypothesis:

(1) *Identifying the correlation among evolving legal challenges, socio-economic needs, public, qualified applicants, business and academic sector and their interest, as well as supporting positive attitude to cross—border living in the time of 2014–2020.*
(2) *Interpreting the disparities in the level of changed socio-economic behaviour in the priority sectors of Next Generation policy Joint Risk and Social Cooperation.*

(3) *Identifying the relationship between changing reactions to crisis in NextGenerationEU 2021–2027 and crisis Joint Risk Policies in cross—border territorial strategy living during 2014–2020.*

The hypothesis will prove whether the region is ready to handle the crisis in the field of project implementation and whether the participation on project willingness to participate has changed. It also includes strategies which can be evitable to deal with Crisis in NexGenerationEU based on INTERREG HUSK experience under Joint Risk Management Priority Axis.

4 Methods

Regarding research methodologies, surveys with a quantitative approach were disseminated. We examined a hypothesis in the following manner. This represents one of the structured scientific approaches to analyze and depict a particular phenomenon or issue meticulously. It involves the precise description and classification of the phenomenon, organization of information, exploration of its connections with other phenomena or issues, and drawing conclusions that contribute to the advancement of the subject under investigation. The investigation took place in the year 2023.

Project impact assessment derived from the present Priority Axes outlined in the ongoing 2023 call aims to make meaningful contributions to three pertinent themes:

1. Protection and preservation of Nature,
2. Social Cooperation and
3. Joint Risk Management (Table 1).

The research carried out within the local community encompasses individuals from the cross-border region who are applicants residing across borders including the Western part of the cross-border area namely the area of INTERREG projects (Bratislavský kraj-Bratislava Self-governing region, Trnavský kraj-Trnava Selfgoverning Region, Gyor-Moson-Sopron Megye-GYMS County). The study employed purposive sampling, comprising 188 individuals *(N=188)*. Subsequently, an online questionnaire using the Google Forms application was developed as the research tool.

Table 1 Correlation of cross-border policies with the recognition of Next Generation priorities within the financial framework

Priority axis	Million euro	Size of impact	Pearson correlation coefficient (R)
Protection and preservation of nature	**13.08**	**1.6**	**0.35**
Joint risk management	**15.4**	**0.06**	**0.28**
Social cooperation	**21.7**	**0.6**	**0.35**

Source Author 2023

To ensure the questionnaire's psychometric soundness, it underwent evaluation by a panel of experts in scientific research and sociology. The tool's apparent honesty was affirmed through this process. Internal consistency was assessed by calculating correlation coefficients between each phrase's grade and the total questionnaire score. All scale phrases exhibited statistically significant correlation coefficients with the total score at a significance level of *(0.01)*, indicating a high level of internal consistency. The research tool's stability was confirmed using the Cronbach Alpha standard, yielding a reliability coefficient of *(0.82)*. These elevated reliability values affirm the tool's validity, reliability, and the dependability of its results. For data analysis and achieving research objectives, One-Sample T Test and One-Way ANOVA were employed to determine sample effect size.

5 Results

The study outcomes indicated that the tested hypotheses were validated. The first question: What is the level of willingness to carry out INTERREG projects in one settlement (SK) or other (HU) in the time of the programme period 2014–2020 during COVID-19? To address this inquiry, the statistical analysis employed the arithmetic average, and the One-Sample T Test was used at the level of significance ($\alpha = 0.05$)

Another research question was: Is there a statistically significant correlation between the implementation of Joint Risk Management type if INTERREG projects and the period chosen from old and new programme period? (Table 2).

The test has showed that Business sector is focusing more on Joint Risk management opportunities from private and public sources as well, as second are municipalities and educational segment, who doesn't show big interest or have limited sources to deal with this issue of Next Generation.

A positive and medium-strength correlation exists between the interest in and realisation of projects within NGOs, schools, and companies, particularly in response to the impact of the COVID-19 crisis. This implies that larger institutions and those located in more densely populated cross-border areas tend to experience heightened concerns about the pandemic. Consequently, there is an increased inclination to engage in joint risk management projects and implement crisis-related policies to a certain extent, and vice versa.

The effect size, quantified at *(0.154)*, indicates that the willingness to participate in Joint Risk Management projects during the new program period from 2021 to 2027 has a relatively substantial influence, accounting for 55.4%, on addressing the challenges posed by the COVID-19 crisis. We conducted a comparison of regional development data, including statistics, to analyse the origins of cross-border citizens, with a specific emphasis on the capital and southwestern region. In this regard, we assessed their readiness to undertake projects during the ongoing pandemic.

Table 2 Correlation among cross-border joint risk management policy and project awareness

Relationship between project numbers and the eligible applicant	Pearson correlation (R)	Size of impact	Category	Joint risk management
Relationship between the level of project **awareness within municiplatiies** and implementation of cross-border projects	0.35	1.6	PUBLIC	++
Relationship between **school and educational institution** activities within implementation of cross-border projects	0.28	0.06	EDUCATIONAL	+
Relationship **between crossborder business** and in implementation of cross-border projects	0.45	0.6	BUSINESS	+++

Source Own elaboration 2023, **+++ strong, ++ —medium + —low**

6 Discussion

Based from the definitions of the regional policy instruments. For example, *Vanhove and Klaasen* (1987) include among the regional policy instruments (a) infrastructure support, (b) financial incentives, (c) regulatory instruments, (d) decentralization of state authorities,(e) regional allocation of public investment and state contracts, (f) regional development agencies, and (g) macropolicy instruments. A comprehensive classification of regional policy instruments, combining several of the above aspects, is also offered by *Maier and Tötling* (1997), who took into account not only the nature of the instrument itself but also its target group, interest, whereas, the motivation of applicants was major in the fields of information, investment and infrastructure, promoting joint risk management in all phases of projects SME sector as well. Research has shown that there are correlations between the private sector and their motivation to the emphasis on the joint risk management policy in previous programme period as well as in the future plans.

The upcoming Interreg initiatives, financed by the European Regional Development Fund, provide assistance of development of the European Union's territory at different levels and started their new programme period just currently with new Calls of proposals, strongly reacting on Priority Axis as Joint Risk Management and Social Cooperation, reacting properly on the after-pandemic status of EU crisis. NextGenerationEU represents an opportunity to recover stronger from the pandemic, overhaul the economies of the EU, and generate prospects and employment for the people of

Europe. where the vision to invest together €806.9 billion can make Europe greener, more digital and more resilient.

Slovak and Hungarian cross-border initiatives have become increasingly intricate due to the impact of the pandemic. The demographic growth, population trends, and the influence of COVID-19 present challenges without foreseeable prognoses. Despite the varied administrative, legal, and rural histories between the regions in different programming periods, a shared interest in crisis prevention and proactive Joint Risk Management Policies is evident.

In the region, trade and agriculture sectors, as well as the intensive care of countries, regions, and cities, are experiencing significant growth in peripheries. This dynamic introduces new challenges and objectives for projects under NextGeneration, encompassing sustainable development, regional and strategic management, and document-based approaches. Local authorities play a crucial role in disseminating information and advice to those interested in pivotal projects and the population of suburban areas. This forecast aligns naturally with the European Community framework, a focal point in the European Union's INTERREG programming. Furthermore, within the realm of sustainable innovation and the development of cities, applicants and inhabitants exhibit five distinct characteristics. Despite the majority originating from the Capital City and Southern region, there is no discernible correlation between the applicant's origin and the willingness to execute projects across borders (Sadiq 2020).

7 Recommendations

Due to the recent political developments surrounding the NextGeneration EU, there has been increase in government budgetary allocations in these countries in crisis times. Several studies conducted on the joint risk management and social cooperation indicators the growth relationship showed that there are conflicting outcomes. While some studies concluded that joint risk allocation spending affects the growth of the economy positively, other studies found the relationship to be negative. However, there is a common of research on this subject matter in Next Generation counties, hence the rational for embarking on the present study.

The study examined the relationship between joint risk management, social cooperation spending and regional development willingness in countries using project implementation data which spanned from 2014 to 2020 and at the beginning of next generation from 2021 to 2023. The result of the test suggests that the effect model is the most appropriate for the analysis.

In terms of research methodologies, quantitative survey questionnaires were disseminated. We tested 3 hypotheses about Joint Risk Prevention and Crisis Mitigation.

1. *Identifying the correlation among evolving legal challenges, socio-economic needs, public, qualified applicants, business and academic sector and their*

interest, as well as supporting positive attitude to cross—border living in the time of 2014–2020.

This hypothesis was proved, the cross- border living was challenged with numbers of standing and outgoing projects, with relevant portfolio towards risk management.

2. *Interpreting the disparities in the level of changed socio-economic behaviour in the priority sectors of Next Generation policy Joint Risk and Social Cooperation.*

The socio-economic behaviour of willingness to participate on Joint Risk management projects is high in the sector of business and lower in public, academic and NGO sector. Thereis and evidence s a higher effect in private sector in order to protect companies, SMEs and applicants in project implementation phases which was proved by research data on the sample.

3. *Identifying the relationship between changing reactions to crisis in NextGenerationEU 2021–2027 and crisis Joint Risk Policies in cross—border territorial strategy living during 2014–2020.*

Identifying the reactions to crisis it was stated that the source of data we collected between 2014 and 2020 who were eligible and successful applicants of INTERREG projects is in the majority willing to participate (55.14%) and the correlation between old programme period shows that there are territorial and preference issues, which can be a good policy result for applying into new programme manuals and programme documents involving business sector.

This represents a type of structured scientific examination and portrayal of a particular phenomenon or issue. It involves precise description, classification, and organization of information, aiming to comprehend the connections between the studied phenomenon or problem and others. The objective is to draw conclusions that enhance the understanding of the subject, thereby contributing to the advancement of the studied reality. The analysis of project impact in the ongoing call for 2023 is centred around the Priority Axes, aiming to address three critical areas in crisis prevention and development tools. These include the safeguarding and conservation of the environment, fostering social cooperation, and implementing collaborative risk management strategies and how it influences willingness to participate in other projects. From the long-run and short-run results of the study observed that all the explanatory variables are significant in the long-run but insignificant in the short run except for the past year's inflation rate.

In view of the above findings, the study recommends that spending on Next Generation Europe in INTERREG called as Joint Risk Management Priority Axis expenditure should be encouraged in periods of security threats, as it enhances economic growth through infrastructural and technological facilities. This encourages investment, productivity, and income growth in the cross-border area countries.

In addition, the inflation rate should be maintained within the threshold it can influence economic growth positively. Finally, as capital formation negatively influences growth in the cross-border countries, government should invest in more productive ventures and joint risk management strategies together with EU in line with NextGeneration EU. This will promote economic growth in the long run.

References

Anderson JE (2015) Fitting linear mixed effect models using IME4. J Stat & Softw 67(1):1–48

Bavaud F, Kordi M, Kaiser C (2018) Flow autocorrelation a dyadic approach. In: Annals of Regional Science. ISSN 05701864

BIC Bratislava. Projects. https://www.bic.sk/en/projects

Christie R, Claeys G, Weil P (2021) Next generation EU borrowing: a first assessment' Policy Contribution 22/2021, Bruegel. https://link.gale.com/apps/doc/A682056262/AONE?u=googlescholar&sid=googleScholar&xid=dfb5fb72

Chun Y (2008) Modelling network autocorrelation within migration flows by eige spatial flitterins. J Geogr Syst 10:317–344

Codogno L, van den Noord P (2020) Assessing Next Generation EU (December 15, 2020). Amsterdam Centre for European studies Research Paper No. 2020/09. Available at SSRN https://ssrn.com/abstract=3749255 or https://doi.org/10.2139/ssrn.3749255

European Commision (s.a.) NextGenerationEU. European Commision. Retrieved October 29, 2023, from https://commission.europa.eu/strategy-and-policy/eu-budget/eu-borrower-investor-relations/nextgenerationeu_en

European Commission (2022) Strategy and Policy COVID 19. https://commission.europa.eu/strategy-and-policy_en

European Council (s.a.) A recovery plan for Europe. European Council. Retrieved October 20, 2023, from https://www.consilium.europa.eu/sk/policies/eu-recovery-plan/

Danieles A (2020) Next generation EU—European Commission's new recovery instrument. ECCP. Retrieved October 29, 2023, from https://clustercollaboration.eu/news/next-generation-eu-european-commissions-new-recovery-instrument

Delaneuville F (2022) De l'enthousiasme suscité par le "retour vers l'Europe" à la tenation souverainiste: Comprendre les Etats du groupe de Visegrad January 2022. In book: Europe de l'Est et l'Union européenne. Quelles perspectives? Bruylant, Publisher, pp 93–103

European Commission, DG Region Urban Policy, (2023) Regional and Urban Policy. https://commission.europa.eu/about-european-commission/departments-and-executive-agencies/regional-and-urban-policy_en

Guillermo-Ramirez M (2018) The added value of european territorial cooperation. Drawing from case studies. In: European territorial cooperation. The Urban Book Series. Springer, Cham. https://doi.org/10.1007/978-3-319-74887-0_3

Hoyt Thimothy R (2006) Military policy and regional defense policy. ISBN, 9780203969045

INTERREG Europe: Interreg Europe Programme Manual. (2021). https://www.interregeurope.eu/sites/default/files/2022-01/Interreg_Europe__Programme_manual.pdf

INTERREG Europe. Programme manual (2022). https://www.interregeurope.eu/sites/default/files/2023-02/IR-E_programme_manual_annexes.pdf

INTERREG HU SK Programme Document 2021–2027. https://www.skhu.eu/programme/interreg-husk-2021-2027

Jablonczay T, Grajczár T (2014) Területi identitás stratégiák. ISBN: 978–963–08–6949–2

Kováčik V (2023) Fondy Európskej Únie v Slovenskej republike. 2023 Univerzita sv. Cyrila a Metoda v Trnave, Fakulta sociálnych vied ISBN 978–80–572–0013–0 . [online]. 2023, vol 1 [cit. 2023–12–11]. Dostupné na internete https://fsvucm.sk/wp-content/uploads/2023/01/Fondy-Europskej-Unie-v-Slovenskej-republike_kovacik.pdf

Maier G, Todtling F (1997) Regional and Urbanistic Economy. Localization theory and spatial structure. Elita, Bratislava, SK

Malhorzata F, Liszievszky M (2010) Demographic consequences of suburbanization in selected towns in the eastern December 2010. Bullet Geogr Socio-Econ Ser 14(14). https://doi.org/10.2478/v10089-010-0011-4

Marek D, Kantor T (2009) Příprava a řízení projektu strukturálních fonfu Evropské unie. Barrister & Principal, Brno, 215 s. ISBN 978–80–87029–56–5

Medeiros E (2018) European Territorial Cooperation: Theoretical and empirical approaches to the process and impacts of cross-border and transnational cooperation in Europe, vol 54. Springer, Cham, p 269. https://doi.org/10.1080/00343404.2019.1672937

NGEU tracker (s.a.) What is NGEU? NGEU tracker. Retrieved October 29, 2023, from https://www.ngeutracker.org/4

Pawera et al (2015) Regionálna ekonomická politika a politka súdržnosti EÚ. Univerzita Komenského v Bratislave. ISBN 978–80–223–3549–2

Plan of Recovery, Plán obnovy a odolnosti SR Retrieved https://www.planobnovy.sk

Pyka, Delaneuville R, Giba F, Zummerová LM (2022) De l'enthousiasme s suscité par le "retour vers l'Europe" à la tenation souverainiste : comprendre les Etats du groupe de Visegrad

Rayo AO (2021) The Next Generation EU instrument and the European Union new budget cycle. Rev Catalana Dret Public 63:4–20. https://doi.org/10.2436/rcdp.i63.2021.3710

Regional Policy EU: Regional development and cohesion policy. Retrieved https://ec.europa.eu/regional_policy/en/newsroom/news/2018/05/29-05-2018-regional-development-and-cohesion-policy-2021-2027 Inforegio—Regional Development and Cohesion Policy 2021–2027 (europa.eu)

Reitel B, Wassenberg B, Peyrony J (2018) The INTERREG experience in bridging european territories. A 30-year summary. In: European territorial cooperation. The Urban Book Series. Springer, Cham. https://doi.org/10.1007/978-3-319-74887-0_2

Sadiq M (2020) Policing in pandemic: Is perception of workload causing work–family conflict, job dissatisfaction and job stress? J Public Affairs:e2486. https://doi.org/10.1002/pa.2486

Šuplata M, Delaneuville F, Giba M, Demjanovič D (2022) The European Union on cross-roads: An overview of post-lisbon crises and a way forward. European Studies 9:122–153. https://doi.org/10.2478/eustu-2022-0006

Trend (2023) Eurofondy sú jednou hlavných výziev programe vlády ale namejú ani vlastnú kapitolu [online]. 2023, vol 1 [cit. 2023–12–11]. Dostupné na internete: https://www.trend.sk/spravy/eurofondy-su-jednou-hlavnych-vyziev-programe-vlady-ale-nemaju-ani-vlastnu-kapitolu

Van den Berg L (1987) Urban systems in a dynamic society. Gower.Gower PubCo, Aldershot, UK. ISBN-13: 978–0566052514 ISBN-10: 0566052512

Vanhove-Klaasen LH (1987) Regional policy: A European approach, 2nd ed. Aldershot, Gower. ISBN 0–566–05413–2

Vlachinský M *Why we should be concerned about the recovery plan.INESS.* online]. vol 1 [cit. 2023–12–11], pp 5–21. Retrieved https://iness.sk/sites/default/files/documents/pdf/other/2021/ngeu-why_we_should_be_concerned_about_the_recovery_plan.pdf

From Risk to Reinvention in Europe

The contributions gathered in this volume address the complex transformations facing Europe across economic, environmental, technological, and organizational domains. They reflect a shared understanding that contemporary crises, whether inflationary, ecological, labor-related, or geopolitical, are interconnected. This structural condition of "polycrisis" requires integrated answers that transcend sectoral boundaries and traditional frameworks.

A cross-cutting theme is the redefinition of the firm's role in society. Corporate responsibility is no longer limited to financial performance. The adoption of sustainability standards illustrates a broader shift toward transparency, ethical governance, and climate accountability. These developments are framed not only as regulatory demands but also as strategic opportunities to strengthen competitiveness and stakeholder trust.

Digital transformation, particularly the deployment of artificial intelligence, emerges as both a driver and a challenge. The unequal pace of AI adoption across firms and regions points to persistent structural disparities within the European Union. While technological change offers efficiency gains, it also accentuates risks of exclusion, automation-related job loss, and organizational misalignment. Several chapters emphasize the importance of strategic planning tools tailored to uncertain, fast-evolving environments—what is now conceptualized as the VUCA (Volatile, Uncertain, Complex, Ambiguous) world.

The labor market is another focal point in relation to generational shifts and post-pandemic dynamics. Studies on Generation Z highlight changing expectations around purpose, flexibility, and personal development. Traditional employer value propositions are being reassessed, requiring firms to recalibrate recruitment strategies and workplace culture. The management of human capital, especially digital skills development, is now a key determinant of resilience and performance.

In parallel, trade and economic policy are explored through the lens of environmental sustainability. The case of the relation in between European Union and MERCOSUR, exemplifies tensions between economic liberalization and ecological integrity. Negotiations reflect competing priorities as market access versus biodiversity protection. It underscores the need for policy coherence in a carbon-constrained world.

Taken together, the chapters underscore a central insight: current crises are not anomalies but signals of deeper systemic shifts. Addressing them requires not only technical solutions but also a normative rethinking of growth models, institutional roles, and governance frameworks. While the contexts vary, the convergence around adaptability, interdependence, and long-term vision provides a coherent basis for policy, academic, and strategic engagement.

Sociological Perspective on Strategy Tools for the VUCA Environment: A Systematic Literature Review

Jorge Caldeira and Tawfiq Rkibi

Abstract As most companies operate in extremely turbulent environments and considering that strategic management is referred to as having several limitations in these contexts, the need arises to identify the strategy tools (ST) that can operate effectively in volatile, uncertain, complex and ambiguous environments. A systematic literature review (SLR) was undertaken, covering the period from 1990 to 2021, in the search for scientific articles identifying ST for the VUCA environment. The concept of VUCA first appeared in the 1990s. It is an acronym for volatility, uncertainty, complexity and ambiguity, seeking to describe the turbulent and chaotic environment that most companies operate in. Several authors have highlighted the impact of this environment on strategic management. It is also widely agreed that ST play an essential role in strategic management. Using the right ST could help companies avoid strategic disasters in such challenging environments. The ST toolkit identified appears to be incomplete, unbalanced and conservative. Although the VUCA world and strategic management have been interconnected for at least three decades, information on how companies should select strategy tools in a multidimensional perspective (stages of the strategic process vs. VUCA components) is limited. This SLR is aimed at identifying the strategy tools that are appreciated through the sociological view (researchers) for the present context, have an ST selection support matrix and allow specialists in strategic management to reflect on strategic practices.

Keywords Strategy tools · VUCA · Sociological perspective · Systematic literature review

---

J. Caldeira
Instituto Politécnico de Santarém, Santarém, Portugal
e-mail: jorge.caldeira@esg.ipsantarem.pt

T. Rkibi (✉)
Universidade Europeia, Lisbon, Portugal
e-mail: tawfiq.rkibi@universidadeeuropeia.pt

G. Pellat et al. (eds.), *Data-Centric Business and Applications*, Lecture Notes on Data Engineering and Communications Technologies 258,
https://doi.org/10.1007/978-3-031-95068-1_10

1 Introduction

1.1 Framework

The world has changed and the globalization and hyperconnectivity of people and markets has sharply increased the volatility of our societies and economies and one of the victims of this is conventional strategic planning (Schühly et al. 2020). According to Robinson et al. (2017), strategic decisions are being made in contexts characterized by the impact of the effects of volatility, uncertainty, complexity and ambiguity, which Johansen and Euchner (2013) call the VUCA environment.

It is widely agreed that strategy tools (ST) play an essential role in the strategic management of organizations (Rengarajan et al. 2021; Vaitkevičius et al. 2006; Vuorinen et al. 2018) and so it is fundamental to understand how specialists are selecting the models, methods and procedures to adapt to the context of turbulent environments (Jarzabkowski and Kaplan 2015; Rengarajan et al. 2021).

On the other hand, the result of the increase in the number of ST is that it has become more difficult for managers to select ST (Vaitkevičius et al. 2006). Just using certain ST will allow companies to increase their probability of successfully defining their strategies in such challenging environments (Rengarajan et al. 2021). This is why managers are looking for ST that are capable of supporting planning and assuring a competitive advantage (Afonina 2015). By choosing appropriate ST, organizations can avoid strategic disasters (Courtney et al. 1997), so it is important to learn about these ST because, according to De Waal and Knott (2013), ST can only be adopted if the users are aware of their existence.

1.2 Goals

Strategic management developed in a context of the interaction of three perspectives (Anastassopoulos et al. 1988). According to the authors, the first, defined as the sociological perspective, is made up of researchers and academics (e.g. Harvard Business School, HEC and M. Porter) concerned about the universality of the models. The second perspective, called the economic perspective, consists of consultants (e.g. BCG, Arthur D. Little and McKinsey), focusing on the essentials of strategic management, particularly facilitating strategic choices. Finally, the third perspective, called the business perspective, consists of company/business managers (e.g. at General Electric and Shell) (Ibid, p. 107). This last case is somewhat different from the previous approaches insofar as the companies were supported by consultants (Ibid, p. 107). Laamanen (2017) also emphasizes academics and consultants in the development of approaches, theories and frameworks.

Navas-López and Guerras-Martín (2018) underlined the consolidation of knowledge of strategic management through the confluence of several sources, from academic research and the practical experience of consultants and companies. The authors state that this confluence is very often called the ABC (Academics, Business, Consultants) model and has contributed to the growth of the area. For the authors, strategic management has been developed through a concept of balance between theory and practice, where research produces knowledge and company managers generate experience through practice.

Several authors have referred to the production of knowledge of strategic management in the context of different sources. Jarzabkowski and Kaplan (2015) also highlight development capacity and the dissemination of strategic management through academics, company managers and consultants. In line with Jarzabkowski, Qehaja et al. (2017) state that researchers and consultants are responsible for the development of several ST. Academics are an important source of dissemination of knowledge of strategic management (Jarzabkowski and Kaplan 2015), while company managers adopt this knowledge and adapt it to their needs (Jarzabkowski et al. 2013).

Two research questions were defined: Q1. What is the ST toolkit for the VUCA environment? Q2. To what extent is the strategic process covered by ST in a VUCA environment?

The ST toolkit could help specialists in strategic management to learn about and select ST in a VUCA environment. The degree of coverage is directed at identifying how far the points of contact of the effects of the components of the VUCA (volatility, uncertainty, complexity and ambiguity) environment in the different stages of the strategic process (analysis, formulation, action, monitoring and adaptation) are in some way protected by ST. This SLR is aimed at gaining greater understanding of the ST that are appreciated by the sociological view and generating knowledge contributions for another three perspectives, professors (teaching ST), consultants (applying ST) and managers responsible for the strategic management of businesses (using ST).

1.3 Structure

The article is structured in six parts, which are preceded by the introduction, and ends with the final considerations, followed by the references. The central parts of the paper are made up of four points: theory concepts, methodology, tools for VUCA, discussion and conclusions, where the results of the SLR are discussed.

2 Concepts

2.1 Strategic Management

2.1.1 Strategy

Strategic management has been considered to be a fundamental question that explains the success or failure of a company (Rumelt et al. 1991). For Ansoff et al. (2019), although complex and costly, strategy is a powerful tool for dealing with the contexts of change and there is no doubt that it ends up paying for itself.

Until the mid-1950s, organizational management was essentially based on financial planning (Ansoff et al. 2019). In the 1950s, organizations were faced with challenges of competitiveness and expansion that surpassed the capacity of the planning process, thus far focused on budgeting, and it is within this framework that long-range planning appeared (Grant 2020). With long-range planning, the future is expected to be predictable through extrapolation of the historical growth (Hussey 1998). With the increase in turbulence and discontinuity in the environment, the future is no longer predictable, making it vital to analyze the external environment of the organization (Ansoff et al. 2019). Strategic planning first appeared in the 1960s, ten years after long-range planning (Bolland 2020) and replaced extrapolation with analysis of company environments. It was in the early 1980s and 1990s that strategic planning came in for criticism (Mintzberg et al. 2009; Ocasio and Joseph 2008). The criticism focused on its inability to improve company performance (David 2011). It was in this context that strategic management emerged. Grant (2020) believes that the transition from strategic planning to strategic management comes when competition in the environment and the maximization of performance become the main goal of companies.

There are many external factors (see Table 1) that have influenced the changes in strategic management over the years (e.g. the post-World War II recovery period, globalization, competition, internationalization, inflation, the oil crisis, technological changes, etc.).

2.1.2 Strategy Process

The strategy process is different from the strategy content. The goal of the former is to try to answer the question of how strategies are formed, implemented and modified, while the latter describes the strategy itself (Whittington et al. 2020).

According to Grant (2020), the strategy process can be defined as a systematic approach that makes it possible to obtain information, share insights, conduct analyses, reach decisions, guarantee consistency between the different decisions and commit managers to actions and performance goals. The author also states that the dialog inherent to the strategy process is extremely important in that it makes it possible to share knowledge, communicate ideas and reach a consensus. Several

Table 1 Changes in strategic management and external factors

Period	Conceptual changes	External factors
Up to the mid-1950s	Financial planning	Post-World War II recovery period, market stability and increased demand (Mele and Parra, 2006). Large-scale production in American industry (Bolland 2020)
Late 1950s and 1960	Long-range planning	
1960–1970	Strategic planning	Continuation of the recovery period and prosperity. Post-World War II (Bowman et al. 2002). Increased complexity and dynamism of the environment (Mele and Parra 2006)
1970	Strategic management	Stagnation and inflation (Bowman et al. 2002, p. 33; Hussey 1998). Oil crisis and macroeconomic instability (Grant 2020; Laamanen 2017). Faster pace of change (Ansoff et al. 2019)
1980		Increase in international competition and appearance of the globalization phenomenon (Furrer et al., 2008). Increase in competition pressure from Asian nations (Grant, 2020)
1990		Rapid economic and political discontinuities in the international context (e.g. the end of the Cold War) (Bowman et al. 2002)
		Digital technology, disruptive technology and faster changes in the environment (Grant 2020)
2008		2008–2009 financial crisis (Grant 2020; Iborra et al. 2019)
2018		Turbulent environments (Grant 2020)
Post-2020		Big data, artificial intelligence (Schühly et al. 2020; Talaoui et al. 2023)

authors present the strategy process divided into different, sequential stages that are inter-related in a dynamic and continuous process (Kaplan and Norton 2008a, b; Lynch 2018; Vuorinen et al. 2018; Whittington et al. 2020). Although there are some differences in the way the process is structured, the sequence of stages presented by the different authors is quite similar. To support this contribution, it was decided to use the process devised by Vuorinen et al. (2018) due to its simplicity and timeliness. The substages of the action (translating strategy to operations and organizing activities) were abbreviated for implementation (see Fig. 1).

In the model proposed by Vuorinen et al. (2018), the strategy process begins with the architecture phase, which consists of a strategic analysis and is aimed at preparing the diagnostic of the external and internal environment of the organization (e.g. macrotrends, business environment, sector, competitors, strong and weak points, capacity and resources, etc.) and strategic formulation (e.g. definition of the strategic aims, goals and strategies). The action stage involves translating strategy to operations and organizing activities, particularly through designing detailed actions and identifying the necessary resources. The adaptation phase consists of monitoring and learning, which involves monitoring and discussing the performance of

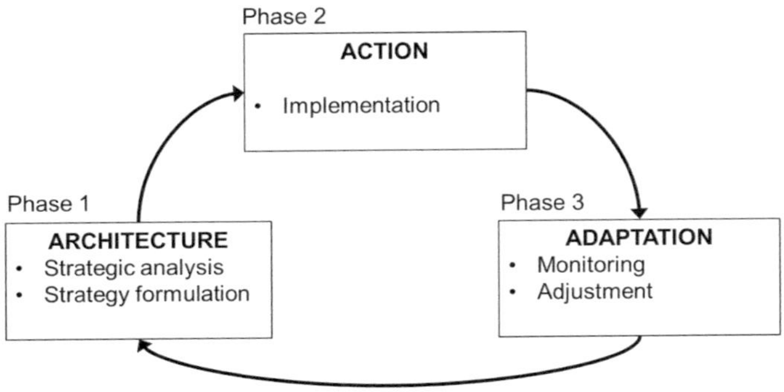

Fig. 1 Sequence of stages and substages in the strategy process (Vuorinen et al. 2018)

the goals, analysis of the effectiveness of the strategies and actions planed, analysis of the impact of the decisions taken, etc. The adaptation stage also includes the organizational learning process, giving rise to decisions with possible adjustments to the planning documents.

2.1.3 Strategy Tools

According to Bowman et al. (2002), a significant part of the strategy field is associated with the development of strategy tools. Hussey (1998) states that the discounted cash flow method was one of the first tools to be used to support planning. The early years of strategic management saw the appearance of some of the most iconic frameworks, such as BCG Matrix, GE-McKinsey Matrix, ADL Matrix—Arthur D. Little (Cummings and Daellenbach 2009), Five Forces Framework (Porter 1990), Generic strategies (Porter 1980), Experience curve (Rumelt et al. 1991), SWOT analysis, PIMS, Scenario analysis, 7-S McKinsey Framework, etc. (Bowman et al. 2002).

There is no one generally accepted term for strategic management models and techniques (Afonina and Chalupský 2013). Of the different terms found, the most common are "Strategy tools" according to Jarzabkowski and Kaplan (2015); Kaplan and Norton (2008a); Knott (2006), "Strategic Management tools" according to Clark (1997), "Models" according to Mintzberg et al. (1998), and "Tools & Techniques" according to Knott (2006). In this article, the decision was made to choose the term strategy tool (ST), as there is some consensus about its use.

The term ST will encompass frameworks, concepts, models and methods (Jarzabkowski and Kaplan 2015) whose aim is to provide support for organizations in the different stages of the strategic process, which is in line with the approach of Knott (2006), who believes that ST are something that structure or influence strategic activities.

According to Hakala and Vuorinen (2020), ST are used in the strategic process to create strategy content. Several functions are referenced by different authors regarding the use of ST: guiding the strategy process (Knott 2006), supporting and improving thought and strategic decision-making (He et al. 2012), increasing the analytical capacity of managers (Webster et al. 1989), solving complex problems (Frost 2003), clarifying (Wright et al. 2013), supporting communication and promoting different perspectives (Wright et al. 2013), providing rationality and objectivity (Kaplan and Jarzabkowski 2006).

According to Hakala and Vuorinen (2020) the classification of ST is not always clear, insofar as there is no general consensus as to whether some tools constitute ST. For example, budgeting, mission and vision declarations, spreadsheets and customer satisfaction surveys are sometimes deemed to be ST. However, in the opinion of the authors, it is not clear if they can be strictly termed ST. In the study, *Management Tools & Trends*, by Bain & Company (Rigby and Bilodeau 2018), the mission and vision declarations are ranked as the tenth most commonly used management tools.

2.2 VUCA

Many authors are unanimous on the fact that the level of turbulence has increased substantially in recent years (Ansoff et al. 2019). Turbulence can also have different levels of intensity. The authors describe the turbulence of company environments by establishing a scale of five levels to describe the level of turbulence: 1. Repetitive (stable environment), 2. Expanding slow incremental, 3. Changing fast incremental, 4. Discontinuous predictable, and 5. Surprising discontinuous unpredictable. According to Kotter (2014), we are crossing a line to a place of unpredictable turbulence, where change is increasing exponentially, something we are not prepared for.

VUCA is an acronym that means: V for volatility, U for uncertainty, C for complexity and A for ambiguity, which seeks to describe the turbulent and chaotic environment in which most organizations operate (Murugan et al. 2020). The VUCA concept first appeared in 1990 at the US Army War College, recognized as an elite school of higher education in leadership and strategy, also known as the school for future American generals (Johansen and Euchner 2013; Kaivo-oja and Lauraeus 2018) and informally known as the "VUCA University" (Johansen 2007). The VUCA environment seeks to describe the world that emerged from the end of the Cold War in 1991, characterized by the appearance of a new global landscape that was diverse, multilateral, asymmetrical and diffuse instead of monolithic and with clearly identified enemies and strength (Millar et al. 2018; Schoemaker et al. 2018).

The acronym itself only appeared in the late 1990s and it was after the terrorist attacks in the USA on September 11, 2001 that the VUCA concept began to gain exposure (Lawrence 2013). The adoption of the term VUCA to describe the chaotic, turbulent and rapidly changing business environment came after the financial crisis in 2008 and 2009 (Kaivo-oja and Lauraeus 2018). The term has now become fashionable and has been widely adopted to describe the "new normal" (Chadha 2017; Du and Chen 2018; Giones et al. 2019; Kaivo-oja and Lauraeus 2018). The VUCA world has reached everybody, irrespective of whether the organization is in the public or private sector, whether it is for profit or not or it is a religious institution or a company listed in Fortune 500 (Kutz et al. 2018). Nowadays, the VUCA world is having an impact on all aspects of life (Johansen and Euchner 2013).

Many authors have tried to define each one of the components of the VUCA environment. Volatility can be described as instability, change or rapid and unpredictable movement (Kaivo-oja and Lauraeus 2018), as opposed to a stable environment. Uncertainty is characterized by a lack of information and indefinition (Bennett and Lemoine 2014), as opposed to a determinable environment. Complexity is associated with a lot of interconnected and interdependent variables (Thorén and Vendel 2018), as opposed to a simple environment. Finally, ambiguity refers to the unknown and has lots of different meanings, with several conflicting interpretations (Kaivo-oja and Lauraeus 2018; van Rijmenam et al. 2019), as opposed to a clear environment.

The VUCA environment created a huge challenge for the strategic planning process in organizations (Baaji and Reinmoeller 2018; Du and Chen 2018; Kaplan and Jarzabkowski 2006; Millar et al. 2018). For Oboreh and Umukoro (2011), the levels of uncertainty relative to the environment have increased significantly in recent decades, which affected strategic planning in organizations. Milliken (1987) argues that uncertainty can "paralyze" the strategic planning process. Grant (2003) also refers to the difficulty in making strategic plans given the increased volatility of the business environment. Strategic decisions are being extraordinarily affected by the VUCA environment (Bennett and Lemoine 2014). For Miles and Snow (2003), the adaption process of organizations to the environment has become too complex. According to Hakala and Vuorinen (2020), organizations have learned to live with the fact that they cannot make meticulous plans as they could in the past. According to the authors, we live in a world where nothing is certain and organizations have to adapt to constant uncertainty. Qehaja et al. (2017) state that managers have to deal with a multitude of issues at the same time, with all of these coming from different directions.

The VUCA environment has become the reality that many senior executives are faced with every day when making strategic or even operational decisions. (Robinson et al. 2017). Several authors state that the increase in turbulence poses a challenge to organizations with regard to how they make plans (Grant 2003; Ramírez and Selsky 2016; Shinkle and Hundley 2013) and implement strategies (Perrott 2008).

Table 2 shows the potential impacts of the components of the VUCA environment in the different stages of the strategy process.

In overall terms, the main impacts can be boiled down to:

a. Accelerated outdating of the strategic diagnosis.
b. Greater need for resources for preparing the strategic diagnosis.
c. Delays in preparing the strategic diagnosis.
d. Increased difficulty in strategic formulation.
e. Greater need of resources for formulation.

Table 2 Matrix of VUCA impacts on the strategy process

			VUCA components			
Strategy process	Phases	Substages	Volatility	Uncertainty	Complexity	Ambiguity
	Architecture	Analysis	Inability to anticipate (Giones et al. 2019; Malewska 2015; Milliken 1987; Sullivan 2012) Information quickly becomes obsolete at the end of the planning process (Malewska 2015; Kaplan and Norton 1996)	Uncertainty about the state of the environment. Forcing time and resources to be spent on analyzing the environment (Milliken 1987)	Not understanding the challenges of the organization (Bennett and Lemoine 2014) Not understanding the interconnections between the elements in the environment (Milliken 1987)	Inability to accurately conceptualize threats and opportunities before they become lethal (Kail 2011; Schoemaker et al. 2018)
		Formulation	Making it more difficult to formulate a strategy (Perrott 2008) The strategic plan quickly becomes obsolete (Reeves and Candelon 2022)	It becomes more difficult to decide which actions to adopt (Şener 2012; Shinkle et al. 2013)	Forcing a careful analysis of the different alternatives possible (Milliken 1987)	
	Action		Plans become obsolete before they have been fully implemented. Implementation of the strategy becomes more difficult (Perrott 2008)			

(continued)

Table 2 (continued)

			VUCA components			
	Adaptation	Monitoring	Rapid and unpredictable changes put pressure on the decision-making and approval process (Perrott 2008)		Information overload, which can "freeze" decision-making (Bennett and Lemoine 2014)	Difficulty in interpreting information (Milliken 1987)
		Adjustment		Forcing constant readjustments to be made to the strategy and the business model (Kaplan and Norton 1996; Schoemaker et al. 2018) Not knowing the answers or options available (Milliken 1987)		Difficulty in organizing responses (Giones et al. 2019)

f. Delays in strategic formulation.
g. Accelerated outdating of the strategic plan.
h. Accelerated outdating of the operational plan.
i. Increased difficulty in implementation of the operational plan.
j. Pressure on the decision-making process.
k. Paralysis in the decision-making process.
l. Weakening of decision-making.
m. Inability to adjust the strategy.

All the components of the VUCA environment have an impact on all the phases of the strategy process, but not all of the components necessarily have a direct impact on all of the phases (e.g. the impact of uncertainty on implementation and monitoring is not identified).

As the strategy process is made up of sequential, interconnected stages, the weakening of one stage may have consequences on the next stage. The strategic diagnosis stage allows essential information to be obtained for the strategic formulation (Whittington et al. 2020), so an unsuitable strategic diagnosis may lead to the strategic formulation being unsuitable. Unsuitable strategic formulation may produce a negative impact on the implementation of the strategy. However, according to (David 2011), correct strategic formulation is no guarantee of good strategic implementation. The assessment of the success of the strategic implementation should be constantly verified and analyzed, and redesigned if necessary (Pietrzak and Paliszkiewicz 2015). An unsuitable analysis may lead to weakened strategic and operational adaptation.

3 Methodology

It was decided to use the SLR in order to identify the ST included in the sociological perspective (Rengarajan et al. 2021; Vuorinen et al. 2018), covering the period from 1990 to 2021, in the search for scientific articles that identified ST for the turbulent environment of organizations. The option to keep the study based on the developments in research in the last 31 years is in line with the introduction to the VUCA concept created in the late 1980s.

The SLR is a research tool capable of responding to specific questions and generating knowledge through the identification and analysis of relevant studies (Torraco 2016). According to Petticrew and Roberts (2006), SLR are literature reviews that closely monitor a set of scientific methods aimed at explicitly limiting systematic error (bias), seeking mainly to identify, assess and synthetize all the relevant studies (of any design) in order to answer a certain question or set of questions.

Torraco (2016) classifies the SLR as a different research method that generates new information and that is justified whenever there is an absence of literature reviews on a certain dynamic topic that has seen rapid developments over an extended period of time. SLR have also been widely used to analyze a variety of complex, contemporary topics and very often controversial issues in the "real world" (Petticrew 2001).

In this SLR, research was carried out into scientific articles in b-on in the area of “business and management” that contained: AB (VUCA or turbulent) AND AB (abstract) (“strategic management” or “strategic planning”) AND TX (text) (tool or technique or model or framework). As suggested by Petticrew and Roberts (2006), criteria can be defined for the exclusion of articles. Four specific exclusion criteria were defined. In this study, articles were excluded when: 1. There are no ST, 2. The stage of the strategy process cannot be identified, 3. The element of the VUCA environment that is minimized cannot be identified, and 4. It is not in English, Portuguese or Spanish. Regarding the fourth criteria, researchers may exclude articles written in languages they are unable to read (Okoli and Pawlowski 2004).

The abstracts of 82 articles were analyzed, which allowed 21 articles to be eliminated. This stage is not based on an in-depth assessment of the quality of the articles, but rather on practical considerations to determine if, for the purpose of the review, the articles warrant more in-depth analysis (Okoli and Pawlowski 2004). The aim behind reading the abstracts of the articles was to eliminate the works that did not present ST in a strategic management context or were not of any significant relevance. Petticrew and Roberts (2006) suggest the tabulation of the features of the articles identified as a means of increasing transparency in the analysis. Excel was used to collect the key aspects related to the research (Durst and Edvardsson 2012).

The 61 articles considered relevant from their abstracts were read and subjected to in-depth analysis, checking to see if they met the exclusion criteria. Following the analysis, 40 were excluded because they met one of the four exclusion criteria, resulting in the identification of 21 relevant articles that present 24 different ST in the strategy process capable of mitigating at least one of the components of the VUCA environment. To prevent research fatigue, it was decided to reduce the ST list (Rengarajan et al. 2021) by excluding these three: (a) “Analytic Hierarchy Process”, because it is covered in Multicriteria Analysis; (b) “Future Studies”, because it is too broad and is already identified in the Scenarios method, and, (c) “Weak Signs and Wild Cards”, because it is covered in the Scenarios method. Following this analysis, the 24 ST were reduced to a final list of 21 ST. The majority of the articles were published between 2010 and 2021, which is in line with the increase in turbulence. This statement is proven by Kaivo-oja and Lauraeus (2018), who state that the adoption of the term VUCA came to the fore after the financial crisis in 2008 and 2009. Figure 2. shows the results obtained in the different stages of the SLR.

4 Results: Tools for VUCA

The analysis of the **21 articles made it possible to identify a total of 21 ST** (see Table 3). Two of the ST are made up of frameworks that use a combination of other ST or adapt existing models (e.g. Framework: Benchmarking & Analytic Hierarchy Process and deconstruction of the value chain). Some of the ST are broad designations that combine several other ST identified in the SLR (e.g. Multiple-criteria decision analysis and Strategic information system). Several of the ST identified

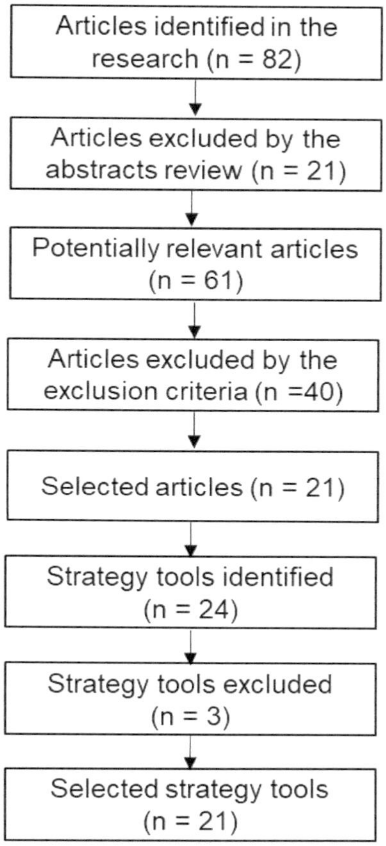

Fig. 2 Selection process for systematic literature review

can be applied though various different approaches (e.g. Scenarios, Frameworks for Mental Spaces, Benchmarking, Statistic forecast models, Crowdsourcing, etc.). For example, Scenarios can be prepared using the following approaches: Global Business Network, Cross Impact Analysis, *La Prospective*, etc.).

The most common ST in the SLR is the Scenarios method, followed by Benchmarking, Stakeholder analysis and Dashboards.

It is important to analyze the toolkit based on evolution over time in the management area, simultaneously comparing the 21 ST identified with the dimensions that "helped" the strategic management to evolve, in particular the external factors and their impacts on the environm ent in terms of turbulence (see Fig. 3).

Following the identification of the ST in the sociological perspective approach, the strategy process coverage matrix was prepared (see Table 4) with the aim of obtaining information to answer the questions in this study: Q1. What is the ST toolkit for the VUCA environment? Q2. To what extent is the strategic process covered by ST in a VUCA environment?

Table 3 Description of strategy tools

#	Strategy tool	Short reference	Purpose of the tool	Year
1	Game theory	(Belt and Bashore 2000)	Analyzing the choice of optimal decisions	1944
2	Delphi method	(Kaivo-oja and Lauraeus 2018)	Obtaining a consensus of opinions in a group of specialists through the intensive use of questionnaires and the resulting controlled opinion feedback	1953
3	Scenarios	(Begun 1999; Belt and Bashore 2000; Da Silva et al. 2012; Kaivo-oja and Lauraeus 2018; Sevaguru and Safa 2005; Vecchiato 2015)	Support for long-term decision-making through the identification of several plausible futures	1950s
4	Statistical forecast models	(Kaivo-oja and Lauraeus 2018)	Supporting forecasts through the use of historical data and statistical models	1950s
5	Stakeholder analysis	(Kaivo-oja and Lauraeus 2018; Mclarney 2003)	Obtaining a holistic view of the level of interest and power of all those around the organization	1963
6	Real Options	(Vecchiato 2015)	Helping to decide on the selection of resources	1977
7	Multiple-criteria decision analysis	(Moslem et al. 2020)	Supporting the resolution of complex problems that involve several criteria (quantitative and qualitative)	1979
8	Benchmarking	(Begun 1999; Hong et al. 2012; Kaivo-oja and Lauraeus 2018)	Comparing strategies, functions, processes, products or services, performances, etc. between the best organizations in their class	1982
9	Backcasting	(Thorén and Vendel 2018)	Approaching the future in the long term, looking back to the present	1982
10	Financial modeling	(Mclarney 2003)	Solving financial problems by using "What if" questions or projections	1980s
11	Strategic information systems	(Yoshikuni and Albertin 2018)	These are the different information collection (computer apps), processing and analysis systems used to support decision-making	1980s

(continued)

Table 3 (continued)

#	Strategy tool	Short reference	Purpose of the tool	Year
12	Lean management	(Hinterhuber 1994)	Obtaining improvements in times and costs in order to add value for the customer	1990
13	Mental spaces	(Leibold et al. 2004)	Helping organizations design business models	1994
14	STREET process	(Kaivo-oja and Lauraeus 2018)	Analyzing the impact of new technology on the organization	1994
15	Gartner hype cycle	(Kaivo-oja and Lauraeus 2018)	Basing the main technological trends on foresight	1995
16	Benchmarking & analytic hierarchy process	(Korpela 1996)	A framework that complementarily uses the AHP method and benchmarking to identify best practices	1996
17	Dashboards	(Batista et al. 2012; Kachaner et al. 2016)	Supporting management, particularly in the area of support for decision-making	1990s
18	Value chain deconstruction	(Dommisse and Oosthuizen 2004)	Anticipating changes in the value chain that could affect the existing business model	2003
19	Crowdsourcing	(Kaivo-oja and Lauraeus 2018)	Taking advantage of the contributions of the "wisdom of crowds" in order to generate the production of collective collaborative ideas	2006
20	CINE matrix for VUCA context	(Sharma 2019)	Analyzing the configuration context the managers or leaders are working in	2019
21	Moon ocean strategy	(Sharma 2019)	The moon ocean strategy leads to the "quantum leap" over current competition spaces	2019

5 Discussion and Conclusions

The aim of this study was to identify and analyze the ST toolkit in the theoretical fields of literature and two research questions were defined: Q1. What is the ST toolkit for the VUCA environment? Q2. To what extent is the strategic process covered by ST in a VUCA environment ?

(a) The first question was aimed at obtaining information for the construction of an ST toolkit to support strategic management in a VUCA environment context:

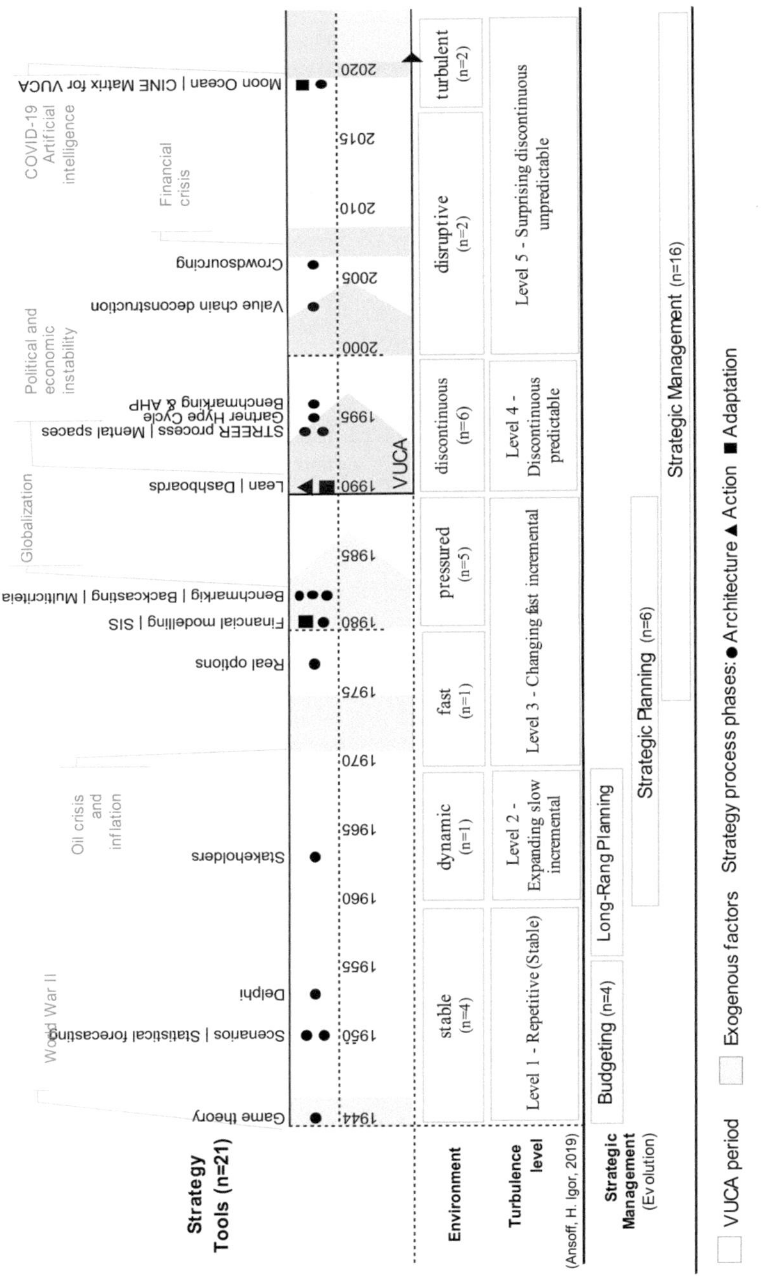

Fig. 3 Changes in strategy tools in the context of turbulence (temporal dynamics)

Table 4 Strategy process coverage matrix using the strategic tools in the environment

			VUCA components			
Strategic process	Phases	Substages	Volatility	Uncertainty	Complexity	Ambiguity
	1. Architecture	Analysis	Stakeholders; Scenarios; Crowdsourcing; Mental Spaces; Gartner Hype Cycle; Financial modeling; Benchmarking & Analytic Hierarchy Process; Statistic forecast models; Real options; STREET Process; Strategic Information Systems; Value chain deconstruction; Game theory	Benchmarking; Scenarios; Moon Ocean Strategy; Gartner Hype Cycle; Delphi; STREET Process; Strategic Information Systems; Value chain deconstruction; Game theory	Delphi; Strategic Information Systems	Stakeholders; Moon Ocean Strategy; Delphi; Real options; CINE Matrix for VUCA Context
		Formulation	Benchmarking; Multiple-criteria decision analysis; Scenarios; Benchmarking & Analytic Hierarchy Process; Strategic Information Systems; Value chain deconstruction	Backcasting; Scenarios; Moon Ocean Strategy; Strategic Information Systems; Value chain deconstruction	Multiple-criteria decision analysis; Backcasting; Strategic Information Systems	Moon Ocean Strategy; Backcasting
	2. Action		Lean management; Strategic Information Systems	Lean management; Strategic Information Systems	Strategic Information Systems	
	3. Adaptation	Monitoring	Benchmarking; Dashboards; Strategic Information Systems	Strategic Information Systems	Strategic Information Systems	
		Adjustment	Dashboards			CINE Matrix for VUCA Context

Note The grey areas identify the VUCA points of contact in the different stages of the strategy process (see Table 2)

The following considerations were used from the perspective of the sociological view:

(a.1) **Number of ST**: The toolkit is made up of 21 ST.

(a.2) **Concentration of ST**: The concentration of ST on two axes is highlighted. The first refers to the architecture axis (analysis and formulation)/volatility, which could denote greater concern about the volatility component and its effects: inability to anticipate (Giones et al. 2019), need to obtain information continuously (Mason 2007) and the difficulty in formulating strategies (Perrott 2008). The second refers to the architecture axis (analysis and formulation)/uncertainty, which could denote greater concern about the uncertainty component and the need to use more resources and time to analyze the environment (Milliken 1987), with the process deciding which strategies to adopt (Şener 2012; Shinkle et al. 2013).

(a.3) **Temporal dynamics**: After analysis of the period when the ST appeared, classifying these as before 1990 or after 1990, it was found that a little over half of the ST identified (52.3%) appeared before 1990, which could indicate a user preference for more tested ST rather than more current

ST. The toolkit only includes one ST identified in the study by Vuorinen et al. (2018)—Balanced Scorecard. Vuorinen's study identified 88 new ST that appeared between 1990 and 2015. The use of ST from before 1990 rather than more current ST could also suggest that researchers are not interested in investigating less conventional ST.

(a.4) **ST versus Turbulence context**: The results of an analysis of Fig. 3 suggest that the ST toolkit, despite being spread across the timeframe from 1944 to the present day, appears to arise during the pre-VUCA period, advancing a little in the early years of the VUCA period, up to level 5 of surprising discontinuous unpredictable. Although countless ST have appeared, they seem not to have been able to win over specialists in strategic management. It is in the strategic planning and strategic management period that there is a higher concentration of ST.

(b) To answer the second question, the results obtained when the ST were placed on the strategy process coverage matrix for the VUCA environment were analyzed (see Table 3):

(b.1) **Quadrants without ST**. The ST do not cover all the quadrants in the matrix. There are four quadrants with no ST (e.g. action/ambiguity, monitoring/ambiguity, adjustment/uncertainty and adjustment/complexity).

(b.2) **Incomplete**. There are nine quadrants with only one or two ST.

(b.3) **Exposed**. Cross-referencing with the information on the impacts of VUCA on the strategy process (see Table 2), thirteen quadrants were identified where the VUCA components have an impact on the strategy process—the strategy process is exposed in two quadrants: (i) in the monitoring/ambiguity quadrant—the impact could arise from the weakening of decision-making, insofar as it is difficult to interpret more ambiguous information (e.g. conflicting information) to support decision-making; (ii) in the adjustment/uncertainty quadrant—the impact may arise from not knowing the answers or options available (Milliken 1987).

In conclusion, the strategy process **is not absolutely covered** by the ST, as it appears to be exposed to the effects of some of the components of the VUCA environment (exposure in two quadrants). The **coverage is incomplete** [there are quadrants without any ST (n = 4)], **unbalanced**, where some quadrants have lots of ST [(n = 5), five or more ST] and some quadrants have few ST [(n = 9), one or two ST]. The ST toolkit can be considered to be **conservative** in that 52.3% of the ST appeared before 1990.

6 Limitations and Future Directions

6.1 Limitations

This SLR may have some limitations, mainly the following:

a. It does not take into account scientific articles that do not highlight VUCA in their texts, but could individually highlight one or more of its components: volatility, uncertainty, complexity and ambiguity.
b. It does not take into account articles that were excluded by an analysis of the abstract and which could perhaps contain relevant information.
c. It does not take into account the opinion of other researchers who may have produced scientific articles that could contemplate different opinions relative to the ST identified.
d. Difficulty in precisely determining to what extent an ST can minimize or eliminate the effects of the components of the VUCA environment, which means that although there are several ST in certain quadrants, this does not necessarily mean that this quadrant is more protected than a quadrant that has only one ST.
e. The matrix positioning (stage of the strategy process versus VUCA components) is not always strictly based on the scientific article, so the analysis was based essentially on a qualitative rather than a and quantitative approach.
f. Some of the scientific articles do not present evidence of the effectiveness and degree of generalization of the application of the ST.

6.2 Future Directions

Looking at the limitations reveals some important directions for future research:

a. Identification and analysis of scientific articles focusing on the criticism of ST, exposing their limitations in the context of the VUCA environment.
b. Identification and analysis of scientific articles pointing to effective ST, individually specifying each one of the components of the VUCA (volatility, uncertainty, complexity and ambiguity) environment.
c. Comparing the sociological perspective with the business (company managers) and economic (consultants) perspectives.

Funding This research did not receive any specific grant from funding agencies in the public, commercial, or not-for-profit sectors.

Declaration of Competing Interest None.

References

Adcroft A, Mason R (2007) The external environment's effect on management and strategy: A complexity theory approach. Manag Decis 45(1):10–28. https://doi.org/10.1108/00251740710718935

Afonina A (2015) Strategic management tools and techniques and organizational performance: Findings from the Czech Republic. J Compet 7(3):19–36. https://doi.org/10.7441/joc.2015.03.02

Afonina A, Chalupský V (2013) Investigation of strategic management tools and techniques. Acta Univ Agric SilvUlturae Mendel Brun 61(4):833–840. https://doi.org/10.11118/actaun201361040833

Anastassopoulos J (1988) Strategor: Stratégie, structure, décision, identité, Interéditions (ed.)

Ansoff I, Kipley D, Lewis A, Helm-stevens R, Ansoff R (2019) Implanting strategic management, 3rd ed. Palgrave Macmillan

Baaji M, Reinmoeller P (2018) Mapping a winning strategy, 1st ed. Emerald Publishing Limited

Batista C, Souza E, Neto J, Dornela J (2012) Proposta de data mart para análise de faturamento de empresa de varejo utilizando software livre. Revista Brasileira De Administração Científica 3(2):163–180

Begun J (1999) Strategic cycling: Shaking complacency in healthcare strategic planning. J Healthc Manag 44(5):339–351

Belt J, Bashore E (2000) Managed care strategic planning: the reality of uncertainty. Healthc Financ Manage 54(5):38–42

Bennett N, Lemoine GJ (2014) What a difference a word makes: Understanding threats to performance in a VUCA world. Bus Horiz 57(3):311–317. https://doi.org/10.1016/j.bushor.2014.01.001

Bolland EJ (2020) Evolution of strategy: Origin, planning, strategic planning and strategic management. In: Strategizing. https://doi.org/10.1108/978-1-78973-697-720201003

Bowman EH, Singh H, Thomas H (2002) The domain of strategic management: History and evolution. In: Pettigrew A, Thomas H, Whittington R (eds) Handbook of strategy & management, 1st ed, pp 31–51. SAGE Publications Ltd

Chadha S (2017) VUCA provoking world: The future. Human Capital 20(8):14–19

Chakravarthy BS, White RE (2012) Strategy process: Forming, implementing and changing strategies. In: Pettigrew A, Thomas H, Whittington R (eds) Handbook of strategy and management, 1st ed, pp 182–205. SAGE Publications Ltd. https://doi.org/10.4135/9781848608313.n9

Clark DN (1997) Strategic management tool usage: A comparative study. Strateg Chang 6(7):417–427. https://doi.org/10.1002/(sici)1099-1697(199711)6:7%3c417::aid-jsc281%3e3.0.co;2-9

Courtney H, Viguerie P, Kirkland J (1997) Strategy under uncertainty : Scenario planning. Harv Bus Rev, December:1–14

Cummings S, Daellenbach U (2009) A guide to the future of strategy?: the history of long range planning. Long Range Plan 42(2):234–263. https://doi.org/10.1016/j.lrp.2008.12.005

Da Silva ATB, Spers RG, Wright JTC (2012) A elaboração de cenários na gestão estratégica das organizações: um estudo bibliográfico. Revista de Ciências Da Administração 14(32):21–34. https://doi.org/10.5007/2175-8077.2012v14n32p21

David FR (2011) Strategic management: Concepts and cases, 13th ed. Pearson Education Limited

Dommisse W, Oosthuizen H (2004) Value chain deconstruction in the United Kingdom retail life insurance industry and resulting strategic options for new entrants to the market. S Afr J Bus Manage 35(4):11–20. https://doi.org/10.4102/sajbm.v35i4.664

Du J, Chen Z (2018) Applying Organizational Ambidexterity in strategic management under a "VUCA" environment: Evidence from high tech companies in China. Int J Innov Stud 2(1):42–52. https://doi.org/10.1016/j.ijis.2018.03.003

Durst S, Edvardsson IR (2012) Knowledge management in SMEs: A literature review. J Knowl Manag 16(6):879–903. https://doi.org/10.1108/13673271211276173

Frost FA (2003) The use of strategic tools by small and medium-sized enterprises: An Australasian study. Strateg Chang 12(1):49–62. https://doi.org/10.1002/jsc.607
Furrer O, Thomas H, Goussevskaia A (2008) The structure and evolution of the strategic management field: A content analysis of 26 years of strategic management research. Int J Manag Rev 10(1):1–23. https://doi.org/10.1111/j.1468-2370.2007.00217.x
Giones F, Brem A, Berger A (2019) Strategic decisions in turbulent times: Lessons from the energy industry. Bus Horiz 62(2):215–225. https://doi.org/10.1016/j.bushor.2018.11.003
Grant RM (2003) Strategic planning in a turbulent environment: Evidence from the oil majors. Strateg Manag J 24(6):491–517. https://doi.org/10.1002/smj.314
Grant RM (2020) Contemporary strategy analysis, 10th ed. Wiley & Sons
Hakala H, Vuorinen T (2020) Tools for strategy (J.-C. Spenser (ed.); 1st ed.). Cambridge University Press. http://dx.doi.org/10.1017/9781108883757
He H, António N, Rosa Á (2012) Strategic tools in China/strategic tools: An investigation into strategy in practice in China. Afr J Bus Manage 6(26):7823–7832. https://doi.org/10.5897/ajbm11.2520
Hinterhuber HH (1994) The European way to lean management. Int Exec 36(3):275–290. https://doi.org/10.1002/tie.5060360303
Hong P, Hong SW, Jungbae J, Kihyun R, Hong P (2012). Evolving Benchmarking Practices: A Review for Research Perspectives. https://doi.org/10.1108/14635771211257945
Hussey D (1998) Strategic management: from theory to implementation, 4th ed. Butterworth-Heinemann
Iborra M, Safón V, Dolz C (2019) What explains resilience of SMEs? Ambidexterity capability and strategic consistency. Long Range Plan, January:101947. https://doi.org/10.1016/j.lrp.2019.101947
Jarzabkowski P, Kaplan S (2015) Strategy tools-in-use: A framework for understanding "technologies of rationality" in practice. Strateg Manag J 36(4):537–558. https://doi.org/10.1002/smj.2270
Jarzabkowski P, Giulietti M, Oliveira B, Amoo N (2013) "We don't need no education"-or do we? Management education and alumni adoption of strategy tools. J Manag Inq 22(1):4–24. https://doi.org/10.1177/1056492612460588
Johansen B, Euchner J (2013) Navigating the VUCA world. Res Technol Manag 56(1):10–15. https://doi.org/10.5437/08956308x5601003
Johansen B (2007) Sensing the future to compete in the present, 1st ed. Berrett-Koehler Publishers, Inc
Kachaner N, King K, Stewart S, Kachaner N, King K, Stewart S (2016). Four Best Practices for Strategic Planning. https://doi.org/10.1108/SL-06-2016-0046
Kail EG (2011) Leading in a VUCA environment: A is for Ambiguity. Harv Bus Rev:1–5
Kaivo-oja JRL, Lauraeus IT (2018) The VUCA approach as a solution concept to corporate foresight challenges and global technological disruption. Foresight 20(1):27–49. https://doi.org/10.1108/FS-06-2017-0022
Kaplan S, Jarzabkowski P (2006) Using strategy tools in practice—how tools mediate strategizing and organizing. Advanced Institute of Management Research Paper No. 047. https://doi.org/10.2139/ssrn.1309556
Kaplan R, Norton D (1996a) The balanced scorecard: translating strategy into action. Havard Business Review
Kaplan R, Norton D (2008b) The execution premium: Linking strategy to operations for competitive advantage. Harvard Business Review Press
Kaplan R, Norton D (2008a) Balanced scorecard: Integrating strategy planning and operational execution : A six-stage system. Harverd Business School 10(3):1–16
Knott P (2006) A typology of strategy tool applications. Manag Decis 44(8):1090–1105. https://doi.org/10.1108/00251740610690630
Korpela J (1996) Benchmarking Logistics Performance with an Application of the Analvtic Hierarchv Process. 43(3):323–333

Kotter J (2014) Accelerate: building agility for a faster-moving world. Harvard Business Review Press
Kutz MR, Ball DA, Carroll GK (2018) Contextual intelligence behaviors of female hospital managers in the United States. Int J Healthc Manag 11(3):155–163. https://doi.org/10.1080/20479700.2017.1309819
Laamanen T (2017) Reflecting on the past 50 years of long range planning and a research agenda for the next 50. Long Range Plan 50(1):1–7. https://doi.org/10.1016/j.lrp.2017.02.001
Lawrence K (2013) Developing leaders in a VUCA environment. UNC Executive Development, pp 1–15
Leibold M, Voelpel SC, Tekie EB (2004) Managerial levers in cultivating new mental space for business innovation. South African J Bus Manag 35(4):61–72. https://doi.org/10.4102/sajbm.v35i4.669
Lynch R (2018) Strategic management, 8th ed. Pearson Education Limited
Malewska K (2015) Intuition in managerial decision-making: The results of the empirical study. In: Proceedings of the European conference on management, leadership & governance, pp 254–261
Mclarney V (2003) Strategic planning processes in chaotic environments. Vikalpa 28(1):27–47. https://doi.org/10.1177/0256090920030103
Mele D, Parra MG (2006) Intellectual evolution of strategic management and its relationship with ethics and social responsibility. IESE Business School Working Paper No. 658. https://doi.org/10.2139/ssrn.960663
Miles R, Snow C (2003) Organizational strategy, structure, and process. Stanford University Press
Millar CCJM, Groth O, Mahon JF (2018) Management innovation in a VUCA world: Challenges and recommendations. Calif Manage Rev 61(1):5–14. https://doi.org/10.1177/0008125618805111
Milliken FJ (1987) Three types of perceived uncertainty about the environment: State, effect, and response uncertainty. Acad Manag Rev 12(1):133–143. https://doi.org/10.5465/amr.1987.4306502
Mintzberg H, Ahlstrand B, Lampel J (1998) Strategy safari wilds of strategic management. In: Free Press
Mintzberg H, Ahlstrand B, Lampel J (2009) Strategy safari, 2nd ed. Pearson Education Limited
Moslem S, Mirzaee S, Mirfakhredini SH, Nasser A, Sharifabadi AM (2020) Compiling the strategies of alfa mining company and prioritizing them by WASPAS method. Sci J Logist 16(3):447–463. https://doi.org/10.17270/J.LOG.2020.473
Murugan S, Rajavel S, Aggarwal AK, Singh A (2020) Volatility, Uncertainty, Complexity and Ambiguity (VUCA) in context of the COVID-19 pandemic: Challenges and way forward. Int J Health Syst Implement Res 4(2):10–16
Navas-López JE, Guerras-Martín LA (2018) Fundamentals of strategic management, 2nd ed. Thomson Reuters Civitas
Oboreh JS, Umukoro FG (2011) Determining the impact of environmental characteristics and uncertainty on the strategic plans of corporate organizations. Sacha J Environ Stud 1(1):31–40
Ocasio W, Joseph J (2008) Rise and fall— or transformation? Long Range Plan 41(3):248–272. https://doi.org/10.1016/j.lrp.2008.02.010
Okoli C, Pawlowski SD (2004) The Delphi method as a research tool: An example, design considerations and applications. Inf Manag 42(1):15–29. https://doi.org/10.1016/j.im.2003.11.002
Perrott BE (2008) Managing strategic issues in a turbulent environment. J Gen Manag:21–30
Petticrew M, Roberts H (2006) Systematic reviews in the social sciences: A practical guide. In: Systematic reviews in the social sciences: A practical guide, 1 ed. Blackwell Publishing. https://doi.org/10.1002/9780470754887
Petticrew M (2001) Petticrew_2001_Myths_Misconceptions, vol 322
Pietrzak, Paliszkiewicz (2015) Framework of strategic learning: The PDCA cycle. Management 10(2):149–161

Porter M (1980) Competitive strategy: Techniques for analyzing industries and competitors. Free Press, New York, NY
Porter ME (1990) How competitive forces shape strategy. Harv Bus Rev 68(2):73–93
Qehaja AB, Kutllovci E, Pula JS (2017) Strategic management tools and techniques: A comparative analysis of empirical studies. Croat Econ Surv, 19(1), 67–99. https://doi.org/10.15179/ces.19.1.3
Ramírez R, Selsky JW (2016) Strategic planning in turbulent environments: A social ecology approach to scenarios. Long Range Plan 49(1):90–102. https://doi.org/10.1016/j.lrp.2014.09.002
Reeves M, Candelon F (2022) Dynamic business strategy: Competing in a fast-changing, uncertain context (M. Reeves & F. Candelon (eds.); 1ed ed.). De Gruyter. https://doi.org/10.1515/9783110755381
Rengarajan S, Moser R, Narayanamurthy G (2021) Strategy tools in dynamic environments—An expert-panel study. Technol Forecast Soc Chang, 165(December 2020):120560. https://doi.org/10.1016/j.techfore.2020.120560
Rigby D, Bilodeau B (2018) Management tools & trends
Robinson J, Sinclair M, Tobias J, Choi E (2017) More dynamic than you think: Hidden aspects of decision-making. Adm Sci 7(3):23. https://doi.org/10.3390/admsci7030023
Rumelt RP, Schendel D, Teece DJ (1991) Strategic management and economics. Strateg Manag J 12(S2):5–29. https://doi.org/10.1002/smj.4250121003
Schoemaker PJH, Heaton S, Teece D (2018) Innovation, dynamic capabilities, and leadership. Calif Manage Rev 61(1):15–42. https://doi.org/10.1177/0008125618790246
Schühly A, Becker F, Klein F (2020) Real time strategy: When strategic foresight meets artificial intelligence, 1st ed. Emerald Publishing Limited
Sevaguru N, Safa M (2005) Scenario planning approach to strategic management of small travel business in Malaysia. Int J Bus Manag Sci 1(1):65–81
Şener İ (2012) Strategic responses of top managers to environmental uncertainty. Procedia Soc Behav Sci 58:169–177. https://doi.org/10.1016/j.sbspro.2012.09.990
Sharma S (2019) New mantras in strategic management. IBA J Manag Lead 10(2):23–28
Shinkle G, Hundley G (2013) Strategy in turbulent environments: Learning from firms in transition economies, pp 1–22
Shinkle G, Kriauciunas A, Hundley G, Goudsmit M (2013) Strategy in turbulent environments: Learning from firms in transition economies, pp 1–22
Sullivan J (2012) VUCA: The new normal for talent management and workforce planning. Dr John Sullivan: Talent Manag Thought Lead, 1–5
Talaoui Y, Kohtamäki M, Ranta M, Paroutis S (2023) Recovering the divide: A review of the big data analytics—strategy relationship. Long Range Plan 56(2):1–40. https://doi.org/10.1016/j.lrp.2022.102290
Thorén K, Vendel M (2018) Backcasting as a strategic management tool for meeting VUCA challenges. J Strateg Manag 12(2):298–312. https://doi.org/10.1108/JSMA-10-2017-0072
Torraco RJ (2016) Writing integrative literature reviews: Using the past and present to explore the future. Hum Resour Dev Rev 15(4):404–428. https://doi.org/10.1177/1534484316671606
Vaitkevičius S, Merkys G, Savanevičienė A (2006) Model of strategic analysis tools typology. Strat Anal Instrum Tipologinis Modelis 47(2):99–109
van Rijmenam M, Erekhinskaya T, Schweitzer J, Williams MA (2019) Avoid being the Turkey: How big data analytics changes the game of strategy in times of ambiguity and uncertainty. Long Range Plan 52(5):101841. https://doi.org/10.1016/j.lrp.2018.05.007
Vecchiato R (2015) Strategic planning and organizational flexibility in turbulent environments. Foresight 17(3):257–273. https://doi.org/10.1108/FS-05-2014-0032
Vuorienen T, Hakala H, Kohtamaki M, Uusitalo K (2018) Mapping the landscape of strategy tools: A review on strategy tools published in leading journals within the past 25 years. Long Range Plan 51(4):586–605. https://doi.org/10.1016/j.lrp.2017.06.005
De Waal GA, Knott P (2013) Innovation tool adoption and adaptation in small technology-based firms. Int J Innov Manag 17(3). https://doi.org/10.1142/S1363919613400124

Webster JL, Reif WE, Bracker JS (1989) The Manager's guide to strategic planning tools and techniques. Plan Rev 17(6):4–48. https://doi.org/10.1108/eb054273

Whittington R, Regnér P, Angwin D, Johnson G, Scholes K (2020) Exploring strategy text & cases, 12th edn. Pearson Education Limited

Wright RP, Paroutis SE, Blettner DP (2013). How useful are the strategic tools we teach in business schools? J Manag Stud 50(1):92–125. https://doi.org/10.1111/j.1467-6486.2012.01082.x

Yoshikuni A, Albertin A (2018) Effects of strategic information systems on competitive strategy and performance article information. Int J Product Perform Manag 67(9):2018–2045. https://doi.org/10.1108/IJPPM-07-2017-0166

Towards a "More Sustainable" Europe to Overcome the Polycrisis: The Role of Companies and the Sustainability Report

Elisabetta Calvo and Dante Alpi

Abstract The historical period we are living through is characterized by the coexistence of multiple crises developing in the climatic, environmental, health, socio-economic and geopolitical fields; the possibility of new conflicts, the aggravation of existing conflicts, mass migrations, financial crises and potential new pandemics may constitute new future threats, requiring the implementation of transitions likely to condition the social and economic aspects of sustainability. Companies are in the process of discovering and evaluating a more ethically responsible and sustainable way of doing business. The Corporate Sustainability Reporting Directive (CSRD) Directive (EU) 2022/2464, aims to achieve sustainability in business and finance, in line with the European Commission's Green Deal and the United Nations' Agenda for Sustainable Growth, by identifying the need for companies to disclose clear sustainability information and encourage the transition to a sustainable economy. Transparent ESG information, through the production of an appropriate sustainability report, proves that companies are actively committed to environmental and social responsibility, with positive effects on their reputation with consumers and investors, as well as the possibility of accessing capital at a lower cost and improving their competitiveness.

Keywords Polycrisis · CSRD · ESG · Corporate sustainability reporting · Climate change · Ecological transition · Health and well-being · Sustainability

---

E. Calvo (✉) · D. Alpi
Member of the ERECO-PGV International Scientific Network, Laboratorio Athena, Terni, Italy
e-mail: elisabettacalvo@outlook.it

D. Alpi
e-mail: alpidante@gmail.com

G. Pellat et al. (eds.), *Data-Centric Business and Applications*, Lecture Notes on Data Engineering and Communications Technologies 258,
https://doi.org/10.1007/978-3-031-95068-1_11

1 Introduction

1.1 Relevance

The historical period we are living through is characterized by the coexistence of multiple crises developing in the climatic, environmental, health, socio-economic and geopolitical fields; the possibility of new conflicts, the aggravation of existing ones, mass migrations, financial crises and potential new pandemics may constitute new future threats, requiring the implementation of transitions of such magnitude likely to condition the social and economic aspects of sustainability (Commission of the European Communities 2023).

This situation has been described by economists as a "polycrisis" (Morin Kern 1999) and, more recently, as a "permacrisis"[1] (Lagarde 2022). Definitions and lists of identified emergencies vary, but the common element of rising inequality remains (Atkinson 2015). The Coronavirus pandemic made communities aware of their vulnerability, underlining the need for an ecological transition towards greater care for the natural ecosystems to which we are intimately linked (Giraud 2022). Current and future socio-environmental upheavals will require financial resources to contain climate change, sustain biodiversity and provide the opportunities needed to achieve sustainable, inclusive growth. Achieving this goal will require the cooperation of many players, including businesses, which account for 70% of global GDP.

Companies will need to intervene in their governance systems to manage and allocate their investments appropriately, developing an operational compliance that must be in harmony with a philosophy of prevention through integrated systems. All this leads to the need to use multiple indicators to assess a company's commitment to environmental, social and organizational dimensions, according to "ESG" criteria, with potential positive spin-offs for investors and consumers. With the "European Green Deal", the European Commission has emphasized the importance of non-financial disclosures for the cost-effective use of resources and for the elimination of greenhouse gas emissions by 2050; in this context, the centrality of defending, maintaining and improving the health and well-being of populations subject to increasing environmental risks becomes apparent. Appropriate communication by companies on aspects of sustainability that are accurate, comparable and reliable is a fundamental condition for effectively achieving the planned objectives; in this context, the European Parliament, with Directive 2014/95/EU has included the obligation, for certain types of companies and large groups, to publish non-financial information (environmental, social, related to workers, attention to human rights and the fight against active and passive corruption). Adequate sustainability reporting enables the various stakeholders to analyze the non-financial performance of companies and provides the opportunity to strengthen a sustainable economic system for the benefit

[1] Collins Dictionary has named "permacrisis" as the word of the year 2022. It defines the term as "a long period of instability and insecurity, especially resulting from a series of catastrophic events".

of citizens, while the lack of shared parameters and methods of analysis for the qualitative, quantitative and managerial assessment of sustainability issues is a limit to its expansion. The Corporate Sustainability Reporting Directive (CSRD) Directive (EU) 2022/2464 of the European Parliament and of the Council aims to achieve sustainability in business and finance, in line with the European Commission's Green Deal and the United Nations' Agenda for Sustainable Growth, by identifying the need for companies to provide clear sustainability information and encourage the transition to a sustainable economy. Today, more than ever, there is a deep awareness that companies, especially the larger ones, play not just a functional but a systemic role. It is essential that companies define their mission, the meaning and importance of the goals they pursue, focusing on the challenges they face in meeting social and environmental needs in order to distinguish themselves in the marketplace and generate long-term value for all stakeholders.

1.2 Methodology

Through an analysis of current conditions linked to the multiple crises and their impact on society and companies, and through an analysis of scientific publications, the article aims to delve deeper into the theme of sustainability arising from the new CSRD directive, which introduces new rules and indicators for corporate sustainability reporting, in order to guarantee relevant information on all the main issues linked to sustainable development.

2 When Crises Intertwine

The succession of crises that have occurred over the last few decades and are still ongoing, has determined a framework defined as "polycrisis", which identifies a set of upheavals that appears greater than the sum of its constituent elements, which accumulate and reinforce each other, and represents a situation of superimposed and interconnected crises, where the boundaries between environmental, social and health problems seem difficult to define, and the resulting need for a response just as complex and multi-polar (Losasso 2022). Each crisis becomes a constituent element of another, and contributes to aggravating it (Tooze 2022), (Fig. 1).

The World Economic Forum's Global Risks Report 2023 (WEF 2023), highlighted the close interconnection between global risks that could potentially compromise the countermeasures implemented to combat them, and in particular those linked to climate change, biodiversity and investment in human capital. The report outlines a global risk scenario for the next two years, with the possibility of the development of multiple adverse conditions (recession, debt, rising cost of living, polarization of societies due to misinformation, halting rapid climate intervention, zero-sum geo-economic warfare) and, over the next ten years to witness increasing global warming,

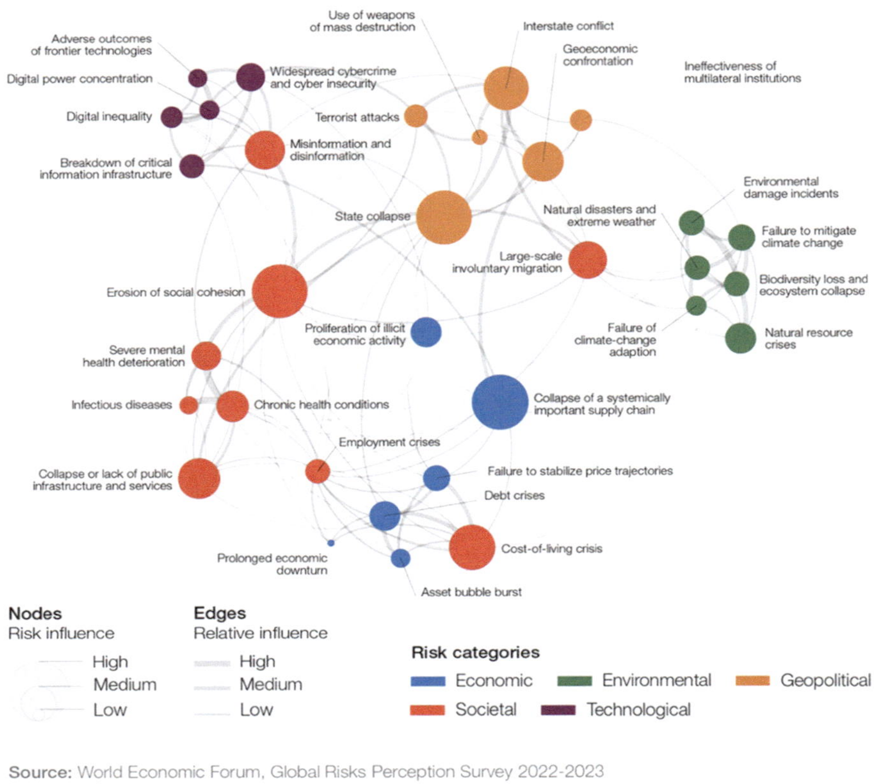

Fig. 1 Global risks: a map of interconnections. *Source* Global Risks Report 2023

leading to ecological disaster if nations don't start cooperating more effectively to mitigate and adapt to climate change (Fig. 2). Geopolitical rivalries are likely to further fuel the disintegration of social cohesion due to reduced investment in health, education and economic development, while heightened tensions are likely to breathe new life into the arms race through new technologies and cyber-attacks.

Managing social, environmental and security conditions simultaneously will be a difficult test for governments, and there is a significant risk that the gap between the "scientifically necessary" and the "politically acceptable" will widen: geo-economic risks threaten efforts to achieve zero emissions, and rising military spending offers fewer budget margins to combat the cost of living, with the consequence that adequate investment in future growth, human development and green technologies will not be achieved. The report calls on leaders to cooperate in decision-making and strike the right balance between short- and long-term perspectives, with urgent climate action and public–private collaboration in favor of financial stability, technological governance, economic development and investment in research, science, education and health.

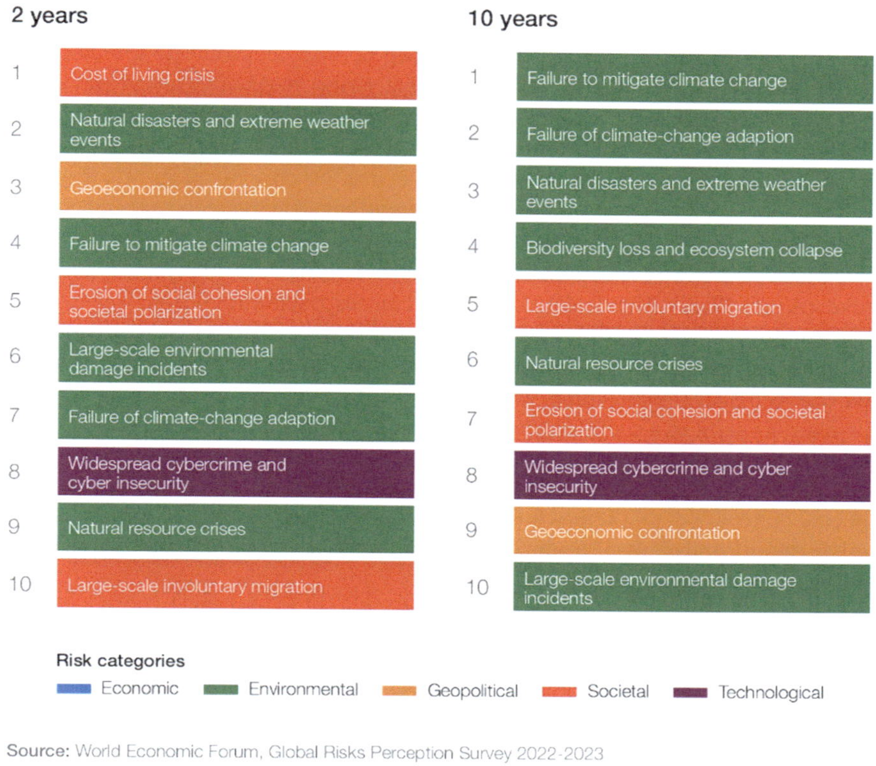

Fig. 2 Global risks ranked by severity in the short and long term. *Source* Global Risks Report 2023

3 Health and Well-Being in a World of Polycrisis

In 2015, as part of the 17 Sustainable Development Goals (SDGs), the United Nations identified Goal 3: "Ensure healthy lives and promote well-being for all at all ages". The Geneva Charter for Well-Being[2] underlines the need for well-being societies to provide the tools to ensure that all people of current and future generations can thrive on a healthy planet, wherever they live. These objectives can only be envisaged on the basis of a positive conception of health, in which the constituent elements of well-being (physical, mental, spiritual and social) are present, and where respect for human rights, social and environmental justice, solidarity, gender and intergenerational equity prevail, in a context of peace. The pandemic has highlighted the close and fragile relationship between man and planet, making us realize just how futile the commitment to developing a safer world can be if the delicate link between people,

---

[2] Based on the results of the Xth World Conference on Health Promotion (Geneva, December 13–15, 2021).

pathogens and climate change is not duly taken into account. There is ample scientific evidence to support the close relationship between exposure to climatic (e.g. global warming) and environmental factors (e.g. rising levels of CO2 and particulate matter) and the increase in cardiovascular, respiratory, neurodegenerative and tumor pathologies, as well as obesity, malnutrition, new infectious and mental illnesses. Particularly alarming is the possibility that the negative consequences of a degraded environment could be passed on to future generations, due to greater genetic vulnerability to diseases induced by epigenetic changes[3]. There is a growing awareness of the need to promote global health and the well-being economy for a sustainable future (ASviS 2022). The depletion of natural resources resulting from overexploitation, combined with the financial and economic crises, the COVID-19 pandemic and the Russian-Ukrainian war, have accentuated inequalities and poverty within and between states, threatening social cohesion. Certain changes are now essential if we are to combat a socio-economic system that is not concerned with the well-being of people, the environment and the planet, but is instead a source of health and socio-economic disparities. In this context, the notion of the Well-Being Economy (an economic model that brings benefits to people and the planet while guaranteeing human dignity and equity) can encourage the spread of new development models for a more sustainable future.

The Well-Being Economy proposes an economic model, as an alternative to the Growth Economy (Marzano 2008), whose main objective is to protect human well-being and the environment. Instead, they are indicators of well-being that include environmental and biodiversity indicators, health indicators (including mental health), social indicators (such as social cohesion), equality indicators, innovation indicators and circularity indicators (Fioramonti 2019). Internationally, the Organisation for Economic Co-operation and Development (OECD) is one of the most supportive bodies of the social economy, affirming the need to reorient economic policies to reduce inequalities; in this context, the OECD encourages the dissemination of indicators that better represent the state of well-being and go beyond the simple estimation of economic performance such as GDP (Llena-Nozal et al. 2019).

Global health and the well-being economy should be the goals towards which to guide the social, economic and environmental choices that underpin the Agenda 2030 Sustainable Development Goals. The growing importance of the influence of climate and the environment on health is underlined in the Intergovernmental Panel on Climate Change (IPCC) 2022 report "Climate Change 2022: Impacts, Adaptation and Vulnerability", which highlights the extent to which the frequency and intensity of climatic events is increasing, with sometimes unpredictable consequences for our society. The WHO highlighted how the main risks resulting from extreme temperatures, particularly heat, are poorly borne by the most vulnerable groups, the over-65 s, children, the chronically ill, athletes, the less affluent population and workers, a category that is increasingly involved. In 2019, the International Labour Organization (ILO 2019) estimated that by 2030, heat stress will reduce total working hours

[3] Epigenetics is the study and analysis of the molecular mechanisms by which environmental situations can influence gene expression without modifying the sequences inscribed in DNA.

worldwide by 2.2% and global GDP by $2,400 billion, with a loss of productivity equivalent to 80 million full-time jobs. Developing countries and those with warmer climates are experiencing even more significant declines in productivity relative to GDP (a reduction in working hours of around 5% is expected by 2030 in Southern Asia and West Africa). Workers in agriculture and construction will be hardest hit, with a 60% and 19% reduction in working hours respectively due to heat stress in 2030.

In January 2023, experts from the World Economic Forum, in the Global risk report 2023 (WEF 2023), warned that the absence of measures to tackle climate change and the disappearance of biodiversity were among the five main risks facing the world in the long term[4]. Between 2030 and 2050, according to WHO estimates[5], as a result of climate change, there could be an additional 250,000 deaths each year from malaria, diarrhoeal syndromes, heatwaves and malnutrition, and health-related costs could reach $2–4 billion a year by 2030.

In the 7th report of the scientific journal "The Lancet" on health and climate change (Romanello et al. 2022), the negative economic repercussions of climate change are highlighted, estimating that in 2021, 470 billion potential working hours were lost (139 h per person) due to heat waves, with an increase of 37% compared to the average for the period 1990–1999 and a potential drop in income equal to 0.72% of global economic output; in addition, in 2021, extreme events caused damage worth 253 billion dollars. According to "The Lancet", it is becoming a priority to introduce health-focused policies capable of limiting the negative effects of climate change, enhancing energy security, creating opportunities for economic recovery and ensuring health benefits. To this end, the use of clean energies could reduce deaths due to exposure to PM2.5[6] which, in 2020, were estimated at 4.2 million, of which 1.2 million were strictly linked to the use of fossil fuels. Another opportunity would be in the agri-food sector, where the transition to a more balanced, plant-based diet would reduce agricultural emissions (linked to meat and milk production) by 55% and prevent up to 11.5 million diet-related deaths every year; eating habits can limit greenhouse gas emissions and promote a healthier lifestyle.

A study (Giosuè et al. 2022) supported by the Foundation of the Euro-Mediterranean Center on Climate Change (CMCC) has demonstrated that the transition to a dietary model based on a balanced weekly diet including all foods and frequencies and quantities useful for the prevention of cardiovascular disease (a higher intake of fruit, vegetables, whole grains, low-glycemic cereals, nuts, legumes and fish, and less beef, butter, high-glycemic cereals and sugar), would reduce the carbon footprint of Europeans' current diet by 48.6%.

---

[4] Climate change is the greatest threat to human health, article by Maddalena Binda published on: https://futuranetwork.eu/focus/533-3980/il-cambiamento-climatico-e-la-maggiore-minaccia-per-la-salute-umana.

[5] Climate change and health, website: https://www.who.int/news-room/fact-sheets/detail/climate-change-and-health.

[6] The term PM 2.5 includes all particles with dimensions of 2.5 microns or less.

The study highlights the extent to which poor-quality food is the greatest risk factor for non-communicable diseases, causing 41 million deaths a year worldwide (71% of all deaths), and that in Europe, cardiovascular disease is responsible for 45% of all deaths (4 million a year). In terms of environmental impact, it is estimated that 37% of anthropogenic greenhouse gas emissions are linked to food production, mainly methane, nitrous oxide and carbon dioxide. Recent estimates show that global greenhouse gas emissions associated with the production of animal-based foods are twice as high as those associated with plant-based foods, and more specifically, the carbon footprint (greenhouse gas emissions per kg of food) is particularly high for beef.

To deal with health risks, it is therefore necessary to implement a One Health approach[7], formalized in 2017 by the WHO, which provides for the collaboration of different disciplines to preserve not only the human species, but also the health of the planet and all its inhabitants in order to create a sustainable, resilient and perdurable ecosystem: the health of each component (human, animal and environmental) is intimately dependent on all the others (global health).

4 The Sustainability Challenge from a Corporate Perspective: The Road to the Goal

For a more sustainable and inclusive future, robust economic growth will be essential, but this alone will be insufficient if it is not accompanied by a profound process of innovation in which businesses, which produce over 70% of the world's GDP, play a key role (McKinsey & Company 2022a, b). Companies will have to change the purpose of their activity, which should no longer depend solely on shareholder profit, but on the benefit of all stakeholders.

In the age of polycrisis, the development of companies towards a[8] purpose is indispensable, and companies attentive to socio-environmental impact will be more robust and resilient in the near future. At a global level, regulations have progressively supported corporate attention to environmental, social and governance (ESG) issues[9] and ESG factors have been perceived by business leaders as increasingly fundamental to their growth.

A sustainable business activity is characterized by the presence of minimal negative or, better still, positive impacts on the environment, society and the economy, a goal that can only be envisaged with a corporate management that cultivates a

[7] https://www.who.int/news-room/questions-and-answers/item/one-health.

[8] The British Academy has conducted a major program of research and engagement examining the purpose of business and its role in society. https://www.thebritishacademy.ac.uk/publications/future-of-the-corporation-principles-for-purposeful-business/.

[9] EVPA Investing for Impact, *Accelerating corporate impact journey is the goal of Business of Impact, Europe's premier corporate impact summit* https://www.businessofimpact.org/venice/about.

goal-oriented strategy, in line with the sustainable development objectives shared by stakeholders.[10]

ESG evaluation criteria are used to analyze the degree of sustainability of organizations and their potential impact on society and the environment. Companies play a key role in the development of sustainable practices, given their influence in terms of socio-environmental impacts, and for this it is necessary that they reduce environmental impact (less waste, energy efficiency, renewable energies), support social responsibility (fair working conditions and worker well-being, support for diversity, inclusion and local communities), implement good governance practices (promoting transparency, accountability, ethical behavior) and involve stakeholders in sustainability and ESG-related projects (cooperation with suppliers, customers and other parties). Achieving ESG objectives can potentially increase revenues, contain costs, attract skilled people and reduce regulatory risk; companies that plan their activities with attention to environmental and social impacts are likely to be more efficient and resilient in the medium to long term (Kurznack et al. 2021).

Companies must implement a holistic and systematic project aimed at corporate sustainability, addressing environmental, social (e.g. quality of life, social cohesion, democracy) and economic (e.g. responsible production and consumption) challenges (Klein et al. 2022).

Capitalism is destined for structural change, following major institutional efforts (UN Agenda 2030), but also due to the pragmatic concept that companies can become a "force for good" capable of triggering a positive impact on society, while continuing to compete and produce value. In this context, the notion of corporate mission, i.e. the ability to distinguish oneself in the marketplace by generating long-term value for all stakeholders, has become a strategic element capable of fostering corporate renewal and prosperity (McKinsey Quarterly 2020).

The purpose and role of business has been analyzed by a British Academy program which has led to a description of the purpose of business which is identified with the ability to "create profitable solutions to the problems of people and the planet, without profiting from problems created to either" (The British Academy 2021); the new conception of purpose should be the origin and raison d'être of a company, and profit should be not the objective of a company, but the product of purpose, through a new relationship of trust with stakeholders and a strong culture of ethics.

Companies should become better "citizens", promoting positive impacts on climate and pollution, workplace safety, worker well-being and local communities, not least because strengthening ESG performance improves the company, and because choosing a sustainable goal increases competitiveness and creates value.

---

[10] CEO Magazine, *How insights from the impact space can empower purpose-driven corporate leaders*. 09 may 2023 https://www.theceomagazine.com/opinion/purpose-driven-corporate-leaders/.

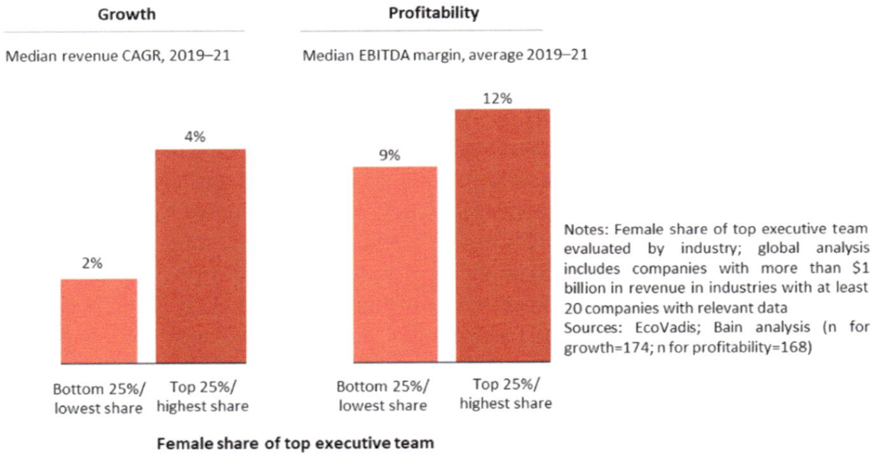

Fig. 3 Companies with more women on their management teams achieve better financial results. *Source* Research by Bain & Company and EcoVadis

A study published in 2023 (Seemann et al. 2023) by Bain & Company[11] and EcoVadis[12], based on an analysis of the revenues of 80,000 private companies and 20,000 listed companies over the period 2019–2021, highlights that the majority of those with a positive socio-environmental impact saw their revenues and profits increase. The study documented that financial performance, in the majority of cases, is positively correlated to ESG activities with notable growth in revenues and EBITDA margins (Earnings Before Interests Taxes Depreciation and Amortization) and, in all cases, with no negative correlations.

The research identifies relationships between corporate performance and four different elements of sustainability:

1. *Diversity, equity and inclusion.* Companies with greater female representation in the management team (high EcoVadis score in the social ESG component) show better financial performance, and those in the top quartile of the sector in terms of gender diversity record around 2 percentage points higher annual sales growth than those in the bottom quartile, and 3 percentage points higher EBITDA profit margins (Fig. 3); other studies have associated this advantage with a broader view of opportunities and risks on the part of a diverse management team.
2. *Employee satisfaction.* Companies with more satisfied employees have 2019–2021 sales growth up to 5 percentage points higher than those with less satisfied employees, and increased EBITDA profit margins up to 6 percentage points (Fig. 4). ESG activities aimed at the workplace include fair compensation, a

[11] Bain & Company is a global consulting firm.

[12] EcoVadis is a company dedicated to providing the world's most reliable corporate sustainability assessments.

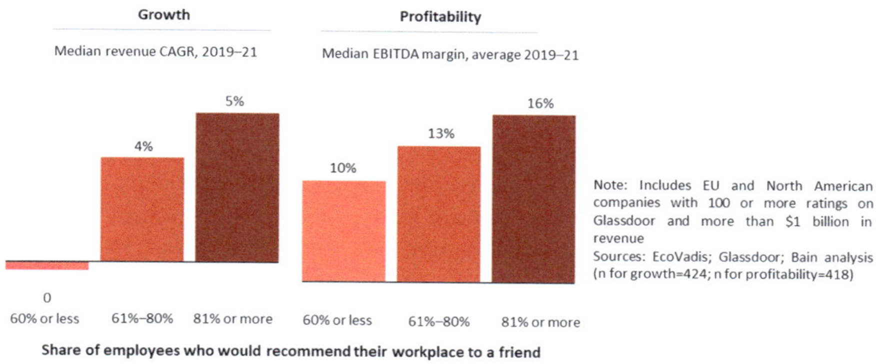

Fig. 4 Companies with the most satisfied employees grow faster and are more profitable. *Source* Research by Bain & Company and EcoVadis

safe working environment, professional training, mental and physical healthcare, childcare and educational opportunities; all of which increase employee satisfaction, productivity, loyalty and the employer's ability to attract top talent.

3. *Renewable energy*. Environmentally aware companies that use the most renewable energy have higher EBITDA margins; in the EU, lower emissions increase profits due to lower carbon tax expenses; and globally, in general, it avoids fluctuations in fossil fuel costs and, sometimes, is simply more economical.
4. *Sustainable supply chain*. Sustainable sourcing increases profitability by 3 percentage points compared to companies with little concern for the ethics, environment and working practices of their suppliers (Fig. 5).

A recent analysis by Mc Kinsey, "The triple play: Growth, profit, and supportability" (McKinsey & Company 2023), pointed out that companies that include ESG initiatives in their growth strategies show superperformance, provided they have a good core operating performance. The message is that not only can you succeed by doing the right thing, you can also do better. The study examined the TSR (Total

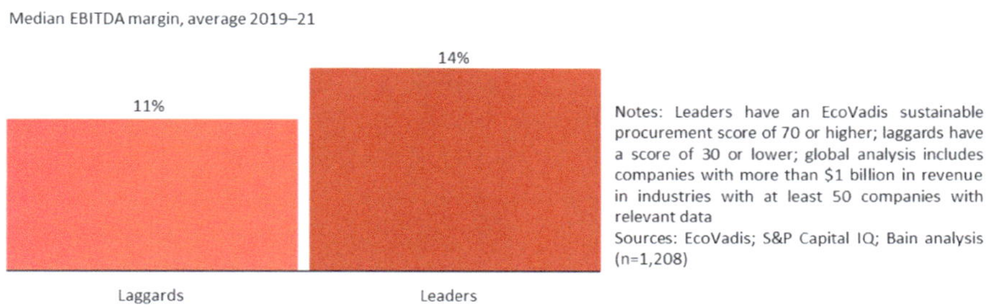

Fig. 5 Sustainable procurement leaders are more profitable. *Source* Research by Bain & Company and EcoVadis

Shareholder Return), financial performance and ESG ratings of 2,269 equity companies over a five-year period 2017–2021, distinguishing them into outperformer and underperformer based on ESG ratings, annual growth, revenues and economic profits; the result is that companies with better growth and profitability, while improving sustainability and ESG, outperform their competitors in terms of shareholder return.

These "tripli outperformers" achieved 2 percentage points more annual TSR than companies that outperformed on financial metrics alone, and 7 percentage points more than the rest of the data set, indicating that strong ESG engagement leads to shareholder value for companies that also outperform their peers in terms of growth and profitability (Fig. 6).

McKinsey's analysis underlines that companies will be able to achieve triple growth if they are guided by the following five principles: (1) include growth, profitability and ESG in their core strategy; (2) innovate on ESG to drive value creation; (3) carry out mergers and acquisitions to rapidly seize ESG growth opportunities; (4)

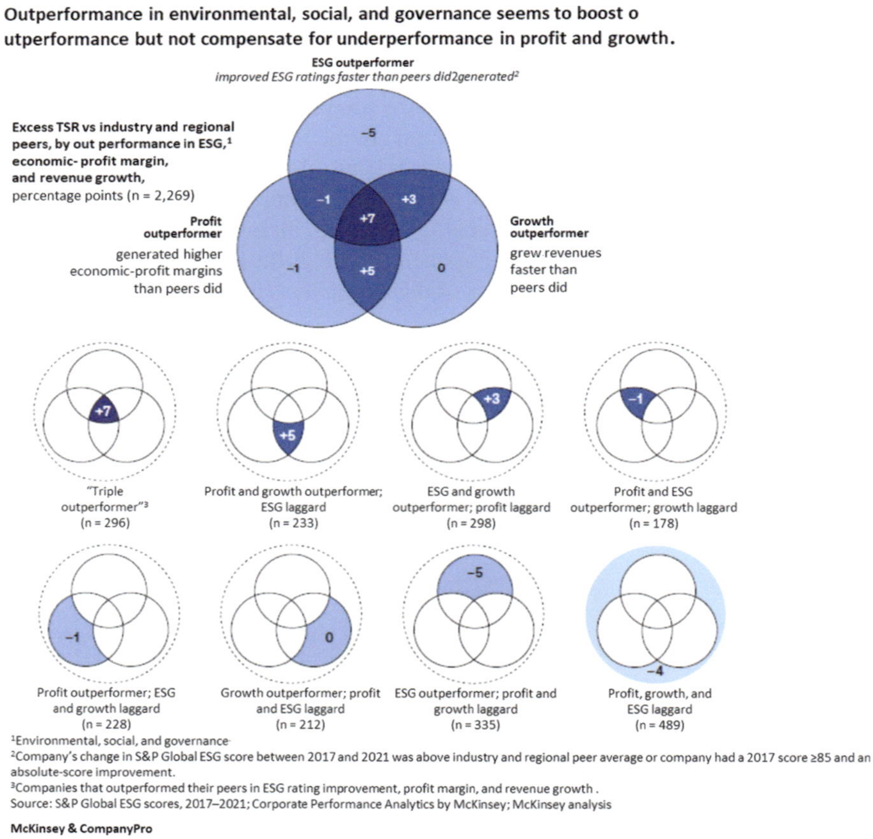

Fig. 6 Results of The triple play: Growth, profit, and sustainability. *Source* McKinsey, 2023

transparently monitor and communicate ESG factors and related data; (5) integrate strategic priorities into the organization's genetic code.

Clear and timely reporting and communication of ESG data is becoming a fundamental priority for companies, also for the purposes of fair recognition by markets and companies that commit to initiatives that require long-term implementation, as is often the case for ESG-related growth projects. Detailed reporting and proactive communication to investors on how ESG initiatives can enhance value, as well as objectives and progress, are essential to gaining such recognition; transparency can accelerate investor recognition of future potential. Companies need to document progress related to implemented initiatives, preferably through external validation, so that these improvements benefit from consumer and investor confidence (McKinsey & Company 2023).

Environmental, social and governance (ESG) objectives and issues represent strategic challenges for companies, and have a fundamental role to play in this context of polycrisis, which leads companies to evaluate their role and responsibilities in society (McKinsey & Company 2022a, b). There is a close correlation between ESG and the objective, which in no way compromises profitability but frequently leads to outperformance (Aydoğmuş et al. 2022), and together they can help company management find possible solutions to the challenges posed by the polycrisis.

Companies are called upon to play an active part in the recovery by becoming increasingly integrated into communities, moving from "purpose" to "sense providing", as it will not be enough for companies to declare their mission and interest in sustainability, but their "humanistic commitment" characterized by new types of solidarity, community relations, sharing and mutualism will be required, with an active role in future communities (IPSOS 2023).

5 The Polycrisis Challenge for a More Sustainable Economy also Involves Sustainability Reporting

A company's sustainability is its long-term viability, and goes beyond social responsibility, corporate reputation and environmental sustainability.

A sustainable company integrates the long-term value of its stakeholders into all dimensions of its activities, including its supply chain, employee relations, community relations, environmental impact and governance practices (Samans and Nelson 2022); the "ESG compliant" company is able to manage both profit-related financial aspects and environmental, social and governance aspects, optimizing its economic value. The company must go beyond profitability, promoting and sustaining benefits for all stakeholders and society at large (Halkos and Nomikos 2021) and, in this context, there is an ever-increasing need for more accurate and clearer information relating to corporate sustainability, and more reliable information about potential risks and present opportunities.

Environmental and social risks have an impact on corporate sustainability and financial health[13]. Sustainability reporting is an important communication method for disseminating management strategies aimed at developing corporate sustainability, with the publication of reports capable of promoting accountability and transparency (Effah et al. 2023).

Sustainability reporting began with the publication of the United Nations' Brundtland Report in 1987, and continued in the late 1990s with the Global Reporting Initiative (GRI) and its efforts to balance economic and environmental issues. It is a fundamental means of communication for reinforcing a company's commitment to sustainable development, and is able to provide financial and non-financial information with greater transparency and reliability, a need increasingly present in the markets and among investors, who require adequate information on the impact of sustainable policies on their investments. Sustainability reporting is increasingly present in the reports published by major companies, not least as a result of the fact that in some countries (European Union, USA and China), listed companies are currently required to disclose information on their ESG performance, the result of requests from both governments and stakeholders sensitive to these environmental issues (Benameur et al. 2023).

The KPMG study "Global Survey of Sustainability Reporting 2022" (KPMG 2022) analyzed the current state of sustainability reporting and assessed that, for the year under review, the world's leading companies showed improvement mainly in the environmental aspect, but need greater progress in the social and governance fields. It is emphasized that sustainability reporting has grown steadily in recent years, and that almost all of the world's 250 largest companies (G250) produce sustainability reports, 96% of them in the ESG context. There has also been a gradual increase in the number of N100s (the top 100 companies in each country surveyed) with sustainability reports, from around 66% ten years ago to 79% today (Fig. 7). In Italy, the companies analyzed that produce sustainability reports reach 94% (86% in 2020); in the USA, 83% of companies registered with the Securities and Exchange Commission (SEC) have published sustainability reports. Even developing countries, albeit to a lesser extent, have seen an increase in sustainability reporting, largely due to new stakeholder expectations influenced by new media and improved quality of life (Ashraf and Nazir 2023).

Sustainability reporting is defined by the European Union as: "the voluntary integration of social and environmental concerns into business operations and stakeholder relations" (Commission of the European Communities 2001); it represents an ethical assertion aimed at ensuring the transparency of organizations and companies towards civil society, through the dissemination of their services and impacts on sustainability in the environmental, social and governance fields. The relevant regulations are currently defining the content of non-financial ESG reporting, with the aim of making it increasingly comparable and capable of describing the strategic

[13] Deloitte *Sustainable Activities—A Need for Transparent Reporting Financial Reporting Brief March 2020*. https://www2.deloitte.com/ie/en/pages/audit/articles/transparent-reporting-for-sustainability-frb.html.

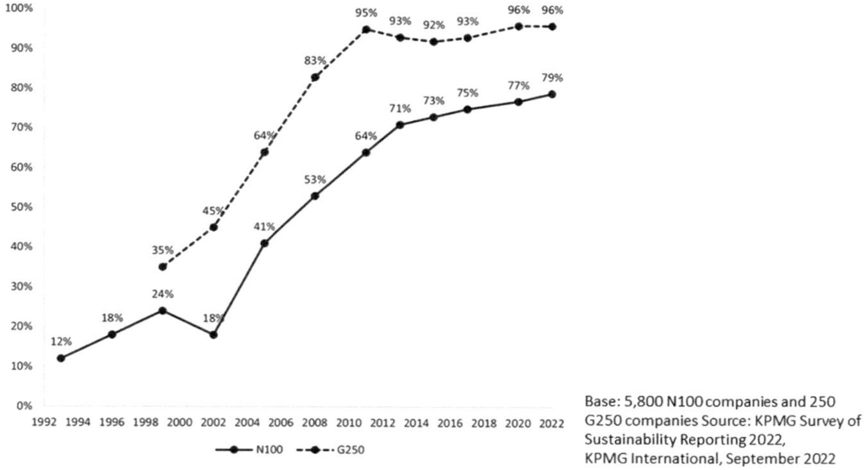

Fig. 7 Sustainability reporting rates worldwide (1993–2022). 2022 *Source* KPMG Survey of Sustainability Reporting

directions taken by companies. On December 16, 2022, Directive 2464/2022, "Corporate Sustainability Reporting Directive" (CSRD) (European Parliament and of the Council 2022), of the European Parliament and the Council, was published in the Official Journal of the European Union. The CSRD is the cornerstone of the EU Sustainable Finance Package, which includes a set of measures to improve the flow of capital into sustainable activities across the EU.

The CSRD is included in the European project that aims for sustainability in finance (European Parliament and of the Council 2019) and in the conduct of business (European Parliament and of the Council 2022), in line with the growth strategy promoted by the European Commission's Green Deal and the sustainable growth agenda set by the United Nations.[14] The CSRD emphasizes the areas that, during a 2018 public consultation on the NFRD (Non-Financial Reporting Directive) and supported by the European Commission, proved to be more fragile, notably the qualitative aspect and in terms of sustainability, comparability and reliability. The CSRD envisages a change in the information framework so that the report includes the information essential for assessing the company's sustainability and its impact on the company's development, performance and positioning. The innovation elements of the CSRD are represented by the "intangibles" produced within the company as part of intellectual, human, social and relational capital, which cannot be highlighted in the accrual balance sheet, but are nonetheless important for assessing the company's ability to produce value over time. The information required by the CSRD, in addition to that set out in the NFRD (environment, social responsibility and workers' rights, respect for human rights, anti-corruption), covers a wide range of areas:

[14] https://sdgs.un.org/goals.

1. presentation of the business model and the strategy inherent in the sustainability objectives;
2. sustainability objectives set by the company and the progress made towards achieving them;
3. administrative, operational and control governance related to ESG issues;
4. company policies and practices with regard to sustainability;
5. corporate impacts linked to the value chain (operations, products and services, commercial relations, supply chain);
6. presentation of the company's main sustainability-related risks, including interdependencies and risk management methods;
7. relevant indicators for ESG information and how they have been identified by the company.

This information must be supplemented by that already required by the NFRD on diversity in the composition of the company's administrative, management and supervisory bodies, with regard to aspects such as age, gender or training and professional experience. The report must set out the objectives of a diversity policy, and any lack of such information must be justified; this information is qualitative, quantitative, retrospective and, above all, forward-looking. With regard to the choice of information to be produced, an innovative component is constituted by the principle of "double materiality" which identifies, on the one hand, the elements of sustainability likely to influence the development, performance and commercial position of the company (inside-out approach) and, on the other hand, the documentable effect of sustainable choices on society and the environment (outside-in approach). One weakness of the NFRD is the absence of any obligation to comply with defined reporting models, which makes it difficult to compare different reports. The CSRD addresses this point by requiring the use of the European Sustainability Reporting Standards[15] (ESRS), developed in harmony with other EU disciplines and policies, in particular: a) *the Sustainable Finance Disclosure Regulation* (SFDR); b) *the Taxonomy Regulation*; c) *the Benchmarks Regulation*. The SFDR introduces innovations both in the preparation of the report (information classified numerically and according to a numerical taxonomy) and in its place in the context of the balance sheet file (mandatory inclusion in the management report). The CSRD provides for the Commission to include the obligation of reasonable assurance in the sustainability report, over and above the limited assurance provided by the NFRD, whereby qualified independent operators (auditors) verify the content by establishing an appropriate relationship (Stabile 2023).

The CSRD directive represents an evolution of current, limited sustainability reporting requirements, with particular emphasis on the fact that the data produced must be considered as "sustainability information" and not simply "non-financial

---

[15] The development of the ESRS has been entrusted by the Commission to EFRAG. https://ec.europa.eu/finance/docs/level-2-measures/csrd-delegated-act-2023-5303_en.pdf. On July 31, 2023, the European Commission definitively adopted the Sustainability Reporting Standards—ESRS (European Sustainability Reporting Standards) with its own delegated regulation. https://eur-lex.europa.eu/legal-content/EN/ALL/?uri=PI_COM:C(2023)5303.

Entreprises concernées par la CSRD

Avec la Directive CSRD, l'obligation de reporting s'étend :

- aux grandes entreprises cotées et non cotées et aux Groupes qui ont à la date de clôture du bilan au moins deux des trois critères suivants:
  1. >250 salariés au cours de l'année financière ;
  2. Actif du bilan >€ 20 millions ;
  3. Produits nets des ventes et des services (chiffre d'affaires) >€ 40 millions ;
- les sociétés cotées, y compris les PME cotées (à l'exclusion des micro-entreprises cotées);
- les Filiales UE de sociétés non UE avec un chiffre d'affaires de >€ 150 millions

Les entreprises devront commencer à publier les informations prévues dans les ESRS selon le calendrier suivant:

- entreprises précédemment soumises à la directive sur la publication d'informations non financières (grandes sociétés cotées, grandes banques et grandes entreprises d'assurance – comptant toutes plus de 500 salariés) et grandes sociétés cotées hors UE comptant plus de 500 salariés: exercice 2024, la première déclaration en matière de durabilité étant publiée en 2025;
- autres grandes entreprises, y compris les autres grandes sociétés cotées hors UE: exercice 2025, la première déclaration en matière de durabilité étant publiée en 2026;
- PME cotées, y compris les PME cotées hors UE: exercice 2026, la première déclaration en matière de durabilité étant publiée en 2027. Toutefois, les PME cotées peuvent décider de se soustraire aux obligations d'informations pendant une période supplémentaire de deux ans. La date ultime à laquelle une PME cotée doit commencer à publier des informations est l'exercice 2028, la première déclaration en matière de durabilité étant publiée en 2029.

En outre, les entreprises hors UE qui génèrent plus de 150 millions d'euros par an dans l'UE et qui ont dans l'UE soit une succursale dont le chiffre d'affaires dépasse 40 millions d'euros, soit une filiale qui est une grande entreprise ou une PME cotée devront rendre compte des incidences sur la durabilité au niveau du groupe de cette entreprise hors UE à partir de l'exercice 2028, la première déclaration en matière de durabilité étant publiée en 2029. Des normes distinctes seront adoptées spécifiquement pour ce cas.

mis à jour au 31 juillet 2023

Fig. 8 Companies covered by the CSRD and timetable for implementation. Revised. *Source* https://ec.europa.eu/commission/presscorner/detail/fr/qanda_23_4043

information", as its relationship to the financial plan is becoming increasingly evident. This conceptual evolution has resulted in a close connection between sustainability reporting and financial reporting, with growing interest on the part of all stakeholders and, in particular, ESG-conscious investors.

Companies that are not required to draw up a CSRD sustainability report could also be involved if they operate as suppliers or producers for companies subject to sustainability reporting under the CSRD, since the latter must also provide certain sustainability indicators relating to their value chain (only some, not all). The new rules contained in the CSRD will make companies accountable to citizens, obliging them to publish regular information on their social and environmental impact in order to curb greenwashing, strengthen the EU's social market economy and consolidate global sustainability transparency standards[16] (Fig. 8).

The CSRD will have a transformative effect on the way business is conducted, forcing companies to take responsibility for sustainability issues by correlating them with the company's financial opportunities and risks, and their effects on

[16] News European Parliament, Sustainable economy: new EU transparency obligations for multinationals, press release https://www.europarl.europa.eu/news/it/press-room/20221107IPR49611/economia-sostenibile-nuovi-obblighi-ue-di-trasparenza-per-le-multinazionali.

society and the environment. The parallel dissemination of strategies and plans for managing sustainability and financial performance will be decisive. CSRD reporting will provide investors with more uniform and comparable information to reward those companies that best demonstrate their ability to compete in the market from a sustainability perspective, by combining sustainability risks and opportunities with accounting projections.

Sustainability reporting will help companies to identify their objectives, estimate their performance and measure the risks and opportunities associated with sustainability, by describing the extent to which companies are able to generate value for their stakeholders, thus enabling a better understanding of performance, beyond economic results (Tettamanzi and Minutello 2021).

Business leaders should not make the mistake of treating CSRD reporting as a perfunctory exercise, thereby risking a potential reduction in economic return when investors realize that the strategy implemented does not correlate with sustainability risks and opportunities; on the contrary, management that convincingly embraces the CSRD philosophy is able to achieve more reliable financial results by leveraging sustainability factors capable of determining value creation.

According to PricewaterhouseCoopers (PWC)[17], to embark on this new path, companies need to implement four management innovations: integrate sustainability into strategy, identify the impact of business on the world, perfect the decision-making process and generate more useful data.

In research from 2023 (Friske et al. 2023), the relationship between voluntary sustainability reporting and corporate value was analyzed by examining the main hypotheses put forward by the signaling theory[18] and the interdisciplinary literature on sustainability reporting in the decade 2011–2020 on a large group of reporting and non-reporting organizations; the results of the analysis indicate that, although reporting may be costly at the outset, it then determines an increase in corporate value in parallel with the growth in the company's ability to better communicate the sustainability initiatives implemented to stakeholders, which translates into an adequate assessment of the reports on the part of investors.

A recent study (Bruna et al. 2022), analyzing the influence of ESG performance on a sample of 350 listed European companies observed from 2014 to 2019, found that in the presence of mandatory non-financial disclosure legislation, there is significant positive evidence of ESG performance on financial performance.

Further research (Carnini Pulino et al. 2022) assessed the impact of ESG disclosure on corporate performance in Italy, where Legislative Decree 254/2016 transposed European Directive 2014/95/EU, obliging large companies to disclose information on their social and environmental activities; the analysis of a sample of leading listed companies from 2011 to 2020, using EBIT and ROA, has led to the conclusion that

[17] PWC, *CSRD is resetting the value-creation agenda* https://www.pwc.com/gx/en/issues/esg/csrd-is-resetting-the-value-creation-agenda.html.

[18] According to signal theory, companies are inclined to provide additional information to reduce information asymmetries, satisfy stakeholder expectations, optimize financing costs and increase corporate value.

ESG disclosure, as tested by the ESG score, has a positive impact on corporate performance, and that companies have implemented a new business strategy combining commitment to business performance with interest in the community, environmental protection and respect for human rights.

In the current context of polycrisis, or "multiple crises" linked to each other, the challenge of sustainability must include the decisive role of companies, towards which a high level of trust is essential for the affirmation of a sustainable global economy. Stakeholders play a decisive role in identifying non-financial risks and opportunities, and the transparency achieved through their involvement in decision-making processes leads to better decisions and greater trust in companies.[19]

The implementation of an ESG program, the guarantee of accountability, the realization of systems and processes to administer a company's impact are closely linked to transparency, documentable by the sustainability commitment, which presents cognitive requirements that are more extensive and articulated than financial requirements alone, and potentially capable of renewing the company's mission (Marchegiani 2021).

6 Conclusion

The interdependent risks of a polycrisis are capable of weakening economic resilience, and this is the potential risk that is occurring in the presence of problems such as global warming, COVID-19 and international conflicts; on the other hand, the principles of environmental protection and respect for people's interests are leading to major social changes. The growing attention paid to the social and economic aspects of sustainability is contributing to the discussion on the need for a new economic model, based on the well-being of people and the quality of the natural environment.[20] Companies are discovering and evaluating a more ethically responsible and sustainable way of doing business over the long term, and the inclusion of environmental, social and governance aspects in their strategies has become a focal point of the capital markets. Transparent ESG information, through proper sustainability reporting, proves that companies are actively engaged in environmental and social responsibility, with positive effects on their reputation with consumers and investors, as well as the possibility of accessing capital at a lower cost and improving their competitiveness.

---

[19] EcoVadis, *Relazione sulla sostenibilità* https://ecovadis.com/it/glossary/sustainability-reporting/.

[20] https://eur-lex.europa.eu/legal-content/IT/TXT/PDF/?uri=CELEX:52023DC0376.

Companies will no longer be able to avoid openly assuming their responsibilities in terms of sustainability, if they don't want to risk being excluded from the present and the future.

"Either they will be involved, or the story will go over their heads".[21]

References

Ashraf Y, Nazir MS (2023) Corporate sustainability and environmental reporting: triggers and consequences. Artic Environ Sci Pollut Res. https://www.researchgate.net/publication/370465447

ASVIS—Quaderni dell'AsviS—Alleanza Italiana per lo Sviluppo Sostenibile (2022) Salute Globale e determinanti Sociali, Ambientali, Economici. Una nuova consapevolezza dopo la pandemia da COVID-19 , Ed. Editron s.r.l. https://asvis.it/public/asvis2/files/Pubblicazioni/QuadernoASVISGoal3.pdf

Atkinson AB (2015) Disuguaglianza. Che cosa si può fare? Raffaello Cortina Editore

Aydoğmuş M, Gülay G, Ergun K (2022) Impact of ESG performance on firm value and profitability. Borsa Istanbul Review 22, Supplement 2: S119–S127. https://www.sciencedirect.com/science/article/pii/S221484502200103X?ref=pdf_download&fr=RR-2&rr=7f918f933b990dc5

Benameur KB et al (2023) Sustainability reporting scholarly research: a bibliometric review and a future research agenda. Manag Rev Q:2

Bruna MG, Loprevite S, Raucci D, Ricca B, Rupo D (2022) Investigating the marginal impact of ESG results on corporate financial performance. Financ Res Lett 47, Part A, 102828. https://www.sciencedirect.com/science/article/abs/pii/S1544612322001283

Carnini Pulino S, Ciaburri M, Magnanelli BS, Nasta L (2022) Does ESG disclosure influence firm performance? Sustain Financ Perform Relat 14:7595. https://www.johncabot.edu/sustainability/does-esg-disclosure-influence-firm-performance.pdf

Commission of the European Communities, Green Paper Promoting a European Framework for Corporate Social Responsibility. Brussels, 18.7.2001 COM(2001) 366 final. https://www.europarl.europa.eu/meetdocs/committees/deve/20020122/com(2001)366_en.pdf

Commission to the European Parliament—Communication from the Commission to the European Parliament, the European Council, the Council, the European Economic and Social Committee and the Committee of the Regions "The Green Pact for Europe", Com/2019/640 final. https://eurlex.europa.eu/legalcontent/FR/TXT/PDF/?uri=CELEX:52019DC0640

Commission to the European Parliament—Report from the Commission to the European Parliament, the Council and the European Economic and Social Committee on the review clauses in Directives 2013/34/EU, 2014/95/EU and 2013/50/EU. Brussels, 21.4.2021 COM (2021) 199 final https://eur-lex.europa.eu/legal-content/FR/TXT/PDF/?uri=CELEX:52021DC0199

Communication from the Commission to the European Parliament and the Council Strategic Foresight Report 2023. Sustainability and human well-being at the heart of Europe's open strategic autonomy. Brussels, 6.7.2023 COM (2023) 376 final. https://eurlex.europa.eu/legal-content/FR/TXT/PDF/?uri=CELEX:52023DC0376

Effah NAA et al (2023) Contributions toward sustainable development: a bibliometric analysis of sustainability reporting research. Environ Sci Pollut Res 30:104–126.https://link.springer.com/article/10.1007/s11356-022-24010-8

European Parliament and of the Council—Proposal for a Directive of the European Parliament and of the Council on corporate sustainability due diligence and amending Directive (EU) 2019/

[21] *"Economy of Francesco—Pope Francis and young people from around the world for tomorrow's economy", Assisi—November 19,20,21, 2020.*

1937. Brussels, 23.2.2022. https://eur-lex.europa.eu/legal-content/FR/TXT/PDF/?uri=CELEX:52022PC0071

European Parliament and of the Council—Directive (EU) 2022/2464 of the European Parliament and of the Council of December 14, 2022 amending Regulation (EU) No 537/2014 and Directives 2004/109/EC, 2006/43/EC and 2013/34/EU as regards the publication of sustainability information by companies. https://eur-lex.europa.eu/legal-content/FR/TXT/PDF/?uri=CELEX:32022L2464

European Parliament and of the Council—Regulation (EU) n. 2088/2019 of the European Parliament and of the Council, of November 27, 2019, on the publication of sustainability information in the financial services sector (SFDR) https://eur-lex.europa.eu/legal-content/FR/TXT/PDF/?uri=CELEX:32019R2088, amended by Regulation (EU) 2020/852 of the European Parliament and of the Council of June 18, 2020, known as the Taxonomy Regulation Taxonomy Regulation https://eur-lex.europa.eu/legal-content/FR/TXT/PDF/?uri=CELEX:32020R0852&from

Fioramonti L (2019) Il mondo dopo il Pil. Economia e politica nell'era della post-crescita. Ed Ambiente, p 262

Friske W, Hoelscher SA, Nikolov AN (2023) The impact of voluntary sustainability reporting on firm value: Insights from signaling theory. J Acad Mark Sci 51:372–392.https://link.springer.com/article/10.1007/s11747-022-00879-2

Giosuè A, Recanati F, Calabrese I, Dembska K, Castaldi S, Gagliardi F, Vitale M, Vaccaro O, Antonelli M, Riccardi G (2022) Good for the heart, good for the Earth: proposal of a dietary pattern able to optimize cardiovascular disease prevention and mitigate climate change. Nutr, Metab & Cardiovasc Diseas 32:2772–2781, 2773, 2777. https://www.sciencedirect.com/science/article/pii/S0939475322003271

Giraud G (2022) La rivoluzione dolce della transizione ecologica. Come costruire un futuro possibile, Ed. Libreria Editrice Vaticana, p 9

George H, Stylianos N (2021) Corporate social responsibility: Trends in global reporting initiative standards. Econ Anal Policy 69:106–117

ILO - Organizzazione Internazionale del Lavoro (2019) Lavorare su un pianeta più caldo - L'impatto dello stress termico sulla produttività del lavoro e il lavoro dignitoso. L'impatto dello stress termico sulla produttività del lavoro e il lavoro dignitoso. Sintesi del rapporto, p 3. https://www.ilo.org/wcmsp5/groups/public/---europe/---ro-geneva/---ilo-rome/documents/publication/wcms_712842.pdf

IPSOS, Italia (2023) Catenaccio all'italiana. Un Paese in difesa, pronto al contrattacco e alla ricerca di un futuro. Ipsos Flair 2023 Volume online DEF, p 15

Klein F, Josten F, Grundmann G (2022) Sustainability in business: Staying ahead of the curve. How strategic foresight can strengthen organizational. Article on Deloitte Insights. https://www2.deloitte.com/xe/en/insights/topics/strategy/sustainability-in-business-staying-ahead-of-the-curve.html

KPMG International (2022) Big shifts, small steps Survey of Sustainability Reporting. https://assets.kpmg.com/content/dam/kpmg/se/pdf/komm/2022/Global-Survey-of-Sustainability-Reporting-2022.pdf

Kurznack L, Schoenmaker D, Schramade (2021) A model of long-term value creation. J Sustain Financ & Invest:2. https://www.tandfonline.com/doi/epdf/10.1080/20430795.2021.1920231?needAccess=true&role=button

Lagarde C (2022) Macroprudential policy in Europe: building resilience in a challenging environment, Welcome speech by Christine Lagarde, President of the ECB and the European Systemic Risk Board, at the sixth annual ESRB conference. Frankfurt am Main. https://www.ecb.europa.eu/press/key/date/2022/html/ecb.sp221208~f83c604a23.en.html

Llena-Nozal A, Martin N, Murtin F (2019) The economy of well-being: Creating opportunities for people's well-being and economic growth. Oecd Statistics Working Paper Series. https://one.oecd.org/document/SDD/DOC(2019)2/En/pdf

Losasso M (2022) Crisi interconnesse e complessità del progetto, Techne 23, pp 7–9. https://www.proquest.com/docview/2669112667

Marchegiani L (2021) Riflessioni su informazione non finanziaria, comunicazione di impresa e fiducia degli stakeholders. Osservatorio del diritto civile e commerciale Fascicolo 1/2021, gennaio, pp 105–136 https://www.rivisteweb.it/doi/10.4478/101967

Marzano F (2008) L'economia della crescita. Dalla teoria classica alle controversie tra contemporanei Ed. Mondadori Università-Collana Minerva-Manuali, p 352

McKinsey & Company (2022a) The role of ESG and purpose. https://www.mckinsey.com/~/media/mckinsey/business%20functions/strategy%20and%20corporate%20finance/our%20insights/the%20role%20of%20esg%20and%20purpose/the-role-of-esg-and-purpose.pdf

McKinsey & Company (2022) Toward a sustainable, inclusive, growing future: The role of business. https://valored.it/wp-content/uploads/2022/11/2022-McKinsey-Toward_a_sustainable_inclusive_growing_future_the_role_of_business-noV2022.pdf

McKinsey & Company (2023) The triple play: Growth, profit, and sustainability, p 2. https://www.mckinsey.com/~/media/mckinsey/business%20functions/strategy%20and%20corporate%20finance/our%20insights/the%20triple%20play%20growth%20profit%20and%20sustainability/the-triple-play-growth-profit-and-sustainability-vf.pdf

McKinsey Quarterly (2020) More than a mission statement: How the 5Ps embed purpose to deliver value. https://www.mckinsey.com/~/media/mckinsey/business%20functions/strategy%20and%20corporate%20finance/our%20insights/more%20than%20a%20mission%20statement%20how%20the%205ps%20embed%20purpose%20to%20deliver%20value/more-than-a-mission-statement-how-the-5ps-embed-purpose-to-deliver-value.pdf

Kern MAB (1999) Homeland Earth: A manifesto for a new millennium. Hampton Press, Cresskill, pp 73–74

Romanello M et al (2022) The 2022 report of the Lancet Countdown on health and climate change: health at the mercy of fossil fuels. Lancet 2022 400: 1619–1654

Richard S, Jane N (2022) Sustainable enterprise value creation. Palgrave Macmillan, Implementing stakeholder capitalism trough full ESG integration, Cambridge, p vii

Seemann A, Guyoton S, Bianchi A, Han J (2023) Do ESG efforts create value? Research by Bain & Company and EcoVadis. Bain & Company. https://www.esghoy.cl/wp-content/uploads/2023/04/bain-brief-bain-ecovadis-do-esg-efforts-create-value.pdf

Stabile Emanuele (2023) Gli standard di rendicontazione tra Dichiarazione Non Finanziaria e "informazioni sulla sostenibilità". Imprese, mercati e sostenibilità: nuove sfide per il diritto commerciale, xiv convegno annuale dell'associazione italiana dei professori universitari di diritto commerciale "orizzonti del diritto commerciale", Roma, 26–27 maggio, p 9

Tettamanzi P, Minutello V (2021) Il Bilancio di sostenibilità come strumento di rendicontazione sociale. Ed. Università Cattaneo Libri 2021:26

The British Academy, Policy & Practice for Purposeful Business, The final report of the Future of the Corporation programme, Published September 2021. p. 6 https://www.thebritishacademy.ac.uk/programmes/future-of-the-corporation/

Tooze A (2022) Welcome to the world of the polycrisis. Financial Times. https://www.ft.com/content/498398e7-11b1-494b-9cd3-6d669dc3de33

WEF—The global risks report 2023, 18th edn. https://www.weforum.org/reports/globalrisks-report-2023/

Trade Cooperation Between the EU and MERCOSUR in the Context of the Climate Crisis

Sára Smolková and Zuzana Kittová

Abstract Intensified trade cooperation between the EU and MERCOSUR has become one of the priorities of the EU's foreign trade policy in recent times. In response to the Russian Federation's invasion of Ukraine and the energy crisis, the EU is seeking to diversify its imports and cooperate with Latin America, which is one of the richest regions in renewable resources and is considered to have the greatest potential for the production and export of green hydrogen. Therefore, it is in the interest of the EU to conclude a free trade agreement with the MERCOSUR. Environmental protection is a problematic area of the negotiations. Even though the agreement would facilitate bilateral trade and promote development in various areas, there are concerns that increased trade cooperation could lead to intensified deforestation of the Amazon rainforest and have an overall negative impact on the biodiversity of this region. The aim of this paper is to quantify and estimate through econometric modelling the impact that more intensive trade cooperation between the EU and MERCOSUR would have on the environment. Econometric modelling involves regression analysis between environmental and macroeconomic indicators, where the dependent variable represents environmental indicators such as deforestation rate, greenhouse gas emissions, and the independent variable consists of economic indicators reflecting mutual trade between the EU and MERCOSUR. The contribution of this paper is to provide recommendations that can be used in negotiations on the free trade agreement between the EU and MERCOSUR in the context of environmental protection and the fight against the climate crisis.

Keywords Climate crisis · Environmental sustainability · Trade cooperation · European union · MERCOSUR

S. Smolková (✉) · Z. Kittová
Department of International Trade, Faculty of Commerce, University of Economics in Bratislava, Bratislava, Slovakia
e-mail: sara.smolkova@euba.sk

Z. Kittová
e-mail: zuzana.kittova@euba.sk

G. Pellat et al. (eds.), *Data-Centric Business and Applications*, Lecture Notes on Data Engineering and Communications Technologies 258,
https://doi.org/10.1007/978-3-031-95068-1_12

1 Introduction

The climate crisis, caused by climate change, is currently one of the most serious and urgent problems facing our planet. The United Nations (2023) characterizes climate change as long-term fluctuations in temperature and weather, which may occur naturally (due to changes in solar activity or large volcanic eruptions). However, since the nineteenth century, human activity has been the primary driver of climate change, particularly through the burning of fossil fuels such as coal, oil, and natural gas. The Intergovernmental Panel on Climate Change (IPCC), comprised of hundreds of scientists and experts, defines climate change "*as a change in the state of the climate that can be identified (e.g., by using statistical tests) by changes in the mean and/or the variability of its properties and that persists for an extended period, typically decades or longer*" (IPCC 2018). In recent years, environmental sustainability has become a global challenge for humanity. Studies have shown that the majority of environmental problems are caused by anthropogenic factors rather than natural factors (Lee et al. 2023).

In the context of the climate crisis, the European Union (EU) plays a leading role, committing itself to ambitious climate and sustainable development goals. The EU has emerged as a longstanding global leader in the battle against climate change. Its climate policy evolved as an integral part of its comprehensive environmental policy, firmly grounded in the Single European Act, which came into effect in 1987 (Delbeke and Vis 2016). Over time, the ambitions of the EU in the field of environmental sustainability have increased, and its primary goal now is to achieve climate neutrality by the year 2050, fulfilling its commitments under the Paris Agreement. To accomplish this primary goal, the EU has also set a partial target to reduce greenhouse gas emissions by at least 55% by the year 2030 compared to the year 1990 levels. The European Green Deal serves as the key strategy for attaining climate neutrality within the EU. This agreement represents a comprehensive package of policy measures designed to facilitate a green transformation and attain climate neutrality, meanwhile promoting a prosperous society built on a modern and competitive economy (European Council—Council of the European Union 2022).

With the increasing share of international trade in the global economy, the linkages between climate measures and trade have become increasingly significant. Free trade agreements (FTAs) have become a major tool for managing trade policy and diplomacy. For this reason, environmental sustainability measures included in trade agreements hold great importance for EU external trade. Environmental issues in trade agreements began to emerge in the early 1980s, but significant progress in their implementation as part of trade policy was not made until 2010 (Dent 2021).

EU's trade relations with third countries also have a significant impact on environmental sustainability. As mentioned earlier, the EU is constantly striving to make progress in addressing climate change, including through its bilateral relationships. Provisions on trade and sustainable development were first introduced in FTAs through a new chapter on Trade and Sustainable Development (TSD) during the negotiations between the EU and South Korea in the year 2010. The European Green

Deal also emphasized the EU's commitments to "greening" its trade policy, with a promise to enhance the incorporation of social and environmental sustainability issues into EU trade policy (Blot et al. 2022).

Latin America and Caribbean (LAC) region represents an opportunity for the EU to diversify its energy industry and reduce dependence on the Russian Federation, particularly considering the ongoing Ukraine invasion. Latin America possesses abundant renewable resources and is considered to have a potential for the production and export of competitively priced green hydrogen (resource in high demand in the EU) (Nolte 2023).

FTA between the EU and the Southern Common Market (MERCOSUR), which negotiations were initiated in the year 2000, have traversed various evolutionary phases. In May 2016, both the EU and MERCOSUR reinvigorated the negotiation process. This renewal entailed the exchange of updated market access offers and a heightened intensity in negotiations, evidenced by the frequent convening of negotiation rounds and scheduled meetings. A pivotal milestone was reached in June 2019, when the EU and four MERCOSUR members (Argentina, Brazil, Paraguay, and Uruguay) achieved a pivotal political agreement (European Comission n.d.). However, after the wildfires in the Amazon rainforest in August 2019, many of which were intentionally set by Brazilian farmers, public awareness and outrage have increased. Critics have highlighted the mismatch between the EU's climate ambitions and the trade deal (Colli 2019). Deforestation of the Amazon rainforest can be caused by several factors, one of them including international trade. The growth of the agricultural sector and the commercial use of natural resources contribute to deforestation, which is often driven by the demand for commodities in the market. Brazil's agricultural sector plays an important role in international trade, especially the commodities like soybeans, beef, and timber. The expansion of agricultural activities, including deforestation, is frequently linked to satisfying the demand on world markets (Pivello et al. 2021).

So far, the agreement has still not been ratified due to several reasons, but the most significant issue is the environmental sustainability and the climate crisis. Critics argue that the deal does not provide sufficient safeguards against deforestation, particularly in the Amazon rainforest. They are concerned that increased trade could contribute to further deforestation and harm the environment. The EU has expressed concerns over the deforestation issue and has called for additional guarantees. There are also concerns about the emissions from increased economic activity and transport that the agreement could bring. The reduction of trade barriers is expected to lead to a significant increase in the transport of goods between the EU and MERCOSUR, which may result in a significant increase in greenhouse gas (GHG) emissions, especially through maritime or air transport (Keating 2023).

The European Commission argues that, on the contrary, this agreement would contribute to the promotion of sustainability and environmental protection thanks to the provision of the TSD chapter. The agreement commits both parties to effectively implement the UNFCC and the Paris Agreement. It contains provisions to protect sensitive EU agricultural products and promotes sustainable agricultural practices. In addition, it focuses on combating climate change and respecting workers' rights under International Labour Organisation standards. The agreement also includes commitments on sustainable fisheries and forest management (European Comission n.d.).

However, this would not be the first agreement dealing with environmental protection adopted by MERCOSUR. The environmental issue is addressed in the preamble of the Treaty of Asunción, which founded the bloc in 1991, with the mention that the expansion of markets must consider the most efficient use of available resources and the preservation of the environment. In 2001, all MERCOSUR's countries adopted the Mercosur Framework Agreement on the Environment, which was a major step in the region's environmental policy. They have also endorsed global environmental frameworks like the Rio Declaration and Agenda 21 from the 1992 United Nations Conference on Environment and Development. Each Member State has some national environmental protection in place and has active non-governmental sectors focused on environmental issues, while Brazil has the broadest and most sophisticated environmental protection at the national level (Hochstetler 2003).

This changed during Jair Bolsonaro's presidency in Brazil (2019–2022), when deforestation rates in the Amazon rainforest increased significantly. His government's policies were often criticized for prioritizing economic development over environmental protection. Bolsonaro reduced regulatory oversight, decreased funding for scientific and environmental bodies, dismissed environmental professionals, and advocated for diminishing the land rights of Indigenous communities. These measures were primarily taken to benefit the agricultural industry. It is estimated that From August 2019 to July 2021, which mostly coincides with the first three years of Bolsonaro's presidency, over 34,000 square kilometres of the Amazon rainforest were lost (Miller 2022).

2 Literature Review: Correlation Between Trade and Environmental Sustainability

Several authors have addressed the mutual relationship between trade and climate and how they can affect each other. The influence of environmental policy on international trade has often been analysed through traditional hypotheses of comparative advantages, such as differences in productivity and factor endowments (Porter and Linde 1995). Authors like Grossman and Krueger (1991) and Copeland and Taylor (1994) have developed a theory that distinguishes three effects of trade and investment policy on the environment. The first is the scale effect, which demonstrates that increased production in a country can lead to higher emissions of pollutants because if there is no change in technological innovations or willingness to adopt more ecologically friendly practices, increased production and consumption usually leads to an increase in pollution. The second is the composition effect, which is manifested in the sectoral structure of the trade, which leads countries to specialize in certain industries, impacting the environment. If these industries are polluting, trade may increase pollution, but if they are clean, it may improve environmental outcomes. The third is the technological effect, which is driven by the adoption of ecologically efficient technological innovations and investments, which can improve the environment. The overall impact on the environment, therefore, results from the interplay of these three effects.

Other theories argue that companies engaged in exporting to foreign markets possess greater technological knowledge and invest in pollution-reducing technologies (Bernard et al. 2011). Similarly, an export-oriented focus enhances competitiveness in the international arena and facilitates the transfer of innovations from abroad. Thus, international trade not only benefits the economy but can also contribute to improving environmental sustainability (Sorroche del Rey et al. 2022).

Conversely, additional theories by authors such as Eskeland and Harrison (2003) contend that international trade exacerbates environmental pollution. According to them, the primary reason is that international trade generally leads to increased industrial production, resulting in higher emissions of undesirable gases. Moreover, transnational corporations frequently relocate their production to countries with weaker environmental regulations. This phenomenon supports the theory that the impact of regulations transfers polluting manufacturing activities to developing countries, given that their environmental regulations are significantly less stringent than, for instance, those in the EU (Li et al., 2018).

The World Trade Organization (WTO) actively addresses this issue and, in its latest report on global trade in 2022, examines the interconnection between climate change and international trade while highlighting how international trade and trade rules can contribute to addressing climate change. The WTO acknowledges the complexity of this interconnection and argues that it is not entirely accurate to claim that international trade worsens the state of the environment. It recognizes that processes associated with international trade, such as production, transportation, or consumption of traded goods, impact greenhouse gas emissions in various ways. These activities stimulate economic activity, leading to increased income and, subsequently, increased emissions. However, trade also affects the types of goods and services that individual countries produce, and therefore, it can have a positive or negative impact on climate change depending on whether a country has a comparative advantage in greenhouse gas-intensive industries. Similarly, international trade provides easier access to low-carbon goods, services, or technologies, increases environmental awareness, and leads to stricter environmental regulations, thereby promoting the adoption of environmental technologies in production processes. It also assists developing countries with abundant renewable energy resources in utilizing their comparative advantage in clean energy production and contributing to the transition to a low-carbon economy in developed countries. The WTO supports the fight against climate change by helping to ensure effective and efficient climate-related trade policies. While not all climate change policies have a trade dimension, WTO rules govern subsidies, tariffs, support measures, regulatory measures, and other trade-related instruments that are relevant to the implementation of climate policies (WTO 2022).

One of the most frequently mentioned terms in the literature is environmental Kuznets curve (EKC). The basic idea of the EKC is that environmental degradation initially increases as a country undergoes industrialisation and economic growth. However, at a certain income threshold, environmental degradation starts to decrease and as a country becomes wealthier, its environmental pollution decreases. This inverted U-shaped curve suggests that in the early stages of development, countries tend to prioritise economic growth over environmental problems. As income rises, industrial activity also increases, leading to higher levels of pollution. However, as countries become richer and more technologically advanced, they start to invest in environmental protection and sustainable practices, leading to a decline in pollution and an improvement in environmental quality (Ekins 1997).

Several researchers also tested these hypotheses with various econometric methods. Balogh (2022) has analysed the factors contributing to CO2 emissions in Non-EU countries, with a special focus on economic growth, agriculture, and trade. The study employs Fully Modified Ordinary Least Squares (FMOLS) regression and Dynamic Ordinary Least Squares (DOLS) estimation methods to estimate the relationships between the variables. The results of this analysis were that the higher level of economic development, after the tipping point is reached, can reduce CO2 emissions. The study also revealed the specific role of FTAs in the reduction of the emissions, when NAFTA was found to encourage emission growth, while EFTA, ASEAN, and MERCOSUR reduced emission growth.

Kozul-Wright and Fortunato (2012) also used panel data analysis to find evidence on the effect of the trade on the CO_2 emissions. The sample included 38 countries, of which 26 have ratified the Kyoto Protocol during the period 1995–2005. They revealed that the Kuznets curve does not apply to GHG. On the contrary, once the minimum development threshold is crossed, GHG emissions continue to increase with increasing per capita income. Also, the impact of trade openness on GHGs is significantly different from other air emissions. This study revealed a positive and highly significant relationship between the trade openness and CO_2 emissions. A closer examination of the data suggests that trade openness tends to increase CO_2 emissions, especially in less industrialized or developing countries. The findings of this study suggest that industrialised countries use trade opportunities to transfer high-emission activities to their less industrialised and less regulated partners, which leads to carbon leakage.

Mahmood (2022) investigated the environmental effects of trade, renewable energy consumption (REC), and industry value-added (IVA) in South America from 1990 to 2018. In this study, the author distinguished between TCO_2 (territory-based CO_2 emissions, which includes all CO_2 emissions produced within a specific region or country) and CCO_2 (consumption-based CO_2 emissions, which considers the CO_2 emissions associated with the production and transportation of goods and services consumed within a specific region or country). By using regression analysis, the author found that the economic growth within South American countries is positively associated with local TCO_2 and CCO_2 emissions, meaning that as economies expand, their production and consumption activities result in more CO_2 emissions. Economic growth also leads to increased CCO2 in neighbouring countries, likely through increased demand for imports from those countries, which increases their production-based emissions. However, it does not affect TCO_2 in those neighbouring countries.

Locally, exports contribute to TCO_2 because they represent production that generates CO2 emissions. Exports help to reduce TCO_2 in neighbouring countries, possibly because when a country exports goods, the production emissions are accounted for in the exporting country's TCO_2, not in the importing neighbour's TCO_2. This is a positive environmental spillover for the neighbouring countries. However, exports negatively affect CCO_2 both locally and in neighbouring countries, indicating that the production for export is adding to the consumption-based emissions footprint of the region as a whole.

Imports have statistically insignificant effects on the local and neighbouring countries' TCO_2, meaning they do not substantially alter the emissions of the importing or neighbouring countries. Imports increase the local CCO2 emissions since they add to the emissions embedded in consumed goods and services, but do not affect neighbouring countries' TCO_2 or CCO2.

IVA does not significantly affect local or neighbouring TCO_2 emissions and has no effect on local CCO2 emissions. However, IVA has a positive effect on neighbouring countries' CCO2, possibly indicating that increased industrial production in one country can increase the CCO2 in neighbouring countries through trade and economic integration.

Moreover, the use of REC has positive environmental impacts by reducing CO2 emissions, both in local and neighbouring countries.

Based on these findings, the study recommends the following for South American countries:

- Increase the share of renewable energy in the energy mix to promote a sustainable environment. This could help mitigate local and regional CO2 emissions.
- Growing exports might have beneficial effects on the environment, as they can help reduce TCO_2 in neighbouring countries.
- Discourage imports to avoid an increase in consumption-based CO2 emissions, focusing on local production and consumption patterns that are more environmentally sustainable.

In conclusion, the study suggests a complex interplay between economic activities and environmental impacts, where trade and energy sources play critical roles in shaping CO2 emissions profiles both within and across South American countries. The promotion of renewable energy is seen as a key strategy for sustainable environmental management in the region.

When it comes to the deforestation, several studies were also dealing with the relationship between trade and the rate of deforestation. For example, Faria and Almerida (2016) used panel data of 732 municipalities in Brazil for the years 2000 to 2007. The main finding of this study is that with the increase of trade openness, there is increase also in the deforestation. The study also supports the EKC hypothesis, as it finds that as the square of GDP goes up, deforestation decreases. This suggests that economic development and increased income can potentially lead to more sustainable practices and a reduction in deforestation.

In summary, several studies confirm the fact that the relationship between economic growth and environmental sustainability is a very complex phenomenon that depends on a multitude of factors and variables that enter into the observation. This is why, it is not possible to derive a general formula that applies to every single research. However, we can say that the EKC effect has been confirmed in many cases, whereby investment in environmental technologies rises with economic growth. In this case, increased trade can serve as an auxiliary factor. Due to the complexity of this issue and the lack of actual observations for MERCOSUR countries, we decided to process the latest data and evaluate econometric modelling that is directly applicable to MERCOSUR countries in the context of bilateral trade with the EU. This paper focuses on the question whether trade cooperation between MERCOSUR and the EU has environmental impacts such as increased GHG emissions or deforestation for MERCOSUR countries. The findings of this study can serve as a basis for decision-making on the conclusion of an FTA between the EU and MERCOSUR, which is increasingly being discussed at a time of an energy crisis that is driving the need to diversify the EU's energy import mix.

3 Methodology

The aim of this paper is to quantify and estimate through econometric modelling the impact that more intensive trade cooperation between the EU and MERCOSUR would have on the environment. To fulfil the aim, we set two hypotheses:

1. Increased trade between the EU and MERCOSUR will lead to an increase in GHG emissions in MERCOSUR.
2. Increased trade between the EU and MERCOSUR will lead to an increase in deforestation in MERCOSUR.

To fulfil the main objective, as a first step, we studied the scientific literature on environmental sustainability and the climate crisis in the context of international trade and economic development. Further, we employed several quantitative scientific research methods in this study. We analysed MERCOSUR's foreign trade data from the European Commission, Directorate-General for Trade (DG Trade) (2023). Furthermore, we have observed the evolution of trade between the EU and MERCOSUR between 2012 and 2022 according to data from the EU statistical database Eurostat (2023). Based on the European Commission's Emissions Database for Global Atmospheric Research (EDGAR) (2023), we analysed the evolution of greenhouse gas emissions and using Curtis (2018, updated through 2023), we also analysed the evolution of deforestation, which is commodity driven.

These data were used to conduct two fixed-effects panel data analysis. Econometric modelling involves regression analysis between environmental and macroeconomic indicators, where the dependent variable represents environmental indicators such as deforestation rate and greenhouse gas emissions and the independent variable consists of economic indicators reflecting mutual trade between the EU and MERCOSUR.

The panel data and fixed effects model is an econometric model that is used to analyse data that have a combination of cross sections and time series. The basic principle of the fixed effects model is to include individual fixed effects for each unit in the panel data. These fixed effects account for individual specific characteristics that do not change over time. The resulting model can be defined as follows (Lukáčik et al. 2011):

$$y_{it} = \alpha_i + \beta_1 x_{it1} + \beta_2 x_{it2} + \ldots + \beta_k x_{itk} + u_{it}, \tag{1}$$

Selection of Variables

Variable	Description
l_Commodity_driven_deforestation_kHa	Extensive deforestation which is primarily associated with the expansion of commercial agriculture
l_Export_million_of_eur	Export to the European union from MERCOSUR countries in millions of euros
l_GDP_pc	Gross domestic product per capita
l_GHG_pc	Green house gas emissions per capita

Source authors own elaboration

Since the selected variables are expressed in different units, we created logarithms from them. The general Eqs. (2) and (3) of the models with logarithms have the following form:

$$l\_GHG\_pc_{it} = \alpha_i + \beta_1 \times l\_Export_{it1} + \beta_2 \times l\_GDP\_pc_{it2} + u_{it}, \tag{2}$$

$$l\_Deforestation_{it} = \alpha_i + \beta_1 \times l\_Export_{it1} + \beta_2 \times l\_GDP\_pc_{it2} + u_{it}, \tag{3}$$

The equation specifications include dependent variable y (GHG emissions per capita in Model 1, Eq. 2 and commodity driven deforestation in Model 2, Eq. 3) and independent variable x (Export in million EUR from MERCOSUR to the EU and GDP per capita in current USD of MERCOSUR countries by World Bank (2023). We used the first independent variable as an indicator of the level of trade cooperation between MERCOSUR and the EU. Increasing value of export indicates a higher degree of economic interaction between the two regions, which can potentially have implications for the climate crisis. The second independent variable GDP per capita was used for a higher objectivity since this macroeconomic indicator is also very often associated with its possible impact on the environment.

We present the statistical probability of null hypothesis rejection concerning insignificant parameter estimates according to the numbers of an asterisk (* 90% probability; ** 95% and *** 99% probability).

4 Results and Discussion

As we mentioned above, LAC countries play a significant role in becoming more important trading partner of the EU. According to the European Commission, the EU is one of the biggest trade and investment partners of MERCOSUR. The following section presents an analysis of the development of MERCOSUR's external trade, the evolution of trade between the EU and MERCOSUR, the evolution of GHG emissions, the evolution of deforestation, and an econometric observation of environmental indicators with macroeconomic ones.

Table 1 captures the main trading partners of MERCOSUR in terms of export, import, and total trade values, measured in millions of euros, and accompanied by the respective percentage of total trade. In terms of MERCOSUR exports, the main trading partners are China, the EU and the USA. When considering imports into MERCOSUR, the main partners are China, the EU and the USA. According to the data of the European Commission, MERCOSUR's exports to the EU amounted to 59,608 million euros, representing 15.7% of MERCOSUR's total exports in 2022. On the import side, MERCOSUR imported goods worth 57,361 million euros from the EU, constituting 17.2% of the region's total imports.

Figure 1 represents the trade flow between the EU and MERCOSUR in millions of euros between 2012 and 2022. Since 2013, both exports and imports have been declining compared to 2012, with the largest drop occurring in 2020, which can be considered as a consequence of the COVID-19 pandemic and the global slowdown in international trade. From 2021 onwards, trade between the EU and MERCOSUR started to recover, reaching an all-time peak in 2022.

Table 1 Top MERCOSUR's trading partners in millions of euros, percentages, 2022

Export			Import			Total trade		
Partner	Value	%	Partner	Value	%	Partner	Value	%
China	95,537	25.2	China	84,843	25.4	China	180,379	25.3
EU	**59,608**	**15.7**	USA	64,005	19.1	**EU**	**116,970**	**16.4**
USA	44,024	11.6	**EU**	**57,361**	**17.2**	USA	108,029	15.1
Chile	14,412	3.8	India	11,036	3.3	India	21,273	3.0
India	10,237	2.7	Russia	8,544	2.6	Chile	20,089	2.8
Mexico	8,220	2.2	Mexico	7,256	2.2	Mexico	15,476	2.2
Singapore	8,155	2.2	Japan	6,780	2.0	South Korea	14,510	2.0
South Korea	8,111	2.1	Saudi Arabia	6,422	1.9	Japan	14,004	2.0
Japan	7,223	1.9	South Korea	6,400	1.9	Canada	11,997	1.7
Vietnam	6,526	1.7	Canada	5,882	1.8	Russia	11,259	1.6
World	379,027	100.0	World	334,337	100.0	World	713,364	100.0

Source European Commission, DG Trade (2023)

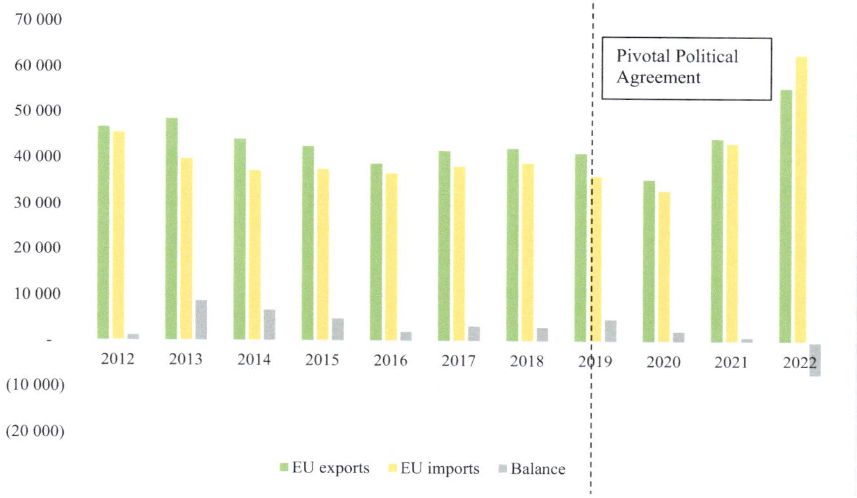

Fig. 1 EU—MERCOSUR trade in millions of euros, 2012–2022. *Source* Eurostat (2023)

Exports increased from 46,759 million euros in 2012 to 55,768 million euros in 2022. Imports have also increased over the years, from 45,508 million euros in 2012 to 63,114 million euros in 2022. During this period, the EU has consistently maintained a trade surplus with MERCOSUR, meaning that it has exported more goods and services than it has imported. The highest surplus was recorded in 2013, amounting to 8,836 million euros. In 2022, there was a significant change from this historical trend, when the EU recorded a trade deficit of 7,346 million euros, meaning that the EU imported more from MERCOSUR than exported.

Total trade between the EU and MERCOSUR is demonstrated in Fig. 2. According to this data, we can conclude that, from 2012 to 2022 the trade relations between the EU and MERCOSUR have experienced significant fluctuations, which are succinctly expressed through observed trade values (in millions of euros). From a trade value of 92,267 million euros in 2012, there has been a marked decline, reaching the lowest value of 68,649 million euros in 2020, potentially reflecting the impact of global economic challenges and the key impact of the COVID-19 pandemic. Interestingly, a remarkable recovery followed, raising the value to 118,882 million euros in 2022.

Analyzing the figures from Table 2, the category "0—Food and live animals" is the leading export category for MERCOSUR to the EU, with its share of the total exports slightly decreasing from 35.75% in 2019 to 33.71% in 2022. Despite a slight reduction in percentage, it continues to be a major export category, suggesting a high demand for these products within the EU market. The category "2—Crude materials, inedible, except fuels" also demonstrates a slight decrease in its share from 28.82% in 2019 to 26.05% in 2022. The category "3—Mineral fuels, lubricants and related materials", although having a lower export volume, shows a significant increase, starting with a smaller share of the total exports at 5.42% in 2019 and growing to

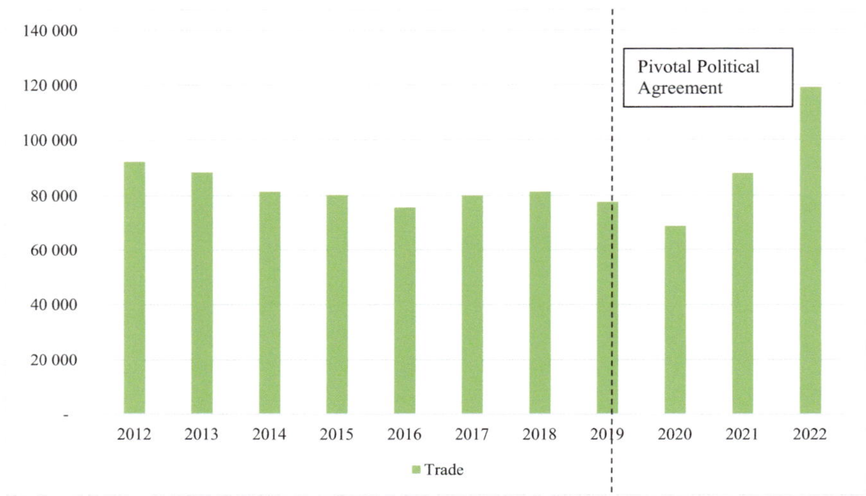

Fig. 2 Total trade EU—MERCOSUR in millions of euros, 2012–2022. *Source* Eurostat (2023)

16.22% in 2022, indicating a sharp increase in export of these commodities from MERCOSUR to the EU. This could be due to a variety of factors, including shifts in global energy markets, changes in the EU's energy mix, or MERCOSUR's increased capacity to export these materials.

The table shows not only MERCOSUR's important role as a supplier of diverse commodities to the EU, but also the growth and potential changes in trade patterns

Table 2 MERCOSUR export to EU by SITCS categories as percentages of total export to EU, 2019–2022

SITC categories	2019	2020	2021	2022
0—Food and live animals	35.75	36.69	32.48	33.71
1—Beverages and tobacco	1.99	2.04	1.55	1.26
2—Crude materials, inedible, except fuels	28.82	28.70	32.64	26.05
3—Mineral fuels, lubricants and related materials	5.42	7.96	8.88	16.22
4—Animal and vegetable oils, fats and waxes	0.59	0.65	0.85	0.64
5—Chemicals and related products, n.e.s	7.13	7.17	7.64	7.43
6—Manufactured goods classified chiefly by material	7.96	6.33	6.11	7.44
7—Machinery and transport equipment	8.45	6.20	5.28	4.86
8—Miscellaneous manufactured articles	1.71	1.60	1.60	1.27
9—Commodities and transactions, n.e.c	1.29	1.68	1.35	0.81
Other	0.89	0.69	1.38	0.32
Total	100.00	100.00	100.00	100.00

Source European Commission, DG Trade (2023)

over the years. The increase in total exports from 36,231 million euros in 2019 to 63,114 million euros in 2022 shows the strengthening of economic ties between the two regions and highlights the importance of MERCOSUR in the EU's external trade. Based on our results, MERCOSUR's exports to the EU have shown composition and technology effects through the change in the composition of exports, especially the growth in "3—Mineral fuels, lubricants and related materials," along with "5—Chemicals and related products, n.e.s.," suggests a shift towards sectors with potentially higher technological complexity and value addition. This indicates a sectoral shift towards more technologically advanced production and a diversification of export profiles, which may have an impact on the environmental and economic development impact within MERCOSUR countries. But also, possibly leading to greater environmental impacts, depending on how these sectors manage their pollution levels and adopt cleaner technologies.

Figure 3 shows the trends in foreign direct investment (FDI) flows from the EU to the MERCOSUR countries individually, but also to the bloc. The data show that Brazil is the main beneficiary, reflecting its dominant economic position within the bloc, followed by Argentina. Paraguay and Uruguay attract significantly less FDI. The peak in FDI across all member countries in 2016 indicates a period of significant EU economic engagement, which is subsequently disrupted by the global economic implications of the COVID-19 pandemic in 2020. The subsequent recovery in 2021, while remarkable, falls short of previous peak levels, suggesting a cautious resumption of investment activity.

When we looked at FDI by sector, we found that Eurostat's statistics are missing a large amount of data per country. The available data showed that the manufacturing

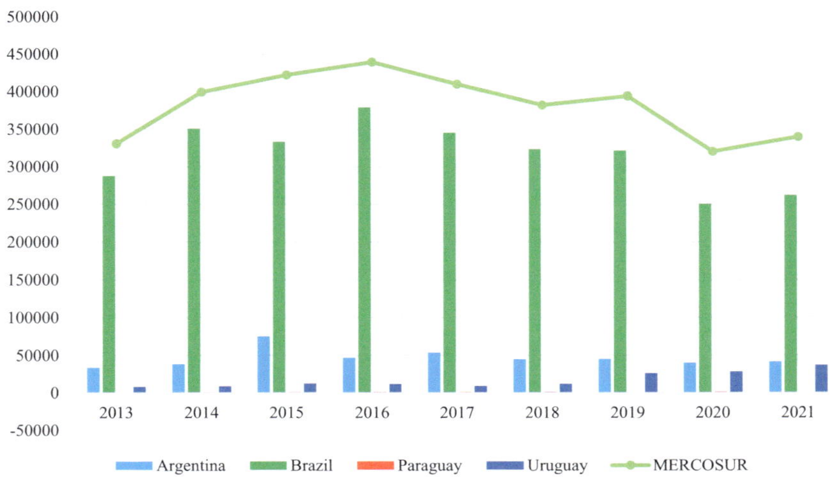

Fig. 3 FDI flows from the EU to MERCOSUR in millions of euros, 2013–2021. *Source* Eurostat (2023)

sector attracted significant investment, with Brazil receiving 87,963.3 million euros in 2013 and reaching a peak of 132,841.7 million euros in 2018. This was followed by financial and insurance activities, whose upward trajectory peaked in 2020 at 101,236.7 million euros. Mining and quarrying investment in Brazil was significant in 2014, reaching 127,466.2 million euros, but showed a declining trend, with only 15,862.1 million euros recorded in 2020. However, the scope of the analysis is limited by the lack of data across sectors and years, suggesting the need for more rigorous data collection to fully understand FDI trends within MERCOSUR.

Figure 4 shows the evolution of GHG emissions in MtCO_2 from 2012 to 2022 in individual MERCOSUR countries, in MERCOSUR as a whole group and in the EU. According to this graph, we can confirm that total GHG emissions in MERCOSUR are lower than in the EU, but this can be attributed to the fact that the EU consists of a larger number of member states than MERCOSUR, thus accumulating more emissions.

Argentina's GHG emissions have remained relatively stable over the past decade. Emissions started at 358.82 Mt CO2 in 2012 and gradually increased to 382.99 Mt CO2 in 2022. While there have been some fluctuations, the overall trend shows a modest increase. Brazil's emissions have also followed a relatively stable trajectory. Emissions began at 1250.58 Mt CO2 in 2012 and reached 1310.50 Mt CO2 in 2022. Based on these data, Brazil is considered the largest GHG emitter among MERCOSUR countries. Paraguay and Uruguay's emissions are the lowest in the group. Paraguay's emissions show a slow increase from 38.86 MtCO_2 in 2012 to 45.57 MtCO_2 in 2022 and Uruguay's emissions have been relatively consistent over the years, ranging from 40.48 MtCO_2 in 2013 to 41.91 MtCO_2 in 2022.

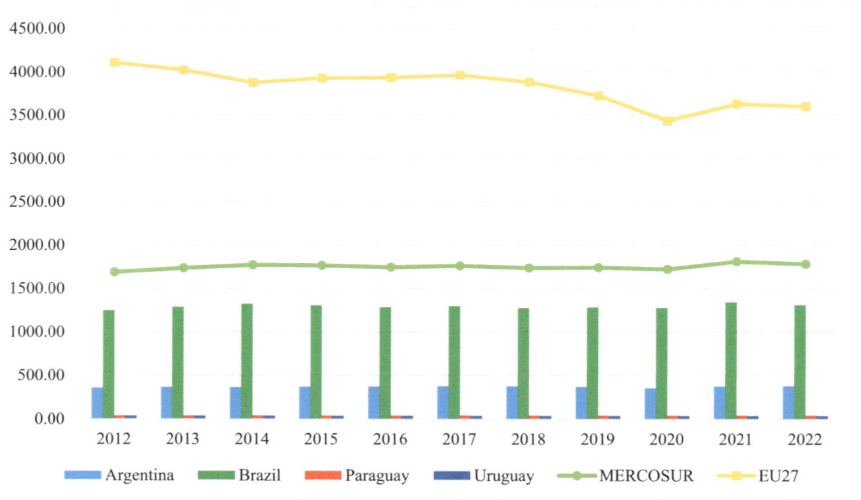

Fig. 4 GHG emissions by country in Mt CO_2, 2012–2022. *Source* EDGAR (2023)

MERCOSUR's total emissions have gradually increased over the years from 1,689.90 $MtCO_2$ in 2012 to 1,780.96 $MtCO_2$ in 2022. By contrast, EU27 emissions are on a downward trend, from 4106.59 $MtCO_2$ in 2012 to 3587.80 $MtCO_2$ in 2022. This represents a significant reduction in emissions, reflecting the EU's commitment to climate action and its success in implementing policies to reduce emissions and transition to cleaner energy sources.

The GHG emissions per capita data for these countries from 2012 to 2022 reveals diverse trends and challenges in addressing climate change. The EU's consistent reduction in emissions per capita stands in contrast to the more fluctuating patterns observed in MERCOSUR countries. These insights provide valuable information for policymakers and underscore the importance of sustainable development and climate action to address global environmental challenges.

Figure 5 demonstrates GHG emissions per capita data by country for the years 2012 to 2022. Per capita GHG emissions in Argentina have remained relatively stable over the last decade with small fluctuations. There was a significant decline in 2020, but emissions began to rise again in 2021 and 2022. Per capita GHG emissions in Brazil have generally followed an upward trajectory, with a few exceptions. There was a decline in emissions in 2018 and 2019, followed by a slight increase in 2020 and a decline in 2022. Paraguay's per capita emissions show some fluctuations, but generally remain in a moderate range. There was a significant increase in 2018, but since then emissions have stabilized. Uruguay has experienced relatively high per capita emissions throughout the period, with fluctuations but no clear trend of increasing or decreasing. Uruguay's emissions remained above 11 tCO_2 per capita throughout the period.

The MERCOSUR bloc, represented by the sum of its member countries, shows a relatively stable trend in per capita emissions, with small fluctuations. Overall, the regional total remains above 30 t CO2 per capita, reflecting a notable carbon footprint for the bloc. In contrast, the EU has shown a more consistent downward

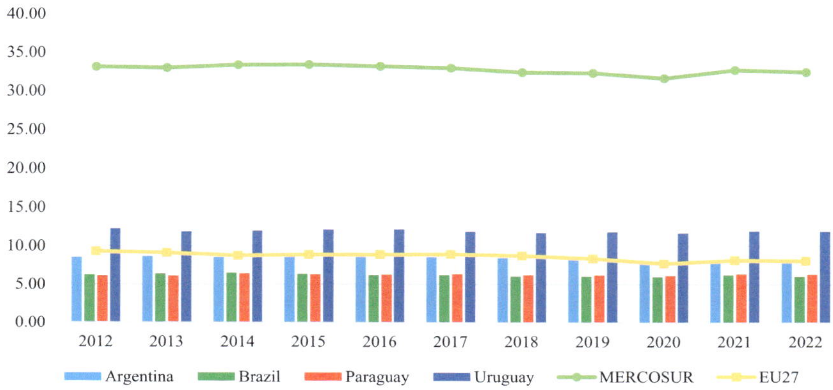

Fig. 5 GHG emissions per capita by country in t CO_2, 2012–2022. *Source* EDGAR (2023)

trend in emissions per capita over the years. The EU has made significant progress in reducing its per capita emissions, with values below 10 t CO_2 in recent years.

Figure 6 represents an analysis of deforestation between 2012 and 2022 in the MERCOSUR countries (Argentina, Brazil, Paraguay, and Uruguay), which reveals remarkable differences and fluctuations. Brazil shows the highest levels, particularly in 2016 at 2180 kHa, while Uruguay maintains consistently minimal levels that barely exceed 1 kHa. Argentina shows a decline from 249 to 71.9 kHa between 2012 and 2019, followed by a subsequent increase to 110 kHa in 2022. Brazil shows considerable variability in deforestation, which may be related to fluctuations in commodity markets or changing regulatory frameworks, which require further investigation. Also, the growth since 2019 is most likely due to President Bolsonaro's policy, which we described at the beginning of this paper. Paraguay reflects a general downward trend, although with occasional increases, however deforestation decreased from 408 kHa in 2012 to 152 kHa in 2022. Uruguay, despite maintaining a generally low deforestation rate, shows a slight increase to 0.54 kHa in 2022. MERCOSUR countries collectively show an increasing trend of deforestation, increasing from 1987.38 kHa in 2012 to 2252.54 kHa in 2022. This suggests the importance of examining the collective impact of international trade, commodity demand, and policy initiatives on regional deforestation.

Since it is difficult to identify and assess the impact of increased trade cooperation between the EU and MERCOSUR countries by simple observation, we decided to conduct two regression models with panel data between 2012 and 2022. The results of the econometric analysis are illustrated in Model 1 and Model 2 below.

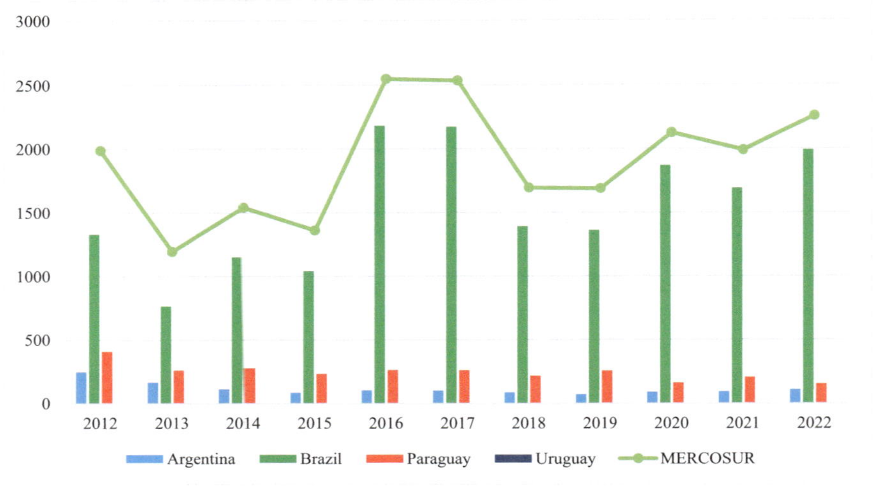

Fig. 6 Commodity deforestation by country in kHa, 2012–2022. *Source* Curtis (2018, updated through 2023)

Model 1: Panel Data Analyses, Green House Gas Emissions and Export from MERCOSUR

Fixed-effects, using 44 observations
Included 4 cross-sectional units
Time-series length = 11
Dependent variable: l_GHG_pc
Robust (HAC) standard errors.

	Coefficient	*Std. error*	*t-ratio*	*p-value*	
const	1.68619	0.0888395	18.98	0.0003	***
l_Export_million_of_eur	−0.0302057	0.00586089	−5.154	0.0142	**
l_GDP_pc	0.0685494	0.00864005	7.934	0.0042	***

Mean dependent var	2.066664	S.D. dependent var	0.273106
Sum squared resid	0.005657	S.E. of regression	0.014214
LSDV R-squared	0.998236	Within R-squared	0.708237
Log-likelihood	134.6655	Akaike criterion	−237.3310
Schwarz criterion	−208.7839	Hannan-Quinn	−226.7443
rho	0.158105	Durbin-Watson	1.487477

Source Authors' own calculations in GRETL, data from EDGAR (2023) and World Bank (2023)

Model 1 represents econometric analysis of panel data of 4 cross-sectional units, which represents 4 MERCOSUR'S countries. These units were observed over an 11-year period (2012–2022). The primary objective of this model is to assess the relationship between GHG emissions and exports from MERCOSUR, considering economic factors like GDP per capita. According to Model 1 the final equitation (Eq. 4) is as follows:

$$l\_GHG\_pc_{it} = 1.686 - 0.030 \times l\_Export_{it1} + 0.068 \times l\_GDP\_pc_{it2} + u_{it}, \quad (4)$$

We interpret the result of this econometric analysis as follows: **with a 1% increase in MERCOSUR exports to the EU, we expect MERCOSUR emissions to fall by 0.03% and with 95% probability, we reject the null hypothesis of insignificance of the parameter estimate of the variable.** With a 1% increase in GDP per capita of MERCOSUR countries, we expect a 0.07% increase in MERCOSUR emissions. With 99% probability, we reject the null hypothesis of insignificance of the parameter estimate of the variable. By using the coefficient of determination R2, we can estimate how much of the variability in the dependent variable is explained by the regression model. In this case, the coefficient of determination has reached the value 0.998236, which means that this model explains 99.82% of the variability of dependent variable l_GHG_pc. In summary, the model results indicate a statistically significant relationship between GHG emissions (l_GHG_pc) and the two independent variables: exports (l_Export_million_of_eur) and GDP per capita (l_GDP_pc). The negative coefficient for exports suggests that an increase in exports is associated with reduced GHG emissions per capita. In contrast, the positive coefficient for GDP per capita indicates that increased economic activity, as measured by GDP per capita, is linked to higher GHG emissions per capita.

Model 2: Panel Data Analyses, Deforestation and Export from MERCOSUR

Fixed-effects, using 44 observations
Included 4 cross-sectional units
Time-series length = 11
Dependent variable: l_Commodity_driven_deforestation_kHa
Robust (HAC) standard errors

	Coefficient	*Std. error*	*t-ratio*	*p-value*
const	6.56527	8.69240	0.7553	0.5050
l_Export_million_of_eur	0.558380	0.262403	2.128	0.1233
l_GDP_pc	− 0.730386	0.856955	− 0.8523	0.4567
Mean dependent var	3.926506	S.D. dependent var	3.474708	
Sum squared resid	4.998613	S.E. of regression	0.422519	
LSDV R-squared	0.990372	Within R-squared	0.422088	
Log-likelihood	−14.58266	Akaike criterion	61.16531	
Schwarz criterion	89.71235	Hannan-Quinn	71.75193	
rho	0.028768	Durbin-Watson		

Source Authors' own calculations in GRETL, data from Curtis (2018, updated through 2023) and World Bank (2023)

Model 2 represents the results of a panel data analysis that explores the relationship between dependent variable, deforestation driven by commodities and independent variable export from MERCOSUR to the EU. For a better objectivity, we added second independent variable which is represented by GDP per capita of MERCOSUR'S countries. Since the p-values are greater than 0.05, we can confirm that we have not demonstrated statistical significance of the variables on commodity-driven deforestation in this model. However, it should be added that this result does not mean that international trade would have no relationship with deforestation, but we could not prove it using this analysis and the selected data.

We have not been able to accurately determine the relationship and correlation between these environmental and macroeconomic indicators by simply observing the evolution of trade between the EU and MERCOSUR in conjunction with the evolution of GHG emissions and deforestation. This was the reason why we decided to apply an econometric method—panel data regression analysis. Before constructing these models, we assumed that an increase in trade cooperation between the EU and MERCOSUR would lead to an increase in GHG emissions as well as that it would lead to increased deforestation. The first model revealed that with a 1% increase in MERCOSUR exports to the EU, we expect MERCOSUR emissions to fall by 0.03%. The second model showed that there is no statistically significant relationship between increased trade cooperation and deforestation. These observations imply that we reject both hypotheses.

Based on these results, we agree with authors who argue that increased trade and economic development do not destroy environmental sustainability, but on the contrary, they may inherently help biodiversity and environmental development, as with increased trade, new technologies and innovations enhancing environmental protection may also flow into the country. We also agree with authors such as Grossman and Krueger (1991) and Copeland and Taylor (1994) who claim that the effect of international trade on the environment is a composite of several effects. Based on our results, in the case of MERCOSUR exports to the EU, the composition and technology effects have been demonstrated and the resulting effect was a positive one.

Our results are in line with findings of Bernard et al. (2011) and Sorroche del Rey et al. (2022) that countries involvement in international trade improves access to innovation, development and welfare, whereby the country is able to invest a greater amount of funds to develop technology and transform to a more environmentally sustainable economy.

Our results differ from that of Mahmood (2022) who found that with increasing exports of South American countries to neighbouring countries both territory-based and consumption-based CO_2 emissions of exporters increase. We attribute the difference in results to the fact that EU countries are sophisticated markets requiring environmentally friendly products.

Our study did not confirm the results of Kozul-Wright and Fortunato (2012), according to which international trade increases emissions in less developed countries as developed countries use trade opportunities to transfer high-emission activities to their less developed trading partners. Our study, where the developed economies are represented by the EU and the less developed economies are represented by the MERCOSUR countries, showed the opposite result—GHG emission reductions in MERCOSUR.

However, despite these theories and our results, it is not entirely correct to dismiss that increased international trade could not also affect the environment negatively. This is a very complex phenomenon that is influenced by a large number of variables. A limitation of our study is the relatively low number of observations that could have influenced the overall result. At the same time, a higher number of independent variables could have been used for greater objectivity.

5 Conclusions

In summary, our research demonstrates that the FTA between the EU and MERCOSUR should not be feared in terms of its impact on the environment. Contrary to the initial hypothesis, our results suggest a decrease in GHG emissions with a rise in MERCOSUR's exports to the EU, while no significant relationship was observed between trade cooperation and deforestation. These findings support the perspective that economic development and environmental sustainability may be complementary, as increased trade is potentially associated with the transfer of innovative technologies that promote environmental protection.

Limitations and Future Research

It is suggested to consider the results of the study within the context of its limitations, including the small sample size and the possibility that other unaccounted variables may influence these results. As indicated in this paper, the other factors influencing the GHG emissions and deforestation in MERCOSUR include the policy of the Brazilian president, international environmental agreements or initiatives aiming at protecting the environment such as the carbon border adjustment mechanism introduced by the EU. Thus, while our study contributes to the ongoing debate regarding the environmental consequences of international trade, it also highlights the need for more comprehensive research. Future research could benefit from a larger dataset and the inclusion of more variables to fully capture the multifaceted nature of trade and its environmental implications.

Acknowledgements This paper is a part of the research under the project: VEGA No. 1/689/23 *Sustainable Growth and the Geopolitics of Resilience in the Context of Crisis Prevalence, 2023-2026, KEGA Nr. 025EU-4/2024: Textbooks on the EU Trade Policy for the Principally Innovated Study Programme International Trade Management, VEGA project no. V4 1/0102/24: Positions and Perspectives of Mutual Trade and Economic Cooperation among the Visegrad Group Countries under the Influence of Geopolitical Changes in the Global Economy.*

All data used in this study were published under the CC BY 4.0 licence and properly cited (terms of use for the World Bank data are available at https://www.worldbank.org/en/about/legal/terms-of-use-for-datasets).

References

Balogh JM (2022) The impacts of agricultural development and trade on CO_2 emissions? Evidence from the Non-European union countries. Environ Sci Policy 137:99–108. https://doi.org/10.1016/j.envsci.2022.08.012

Bernard A et al (2011) The empirics of firm heterogeneity and international trade (NBER Working Paper No. 17627). National Bureau of Economic Research. https://www.princeton.edu/~reddings/papers/NBERWP17627.pdf. Accessed 3 May 2023

Blot E, Oger A, Harrison J (2022) Enhancing sustainability in EU free trade agreements: the case for a holistic approach. Institute for European Environmental Policy, Brussels. https://ieep.eu/wp-content/uploads/2022/12/Enhancing-sustainability-in-EU-Free-Trade-Agreements.-The-case-for-a-holistic-approach-IEEP-2022.pdf. Accessed 1 May 2023

Colli F (2019) The EU-Mercosur agreement: towards integrated climate policy? Egmont Institute. https://www.egmontinstitute.be/the-eu-mercosur-agreement-towards-integrated-climate-policy/. Accessed 17 Oct 2023

Copeland BR, Taylor SM (1994) North-South trade and the environment. Q J Econ 109(3):755–787. https://doi.org/10.2307/2118421

Curtis PG, Slay CM, Harris NL, Tyukavina A, Hansen MC (2018, updated through 2023) Classifying drivers of global forest loss. Science. Accessed through global forest watch on 29 Oct 2023. www.globalforestwatch.org

Delbeke J, Vis P (2016) EU climate policy explained. European Commission. https://climate.ec.europa.eu/system/files/2017-02/eu_climate_policy_explained_en.pdf. Accessed 26 Apr 2023
Dent CM (2021) Trade, climate and energy: a new study on climate action through free trade agreements. Energies 14(4363):1–2. https://doi.org/10.3390/en14144363
EDGAR (Emissions Database for Global Atmospheric Research) Community GHG Database (2023) A collaboration between the european commission, joint research centre (JRC), the international energy agency (IEA), and comprising IEA-EDGAR CO_2, EDGAR CH_4, EDGAR N_2O, EDGAR F-GASES version 8.0. European Commission, JRC
Ekins P (1997) The Kuznets curve for the environment and economic growth: examining the evidence. Environ Plann A: Econ Space 29(5):805–830
Eskeland GS, Harrison AE (2003) Moving to greener pastures? Multinationals and the pollution haven hypothesis. J Dev Econ 70(1):1–23. https://doi.org/10.1016/S0304-3878(02)00084-6
European Council—Council of the European Union (2022) European Green Deal. https://www.consilium.europa.eu/en/policies/green-deal/. Accessed 27 Apr 2023
European Commission (2023) Directorate-General for Trade European Union, Trade in Goods with Mercosur 4. https://webgate.ec.europa.eu/isdb_results/factsheets/region/details_mercosur-4_en.pdf. Accessed 30 Sept 2023
European Commission (2023) The EU-Mercosur Agreement Explained. https://policy.trade.ec.europa.eu/eu-trade-relationships-country-and-region/countries-and-regions/mercosur/eu-mercosur-agreement/agreement-explained_en. Accessed 17 Sept 2023
Eurostat (2023) International trade of EU and Non-EU countries since 2002 by SITC. https://ec.europa.eu/eurostat/databrowser/bookmark/37c3f4eb-3dc0-470b-8109-604e43eaec20?lang=en. Accessed 15 Sept 2023
Faria WR, Almeida AN (2016) Relationship between openness to trade and deforestation: empirical evidence from the Brazilian Amazon. Ecol Econ 121:85–97. https://doi.org/10.1016/j.ecolecon.2015.11.014
Grossman GM, Krueger AB (1991) Environmental impacts of a North American free trade agreement. NBER Working Paper No. 3914
Hochstetler K (2003) Fading green? environmental politics in the mercosur free trade agreement. Lat Am Polit Soc 45(4):1–32. https://doi.org/10.2307/3177129
IPCC (2018) Annex I: Glossary . In: Matthews JBR (ed) Global warming of 1.5 °C. An IPCC special report on the impacts of global warming of 1.5 °C above pre-industrial levels and related global greenhouse gas emission pathways, in the context of strengthening the global response to the threat of climate change, sustainable development, and efforts to eradicate poverty, pp 541–562. https://www.ipcc.ch/sr15/chapter/glossary/. Accessed 23 Apr 2023
Keating D (2023) EU-Mercosur trade deal stalls on climate concerns. Energy Monitor. https://www.energymonitor.ai/international-treaties/eu-mercosur-trade-deal-stalls-on-climate-concerns/?cf-view. Accessed 18 Sept 2023
Kozul-Wright R, Fortunato P (2012) International trade and carbon emissions. Eur J Dev Res 24(4):509–529
Lee CC, Olasehinde-Williams G, Gyamfi BA (2023) The synergistic effect of green trade and economic complexity on sustainable environment: a new perspective on the economic and ecological components of sustainable development. Sustain Dev 31(2):976. https://doi.org/10.1002/sd.2433. Accessed 19 Mar 2025
Li J et al (2018) Developed market or developing market? A perspective of institutional theory on multinational enterprises' diversification and sustainable development with environmental protection. Bus Strat Environ 27(7):858–871. https://doi.org/10.1002/bse.2037. Accessed 19 Mar 2025
Lukáčik M, Lukáčiková A, Szomolányi K (2011) Panelové Dáta v Programe Gretl. http://www.fhi.sk/files/katedry/kove/veda-vyskum/prace/2011/Lukacik-Lukacikova-Szomolanyi2011.pdf. Accessed 24 Oct 2023

Mahmood H (2022) Consumption and territory based CO_2 emissions, renewable energy consumption, exports and imports nexus in South America: spatial analyses. Pol J Environ Stud 31(2):1183–1191. https://doi.org/10.15244/pjoes/141298. Accessed 19 March 2025

Miller M (2022) The fight for the Amazon rainforest is a fight for the planet. Vox. https://www.vox.com/down-to-earth/2022/9/29/23373427/amazon-rainforest-brazil-jair-bolsonaro-lula-deforestation. Accessed 10 Dec 2023

Nolte D (2023) Outlook for the next CELAC-EU summit: is the goal of a strategic partnership still valid and realistic? EU-LAC Foundation. https://eulacfoundation.org/en/outlook-next-celac-eu-summit-goal-strategicpartnership-still-valid-and-realistic. Accessed 9 May 2023

Pivello V, Vieira I, Piratelli A, Ribeiro M, Menezes L, Berlinck C, Overbeck G (2021) Understanding Brazil's catastrophic fires: causes, consequences, and policy needed to prevent future tragedies. Perspect Ecol Conserv 19(3):233–255. https://doi.org/10.1016/j.pecon.2021.06.005. Accessed 19 Mar 2025

Porter ME, Linde Cvd (1995) Toward a new conception of the environment competitiveness relationship. J Econ Perspect 9(4):97–118. https://doi.org/10.1257/jep.9.4.97. Accessed 19 Mar 2025

Sorroche YdR, Piedra-Muñoz L, Galdeano-Gomez E (2022) Interrelationship between international trade and environmental performance: theoretical approaches and indicators for sustainable development. Bus Strat Environ:1–17. https://doi.org/10.1002/bse.3270. Accessed 19 Mar 2025

United Nations (2023) What is Climate Change? https://www.un.org/en/climatechange/what-is-climate-change. Accessed 23 Apr 2023

World Bank (2023) GDP per Capita (Current US$)—Argentina, Brazil, Paraguay, Uruguay. https://data.worldbank.org/indicator/NY.GDP.PCAP.CD?locations=AR-BR-UY-PY. Accessed 25 Sept 2023

World Trade Organization (2022) World trade report. https://www.wto.org/english/res_e/booksp_e/wtr22_e/wtr22_e.pdf. Accessed 28 Apr 2023

Exploring the Relationship Between Employee Value Proposition (EVP) and Organizational Attractiveness in the Context of a Labor Crisis: A Focus on Generation Z

Ghislaine Pellat and Catherine Marty-Peyroux

Abstract In a context of a labor crisis marked by both a recruitment challenge for employers and an individual's crisis in relation to work for employees, the question of recruitment becomes strategic. To attract and retain the best talents, companies are investing in employer branding. Different discrepancies may arise between the expectations of the personnel as assumed by the company and the actual expectations of employees, along with their perceptions of the offered working conditions. This could explain divergences around the Employee Value Proposition (EVP). The study introduces three job offers to participants, selected for their perceived attractive qualities. It is administered to students undergoing professional training, who are ready to enter the job market. We will specify the expectations of this specific audience in terms of EVP, contrasting them with the elements highlighted by employer brands. This study aims to shed light on divergences in expectations and perceptions around the EVP, paving the way for common ground to reduce tensions in the job market and provide insights contributing to the simultaneous resolution of these two crises. It aligns with the European focuses, in which, employment and skills challenge in harmony with the policy of reduction of the carbon footprint.

Keywords Employee value · Value proposition · EVP · Z Generation · Skills · Recruitment · Brand · Employer · Crisis · Skills

G. Pellat (✉)
CERAG, Univ. Grenoble Alpes, Grenoble INP, Grenoble, France
e-mail: Ghislaine.pellat@univ-grenoble-alpes.fr

C. Marty-Peyroux
CORHIS, Université Montpellier-3, Montpellier, France
e-mail: catherine.peyroux@univ-montp3.fr

G. Pellat et al. (eds.), *Data-Centric Business and Applications*, Lecture Notes on Data Engineering and Communications Technologies 258,
https://doi.org/10.1007/978-3-031-95068-1_13

1 Introduction

The labor market is facing a dual crisis, characterized by challenges in recruitment for employers and a shift in employees' relationship with work. These dynamics have elevated the importance of recruitment strategies as a critical organizational priority (Dutta et al. 2021). While this issue is pervasive across Europe, the findings presented in this article are grounded in a study conducted in France. French companies are increasingly struggling to recruit young talent to strengthen their teams. This recruitment crisis can be attributed to several factors, including the retirement of a significant number of employees from the baby boomer generation, a qualitative mismatch between candidates skills and organizational needs, and heightened demand in certain high-growth fields, such as Information Technology. In response, organizations are turning to employer branding to attract and retain the best talent. This approach positions the employer as an active player in human resources marketing, emphasizing the organization's reputation as an employer. Employer branding strategies aim to facilitate a shift in how job seekers perceive the company—from an external viewpoint to an internal experience—during the recruitment process (Grouille 2022). A cornerstone of employer branding is the Employee Value Proposition (EVP), which represents the comprehensive offer made by employers to employees. The EVP encompasses not only financial compensation but also working conditions, workplace environment, and corporate culture. These elements, particularly non-monetary ones, have gained prominence as decisive factors for Generation Z, whose preferences often prioritize personal fulfillment, alignment with personal values, and enriching workplace experiences, in addition to satisfactory remuneration. Despite widespread efforts by major companies to enhance their employer brand capital, the actual effectiveness of these strategies remains unclear. Consulting firms have identified a lucrative opportunity in providing guidance for EVP development, yet critical questions persist:

> *Does a diversified EVP, integrating dimensions aligned with the preferences of Generation Z, enhance organizational attractiveness?*
> *Can such an approach effectively address the dual crises of recruitment difficulties and employee disengagement?*

Addressing these questions requires a closer examination of EVP content and its alignment with the expectations of Generation Z. This exploratory, quantitative study focuses on students in professional training programs, including Professional Bachelor's and second-year Master's degrees. These students are on the cusp of entering the job market, making them an ideal cohort for evaluating EVP perceptions and expectations. By presenting job offers that highlight EVP attributes, this study examines:

- *the "Generation Z candidates" reactions to EVP elements in job postings,*
- *the differences between candidate expectations and employer-provided EVP features.*

The aim is to clarify the relationship between EVP and organizational attractiveness while identifying actionable strategies to reduce tensions in the job market for young graduates. Ultimately, this research aligns with Europe's 2023 economic and political priorities, which emphasize employment and skills development in tandem with carbon footprint reduction. This contribution will be organized with a first part of the paper developing the relationship between EVP and organizational attractiveness. The second section will explain the research methodology and the results of the study will be presented and discussed in a third section.

2 Employer Branding, Employee Value Proposition and Organizational Attractiveness

2.1 The Challenges of Employer Branding

The issues surrounding talent recruitment have led companies to develop an HR marketing approach, giving rise to the concept of employer branding. This concept was first used by consultant Simon Barrow. The employer brand constitutes a "unique employment promise" (Benaiss Noailles et al. 2019) about work environment, culture, benefits, experiences (Indeed Report, 2022). It is, according to Barrow and Ambler (1996), "the set of functional, economic and psychological benefits provided by employment, and identified with the company that employs them."

Employer branding is the result of an investment process that provides a company with intangible capital—employer brand capital. This capital is used to attract and retain customers. According to a study commissioned by Indeed and Glassdoor from Forrester in August 2022 and carried out in the North America, Europe, Middle East & Africa and Asia–Pacific regions, involving 400 employer brand decision-makers, 90% of respondents plan to increase their spending on employer branding over the next few years.

With the development of social networks and specialized sites such as Indeed, employees are now aware of employer brand issues and have access to the opinions of current employees. This makes investing in Employer Brand Capital (EBC) all the more important.

2.2 The Employee Value Proposition: The Most Relevant Link with the Employer Branding

Human resources (HR) practices play a pivotal role in shaping an organization's employer brand. Broadly defined, the Employee Value Proposition (EVP) embodies the company's HR response to attract, retain, and engage talent. A well-crafted EVP is essential for building a robust employer brand (Daubier and Daubier 2022). The EVP represents the comprehensive offer made by a company to its employees, tailored to specific roles and organizational needs. According to Kearney (2022), organizations that actively develop their EVP are better positioned to: foster employee engagement, enhance retention rates, attract critical new talent. However, a critical gap often exists between the employer's promise and the employee's actual experience. Understanding and addressing this disparity is key to optimizing the EVP.

2.2.1 Evolving Expectations and Expanded EVP Components

As candidate expectations evolve, particularly among Generation Z, organizations must adapt their EVP to include elements that extend beyond traditional forms of compensation. While financial remuneration remains a cornerstone, emerging priorities such as skills development and workplace well-being are becoming increasingly influential. Recognizing this shift, Human Ressources consultancies have developed frameworks to structure EVPs effectively. Despite variations in formulation, a consistent typology of EVP components can be identified, encompassing five core dimensions:

- **Compensation**: Direct financial benefits, including salaries, bonuses, and raises.
- **Additional Benefits**: Indirect compensation, such as health insurance, retirement plans, paid leave, and other perks.
- **Career Development**: Opportunities for professional growth through internal promotions and training programs.
- **Work Environment**: Organizational and physical conditions, including team dynamics and work-life balance.
- **Corporate Culture**: The company's mission, values, collaboration dynamics, and sense of belonging fostered among employees.

These dimensions are not only integral to internal and external communications but also directly shape HR practices. For example, recruitment materials such as job descriptions and advertisements often integrate EVP components to highlight the organization's unique value proposition.

2.2.2 EVP's Role in Organizational Attractiveness

Employer branding literature suggests that the EVP significantly influences organizational attractiveness (Jiang et al. 2012). While much of the academic focus has been on employer branding as a holistic concept, discussions around the EVP itself have largely been driven by consulting firms. This gap raises important questions about the measurable impact of EVP on attracting and retaining talent. By addressing these questions, this research seeks to clarify the link between EVP and organizational attractiveness, particularly in the context of Generation Z's evolving workplace expectations.

2.3 *Employer Branding, EVP, and Organizational Attractiveness*

Fray et al. (2015) identify two key research streams concerning organizational attractiveness:

- *the Perceived attractiveness, defined as an attitude unique to a potential candidate.*
- *the Identified attractiveness, understood as a characteristic of the organization itself.*

The latter conceptualization focuses on the organization's ability to generate positive perceptions among candidates, creating the impression that the company is a desirable place to work (Rynes and Barber 1990). Morin's (2006) model highlights four primary factors influencing organizational attractiveness, mentioned by:

Organizational and job-related attributes: these include aspects such as salary, workplace atmosphere, opportunities for promotion, and the organization's social performance.
Organizational reputation: how the company is perceived in terms of brand value and external image.
Compatibility: the alignment between the candidate's skills and values with the company's culture and job requirements.
Interviewer behavior: the role of interpersonal interactions during the recruitment process.

The first of the two factors—organizational attributes and reputation—are integral to employer branding and EVP strategies.

2.3.1 The Evolving Role of EVP in Organizational Attractiveness

The attributes associated with EVP have evolved significantly over time, shaped by changes in HR practices and broader societal shifts. Beyond traditional elements like compensation, EVP now encompasses dimensions such as skill development opportunities, corporate culture, well-being at work, and corporate social responsibility (CSR). These changes reflect the growing importance of environmental factors such as labor market dynamics, public health crises, and climate change, which influence candidate expectations.

Candidates today also value certifications or labels that signify organizational excellence, such as "Great Place to Work" (GPW) or ecological certifications that emphasize an organization's commitment to employee well-being and sustainability (Guillot-Soulez et al. 2019).

2.3.2 The Role of Employee Experience (EE) as a Moderating Variable

Candidates perceptions of an organization's EVP are further influenced by their **Employee Experience (EE)**, which serves as a moderating variable for organizational attractiveness. Borrowing from the concept of **Customer Experience (CX)**, EE is defined as a holistic perception of the employment context. It begins with the candidate's exposure to the job offer, followed by their emotional reaction (e.g., "I like" or "I don't like"), and concludes with a strategic response (e.g., "I will apply" or "I will not apply") (Verhoef et al. 2009).

Similar to CX, EE generates emotional reactions that influence a candidate's decision-making process.

For example: Positive experiences may lead to satisfaction and increased likelihood of recommending the company to others. Negative experiences may deter engagement or applications.

These two parameters—**perception** and **commitment**—are central to this study's analysis. Drawing on the comprehensive literature review by Becker et al. (2020), this research employs the CX framework to better understand how EE shapes candidates' decisions. As with product branding, the employer brand represents a promise that organizations must deliver through well-crafted EVPs (Rahman et al. 2022).

2.3.3 Current Challenges in EVP Implementation

Despite the recognized importance of EVP, many organizations fail to integrate its attributes effectively into their recruitment strategies. For instance, the Job advertisements often neglect key EVP components that could enhance their appeal to potential candidates. It also neglect that the candidates' expectations vary significantly which could complicate the further EVP development.

Generational differences also play a significant role. Research by Mac Gregor (1960) and Schneider (1987) highlights how candidate expectations differ across cohorts. This study focuses specifically on Generation Z, born between 1997 and 2012, who are distinct from their Generation Y predecessors (born between 1981 and 1996).

2.4 Generation Z is a New Paradigm in Workplace Expectations

Generation Z, often referred to as "digital natives," exhibits unique workplace preferences. According to Dalmas (2019), this cohort seeks:

Meaning and ethics *in professional relationships.*
Self-legitimization *through their accomplishments.*

Opportunities for ***socialization and networking****, which are viewed as critical to professional success.*
A willingness to ***innovate and act independently****, often without requiring hierarchical approval.*

Understanding and integrating these preferences into EVP strategies is essential for organizations aiming to attract and retain Generation Z talent effectively.

3 Research Methodology

Our research seeks to explore the tensions between corporate recruitment needs and the expectations of Generation Z candidates. Specifically, it aims to answer the following question:

> *"Can the Employee Value Proposition (EVP) effectively attract young Generation Z graduates, particularly given the presence of perception biases? "*

This study investigates whether biases in employers' perceptions of Generation Z's expectations, as well as Generation Z's own perception biases toward the EVP, influence the attractiveness of recruiting organizations. To address this question, this paper presents an analytical model. Data will be collected from a selected sample of students through a questionnaire survey. The authors have developed an exploratory and quantitative research protocol to examine these dynamics.

3.1 Research Design

Our research aims to enhance the understanding of the tensions between companies' recruitment needs and the expectations of Generation Z candidates. To examine the relationship between the Employee Value Proposition (EVP) and Organizational Attractiveness (OA), this study utilizes an analytical model, a targeted student sample, and an exploratory, quantitative research protocol.

3.2 Analytical Model

The dependent variable in this study is Organizational Attractiveness (OA), defined by Rampl and Kennig (2014) as *"the intention to apply for or accept particular offers from an employer."* Organizational attractiveness is a key determinant in the recruitment process, as effective OA directly impacts candidates' decisions to engage with job opportunities. The independent variable in the model is the Employee Value Proposition (EVP). Efforts to diversify the EVP—by incorporating dimensions such as corporate culture, well-being, and skill development—are hypothesized to have a significant impact on organizational attractiveness. For Generation Z, non-monetary factors, including working conditions and corporate values, often take precedence over compensation. This research is guided by the following hypothesis:

> *A diversified Employee Value Proposition (EVP), which emphasizes well-being, corporate values, and sustainability alongside traditional compensation, positively influences Organizational Attractiveness (OA).*

This can be analyzed through the five-dimensional typology presented in Part 2.2.1 of this article, which includes: Compensation, Additional Remuneration, Career Development, Working Environment, and Corporate Culture.

The literature on EVP and organizational attractiveness highlights the role of perception bias, particularly among candidates. One common bias stems from their perceptions of the company or its activities, which are shaped by personal experience or indirect exposure (e.g., through family, colleagues, friends, or social networks).

As a result, Employee Experience (EE) is selected as the moderating variable in this study.

3.3 *Sample Constitution and Specificity of the Student Population*

This research is based on a questionnaire survey administered to a student population at the end of their professional studies. The sample includes students from Bachelor's programs (third-year of the University Bachelor of Technology—BUT) and Master's programs (fifth-year students).

In the Bachelor's program, students are required to complete either a one-year work-study program or an internship. In the Master's program, students must complete a minimum of two internships: one lasting six weeks in the first year and another lasting six months in the second year. As a result, most of these students have gained professional experience through internships or apprenticeships, allowing them to develop an initial perception of the professional environment based on their early interactions with employers.

These academic programs are explicitly geared towards professional careers, without a research-oriented track. In the short term (less than a year), these students will enter the job market and apply for their first post-graduation position. Consequently, they are particularly concerned with their choice of first employer and, therefore, highly sensitive to organizational attractiveness.

Furthermore, as members of Generation Z, they are particularly attuned to information shared on social networks, which significantly influences their perceptions and career decisions.

3.4 Research Protocol

The research protocol involved interviewing students and asking them to evaluate three job advertisements. These job offers were pre-selected for their alignment with the career paths targeted by the surveyed population. Each advertisement explicitly describes the company's Employee Value Proposition (EVP), allowing for an assessment of the organizational attractiveness generated by reading the ad. All three advertisements are designed for young professionals entering the job market, yet they differ in several aspects: writing style, content (values, salary, mission), and industry sector. One advertisement pertains to the artistic sector, where employment is primarily based on short-term or fixed-term contracts, whereas the other two represent industries offering permanent contracts. This selection was intended to provide respondents with job offers that are comparable in hiring purpose but distinct in key EVP criteria. These three job offers were selected from lesser-known companies to neutralize the effect of corporate reputation and ensure that respondents focused on the content of the job ad (EVP), the sector, and their own perceptions of it. Furthermore, students were asked to react to the advertisements directly, eliminating the influence of the recruiter and their behavior, which is a key factor in Morin's model. The questionnaire aims to measure organizational attractiveness by examining job-related attributes (e.g., salary, work environment, promotion opportunities, social performance) while also considering candidate-company and candidate-job fit.

Each of the three job offers presents a distinct EVP, which can be analyzed using the five-dimensional framework introduced earlier. All job advertisements focus on business development positions.

3.4.1 Description of the Selected Job Offers

Job offer 1 originates from a rapidly expanding start-up specializing in the national and international marketing of an innovative product—a uniquely designed electric toothbrush that differs significantly from existing models. The advertisement incorporates all key elements of an Employee Value Proposition (EVP), though it does not explicitly mention the company's start-up status. Notably, it includes the possibility of acquiring company shares, highlighting an investment opportunity for employees. The corporate culture and values are clearly articulated, alongside standard details concerning compensation and additional benefits. Moreover, the advertisement references the work environment, yet it does not provide information regarding career growth or skill development opportunities.

Job offer 2 is published by a non-profit organization with a nationwide network across France. The advertisement primarily focuses on the organization's mission and the specific responsibilities associated with the position. It places significant emphasis on the personal attributes and competencies required for the role. In terms of EVP, the ad primarily highlights the organization's culture and strong values, yet it omits crucial details such as salary and employee benefits. However, the social

Table 1 Comparative analysis of the three ads. *Source* Generated by Authors

PVE content	Announcement 1 start up International development position	Announcement 2 National Association Sales development position	Announcement 3 Music label Music project manager
Compensation	Salary given CDI, fixed + commission from 40 000 to 70 000	NC*	3-month renewable fixed-term contract, possibly permanent contract
Additional remuneration/benefits	Transport, BSPCE[1], telecommuting	NC*	Ticket restaurants, transport
Skills and career development	NC*	IP indirectly	Open-ended contract possible
Well-being at work	Caring and dynamic team	NC	A passionate and dynamic team
Corporate culture	Values: health, inclusiveness, fair pricing and respect for the environment	– Equal opportunities, non-discrimination, mutual support, social commitment, uniting people around a common cause	Passion, art, music
Job requirements	Major investment Dynamism Impact station First experience	Significant experience in business and associations Dynamic interpersonal skills	Passion, experience of working in a team

impact of the organization and the scope of its network are key components of its attractiveness.

Job offer 3 is issued by an independent music label. The EVP is centered around the concept of "passion", which defines both the work atmosphere and corporate culture. From a material standpoint, the conditions appear less attractive, featuring low compensation and an initial three-month fixed-term contract, with the potential for renewal and eventual transition to a permanent position. Nonetheless, the role presents a notable opportunity to collaborate with well-known artists, which may enhance its appeal despite the contract's constraints.

The following section presents the authors' analysis of these three job advertisements, using the evaluation criteria previously outlined (Table 1).

The questionnaire, which required 20 min to complete, was deemed more than sufficient by respondents. It was structured around four main categories of questions, applied consistently across all three job advertisements: (1) overall attractiveness of the job offer, (2) importance of different EVP components, (3) employee experience,

[1] BSPCE warrants (bons de souscription de parts de créateur d'entreprise) are purchase warrants granted to employees to enable them to buy shares in the company for a fixed period and at a set low price.

and (4) respondent demographics. Overall attractiveness was measured by assessing participants' general perception of the job offer, their level of interest, and their willingness to apply. The objective was to capture both their immediate reaction upon reading the advertisement and the extent to which their initial perception aligned with their intent to apply. These two dimensions—perception and commitment—constitute the employee experience (EE) framework previously discussed in this study.

Additionally, participants were asked to rank the three job offers in order of preference. This comparative ranking question was intentionally placed at the end of the questionnaire to prevent it from biasing respondents' initial perceptions during the advertisement review process.

The Employee Value Proposition (EVP) was evaluated across five key criteria. For each job advertisement, respondents assigned a score to each criterion on a five-point scale (ranging from 1 = least important to 5 = most important). This methodology enabled both an individual ranking of criteria per job offer and an overall prioritization of EVP components. By calculating an average score for each criterion, we conducted a detailed analysis of candidates' decision-making process when applying for a job. This approach allowed us to identify eliminatory factors present in the job advertisements. When compared with the previous rankings of the ads, this data helps determine whether certain EVP components have a decisive impact on a candidate's intention to apply or, conversely, lead them to dismiss a job offer.

The findings reinforce the relevance of the customer experience (CX) model as a framework for analyzing employee experience (EE). In CX, a customer's initial interaction with a company or purchasing situation plays a crucial role in their decision to proceed with a transaction. Key factors such as the transparency of the purchasing process and access to product information or trials influence the decision to buy. When these elements are unclear, certain criteria become deal-breakers (Mishra 2022).

Similarly, in employee experience, we examine the reference points candidates rely on when making decisions—particularly when choosing to apply and engage further in the recruitment process (i.e., personal commitment). Our study assessed both external influences (e.g., social networks, family opinions, personal values) and individual past experiences that may shape a candidate's decision.

Finally, the questionnaire concluded with a section collecting demographic data, including respondents' education level, age, and professional experience at the time of the survey.

4 Results and Discussion

Following the administration of the questionnaire to a cohort of 60 students, we obtained 56 fully completed and usable responses. Consequently, each job offer was evaluated 56 times, yielding a total dataset of 168 responses across the three job offers. This distinction is crucial when interpreting the results—whether drawing overall conclusions based on the entire dataset (N = 168) or conducting comparative analyses between the three job offers using their respective individual response bases (N = 56 per offer).

4.1 Generation Z and EVP: The Impact of Diversity

An analysis of the responses presented in Table 1.2 highlights a high level of consistency between respondents' interest in a job posting and their willingness to work in the advertised position. Across all job offers, 97% of participants indicated that if they found a vacancy appealing, they would also be inclined to pursue recruitment for that position—and vice versa. Only 3% of respondents reported a discrepancy between their interest in the vacancy and their intent to apply, a margin that can be considered a reasonable and acceptable error within the context of human and social sciences research. This finding is critical in validating the reliability of our sample, as it suggests that students' action-oriented decisions (commitment to a position) are strongly aligned with their perceptions of the job opportunity.

Table 2 presents a comparative analysis of the attractiveness ratings assigned to the three job offers.

These findings reflect the initial, instinctive intention associated with discovering a job advertisement upon first reading it. This response serves as an indicator of the emotional reaction experienced by candidates, constituting the first stage of the Employee Experience (EE). Drawing on the Customer Experience (CX) model, which serves as a reference framework for analyzing EE, the next logical step involves evaluating the key components of the job offer as presented in the advertisement, as well as the way these elements are framed and communicated (Mishra 2022).

For this reason, Table 3 presents a two-step analysis: first, respondents were asked to rank the job offers relative to each other, and second, they were asked to identify the criteria that influenced their ranking decisions. The results underscore that a lack of transparency regarding salary and benefits in job advertisements can negatively impact attractiveness. Additionally, some respondents expressed self-doubt regarding their qualifications, citing factors such as insufficient English proficiency or limited commercial skills as barriers to applying.

Table 2 Attractiveness by ad. data supplied by the authors

	Announcement 1	Announcement 2	Announcement 3	Proximity rate between application and recruitment
I would like to apply/I would like to be recruited	64.30%	23.21%	62.50%	97%

Table 3 Ranking of the 3 ads and explanations for the ranking. Data provided by the Authors

Ranking	Announcements by	Preference (%)	Redhibiting factor (%)	Explanations
Announcement 1	First position	68	34	Own experience or skills deemed unsuitable by the candidate (e.g. level of English)
Announcement 3	Second position	46	27	The form of the contract (fixed-term) is not very attractive, and neither is the sector of activity itself (music)
Announcement 2	Third position	41	39	Because the ad doesn't mention any information about salary and benefits

This suggests that the wording of job postings plays a crucial role in shaping candidate behavior. Specifically, when opportunities for growth and training are not clearly emphasized, candidates may hesitate to apply—especially if they do not meet all the stated requirements. Rather than encouraging potential applicants to apply despite minor skill gaps, some advertisements appear to deter applications by not explicitly addressing professional development prospects.

An analysis of the relative importance of different Employee Value Proposition (EVP) components influencing candidates' decision-making reveals two key evaluation criteria: (1) benefits related to compensation and workplace well-being and (2) career development opportunities. These findings are detailed in Table 4.

The results indicate that job advertisements that fail to outline career progression tend to be perceived as less attractive. Similarly, when companies do not disclose salary details or additional financial benefits, candidates are less inclined to consider the position. Respondents express clear expectations that organizations should enhance their daily work experience and actively invest in their well-being. Notably, well-being-related benefits received slightly higher importance scores than salary, although the latter remains a significant factor, particularly when explicitly mentioned (as in Job Offer 1).

Nevertheless, salary transparency remains a critical determinant of job attractiveness and application intent. Candidates value clarity and openness in compensation details, indicating that remuneration should not be treated as a taboo subject in recruitment communications (Table 4).

It is important to highlight that the surveyed participants are young students who will be entering the job market within the next six months. Given their limited professional experience, they lack significant bargaining power when negotiating compensation. As a result, their salary expectations are primarily shaped by personal benchmarks or comparisons with other job offers they have encountered in the market.

However, approximately 20% of respondents indicate that they rely on additional reference points beyond their own experiences when assessing job offers. Among

Table 4 Scores of criteria obtained by offers (5 max score, 1 min score). Data provided by the Authors

Hierarchy of criteria that hold your attention positively in relation to the ad 5 (most important) to 1 (least important)	Score obtained by offer 1	Score obtained by offer 2	Score obtained by Offer 3	Average of 3 bids	Standard deviations from the average of the 3 offers
Corporate culture	*2.80*	***3.22***	*2.93*	*2.98*	*Low*
Career development	*3.04*	*3.05*	*3.07*	***3.05***	*Low*
Working environment	*2.71*	*2.98*	*2.89*	*2.86*	*Low*
Compensation	*3.05*	*2.64*	*2.88*	*2.86*	*Fort*
Additional benefits	***3.39***	*3.11*	***3.25***	***3.25***	*Low*

these, comparisons with their own prior experiences emerge as the most influential factor. Additionally, respondents evaluate similar job offers within the same sector to contextualize their expectations. Another critical consideration influencing their decision-making is the alignment between their personal values and the values they perceive in the job advertisement.

Table 4 presents the demographic and professional profile of the students surveyed, including their work experience. With an average age of 21 (ranging between 20 and 23), these students are still in the process of developing the skills necessary for workforce integration and have a limited long-term vision of their career trajectories. However, this stage marks a pivotal moment in which they begin to actively engage with the world of work, making it essential for companies to play a strategic role in shaping their early professional experiences.

Notably, the perspective of these respondents—many of whom already have some professional exposure—is particularly discerning. They tend to reject organizations that rigidly define employment as the mere fulfillment of qualifications for a specific position (as observed in advertisement 2). Instead, they favor companies that position themselves as enablers rather than gatekeepers, fostering a mutually beneficial, cooperative approach to employment.

The job offer that resonates most with students is one that strikes a balance in its Employee Value Proposition (EVP) by incorporating diverse and appealing elements. Compensation remains a crucial piece of information, and even when monetary remuneration is relatively low, students may still find the offer attractive if it is compensated by other favorable factors—such as a positive work environment or alignment with corporate values that match their expectations and aspirations.

These findings support the hypothesis that a diversified EVP enhances organizational attractiveness, reinforcing the importance of comprehensive employer branding strategies in recruitment efforts (Table 5).

Whether evaluating the criteria for decision-making or the references used to guide those choices, the elements of Employee Experience (EE) are clearly present, much like in Customer Experience (CX). These elements encompass, on the one hand, objective decision-making criteria and, on the other, the context in which

Table 5 Descriptive elements of the sample and participants' professional experience. *Source* Authors

Average age	21 years old
Current level of education	21% Master's, 79% Bachelor's
No professional experience	Only 4% of students say they have had no work experience or internship
Internship student contract, freelance, association, temporary contract	96% of respondents have already had at least one professional experience. Mainly in the form of an internship or work-study contract

these decisions are made, including factors such as previous personal experiences or comparisons with other situations. These choices are influenced by varying degrees by different components of EE, suggesting that EE serves as a moderating variable in the relationship between Employee Value Proposition (EVP) and organizational attractiveness.

This conclusion provides valuable insights for recruiters who wish to attract young talent to their teams. Specifically, these future professionals seek transparency, additional benefits, and the opportunity to envision their future—particularly in terms of career development. These factors align with the development of a cooperative relationship between the candidate and the company, where the employee feels that they are becoming "more educated" through their employee experience.

Given that candidates value the prospect of mutual growth and development with their employers, similar to the customer experience model, companies can enhance the employee experience by clearly highlighting opportunities for career advancement and skills development.

5 Conclusion

In conclusion, the attractiveness of a job offer targeting young students about to enter the professional world is significantly influenced by the Employee Value Proposition (EVP). Organizational attractiveness plays a crucial role in the selection process of candidates, though it is heavily mediated by the employee experience encountered when reviewing the job offer. Based on these insights, we suggest that recruiters should clearly outline the career development opportunities available to candidates, as well as the associated remuneration and benefits. Transparency emerges as the most valued criterion among these students, whose professional references are primarily shaped by personal experience, despite the pervasive influence of social networks.

Recruiters should focus on fine-tuning their employer brand on the social networks most used by young recruits. They should avoid overt self-promotion and instead leverage the sharing of authentic employee experiences, which candidates are likely to appreciate and disseminate. From a strategic standpoint, recruiters must establish clear policies for integrating young talent, emphasizing what the organization

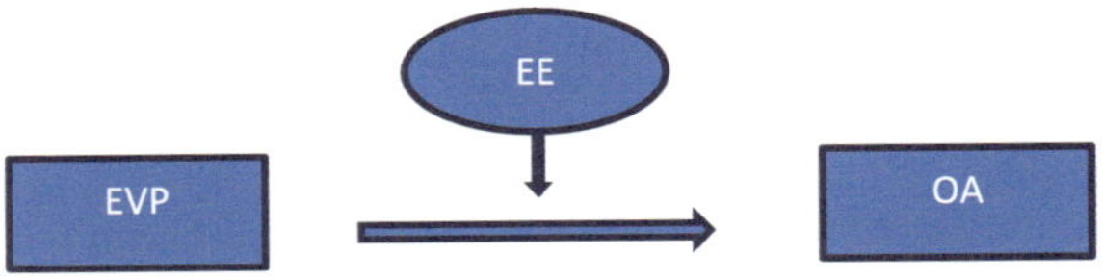

Diagram 1 The explanatory model. *Source Created by the Authors*

EVP: Employee Value Proposition

EE: Employee Experience

OA: Organizational Attractiveness

can offer in terms of career development, while maintaining alignment with the company's strategic goals, including economic performance.

This research aligns with the ongoing trend of integrating employer branding with marketing tools, offering a fresh perspective on Human Resource Management (HRM), particularly recruitment, by incorporating the employee experience (a concept inspired by customer experience). The satisfaction or appreciation of candidates' interactions with the job offer is a valuable metric, one that can be monitored on social media platforms. Similar to customer experience (Mishra 2022), employee experience is influenced by human interaction. In the case of customer experience, the key evaluative criteria include: "interaction with customer service, the entertainment or engagement provided, the outcome of the interaction, and the process of handling the interaction." Drawing from these insights, to enhance the attractiveness of job ads, companies could prioritize early interactions with candidates through social media platforms that facilitate personal engagement.

In marketing, the brand's approach is increasingly focused on individualized responses to potential customers. The findings of this study suggest a similar evolution for recruitment. Depending on the sought profile, job ads should incorporate diverse EVP elements, activating those components most aligned with the candidate's expectations. This research further reveals that for these Generation Z candidates, recruitment is viewed as part of a life project rather than merely an economic necessity. Ultimately, the success of the recruitment process hinges on a shared vision—the company's goals aligning with the candidate's personal aspirations (Diagram 1).

References

Barrow S, Ambler T (1996) The employer brand. J Brand Manag 4(3):185–206

Becker L, Jaakkola E (2020) Customer experience: fundamental premises and implications for research. J Acad Mark Sci 48: 630–648. https://doi.org/10.1007/s11747-019-00718-x

Benraïss L, Herrbach O, Viot C (2019) L'effet du capital-marque employeur sur l'attractivité organisationnelle : le rôle modérateur de la familiarité. Management Et Avenir 1(107):37–59

Dalmas M (2019) Marque employeur et génération Z: une nouvelle approche du recrutement. Revue des Sciences de Gestion 297–298, 113–120. https://doi.org/10.3917/rsg.297.0113

Daubier M, Daubier, S (2022) Outil 53. La marque employeur : 162–163 in M. Daubier & S. Daubier (Dir). Pro en Agilité: 66 outils et 10 plans d'action Vuibert Edition : 208.

Dutta D, Mishra SK, Varma A (2021) Predictors of job pursuit intention across career stages: a multi-phase investigation. Int J Hum Resour Manag 32(20):4215–4252. https://doi.org/10.1080/09585192.2019.1651376

Fray A, Morin L, Renaud S (2015) L'attraction des salariés de la génération Y: Analyse des attentes selon le genre et le niveau de poste. Manag & Prospect 32:117–132

Grouille R (2022) Segmenter les perceptions de la marque employeur chez des recrutés: quel(s) apport(s) RH? Management Et Avenir 131(10):153–175

Guillot-Soulez C, Saint-Onge S, Soulez S (2019) Exploration des liens entre la communication de labels employeurs dans les annonces de recrutement, le mode de gouvernance et l'attractivité des organisations aux yeux des candidats2. Rech Appl Mark 34(3):6–32. https://doi.org/10.1177/0767370119828947

Indeed (2022) L'impact de la marque employeur sur le recrutement et la rétention. Créer une marque dynamique et authentique. April

Indeed et Gassdoor (2023) Hirind and workplace trends. Report 2023

Jiang K, Lepak DP, Han K, Hong Y, Kim A, Winkler AL (2012) Clarifying the construct of human resource systems: Relating human resource management to employee performance. Hum Resour Manag Rev 22:73–85

Kearney M, Bruno M, Field A (2022) Reframing your employee value proposition. Why organizations are finally examining their EVP. Report Ernst et Young (8):13

Mac Gregor D (1960) The human side of enterprise. McGraw-Hill, French Translation : La dimension humaine de l'entreprise. Gauthier-Villars, 1971

Mishra M (2022) Customer experience: Extracting topics from tweets. Int J Mark Res 64(3):334–353. https://doi.org/10.1177/14707853211047515

Morin D (2006) Les déterminants de l'attraction organisationnelle. Léad, Adecco

Rahman SM, Carlson J, Gudergan SP, Wetzels M, Grewal D (2022) Perceived Omnichannel Customer Experience (OCX): Concept, measurement, and impact. J Retail 98:611–632

Rynes SL, Barber AE (1990) Applicant attraction strategies: An organizational perspective. Acad Manag Rev 15(2):286–310. https://doi.org/10.5465/amr.1990.4308158

Schneider B (1987) The people make the place. Pers Psychol 40:437–454

Siebert A, Gopaldas A, Lindridge A, Simoes C (2020) Customer experience journeys: loyalty loops versus involvement spirals. J Mark 84(4):45–66. https://doi.org/10.1177/0022242920920262

Verhoef PC, Lemon KN, Parasuraman A, Roggeveen A, Tsiros M, Schlesinger LA (2009) Customer experience creation: Determinants, dynamics and management strategies. J Retail 85(1):31–41

The European Union Faced with the Return of Inflationary Crises and Corporate Behaviour

Alain Buzelay, Frédéric Delaneuville, Žaneta Lacová, and Anna Jurišová

Abstract Our article addresses the issue of crises in Europe: from the financial crises of 2008 and the health crisis of 2020 to the current inflationary crisis. Firstly, the origins of the return of inflation in Europe are clarified, particularly in its eastern part, including Slovakia. Next, we specify the role of businesses in the inflationary process. Finally, we analyze the specific aspects of anti-inflation policies in Europe. Our ambition is to show that while crises create or break dependencies, they are becoming increasingly intertwined with one another. Regulatory strategies now face not just a single crisis, but a tangle of crises whose consequences become the origin of the subsequent ones.

Keywords Inflationary crisis · Enterprises anti-inflation policy · European union · Slovakia

1 Introduction

A crisis can be defined as the abrupt disruption of a previously stable equilibrium, whether on a human, political, social, economic or financial scale... Economic and financial crises, which are most often closely intertwined, result more specifically from a significant discrepancy between the volume of quantities produced and

A. Buzelay (✉)
Université de Lorraine, Nancy, France
e-mail: alain.buzelay@univ-lorraine.fr

F. Delaneuville · A. Jurišová
Université Comenius, Bratislava, Slovakia
e-mail: frederic.delaneuville@uniba.sk

A. Jurišová
e-mail: anna.jurisova@uniba.sk

Ž. Lacová
Université Matej Bel, Banská Bystrica, Slovakia
e-mail: zaneta.lacova@umb.sk

G. Pellat et al. (eds.), *Data-Centric Business and Applications*, Lecture Notes on Data Engineering and Communications Technologies 258,
https://doi.org/10.1007/978-3-031-95068-1_14

that of quantities consumed, and/or from a profound decoupling between financial magnitudes that soar in relation to the real magnitudes they are supposed to reflect. Economies have always been subject to a large number of crises. The Great Depression of 1929, for example, was triggered by the FED raising its key interest rates to halt the explosion in stock market values following the industrial boom of the time. More recently, three very different crises have forced the ECB to completely reorient its monetary policy.

The 2008 crisis is a financial crisis which, like many others, originated in the United States, the world's most financialized economy. It resulted from the combination of three financing techniques used at the time. Real estate mortgages, used to secure many housing loans at a time when the price of housing was rising steadily, and banks believed they could largely recover the financing granted in the event of non-repayment. Variable rates, which were very low at the start of the loan but became very high over time. Securitization, whereby good loans masking bad debts are integrated into a single medium offered on the financial markets—in order to get rid of them and regain the liquidity needed to grant new loans. But the rapid rise in variable rates suddenly triggered the insolvency of many borrowers, which spread to the banks, who could no longer recover the amount of the loans they had granted by reselling the mortgaged properties, due to the collapse in property prices. The insolvency of the banks blocked the financing of the American economy, triggering a crisis which, through financial and commercial flows, spread worldwide.

The so-called health crisis of 2020 resulted from a generalized worldwide cutback in production supply, following the need to use containment practices to curb the Covid 19 pandemic, while awaiting vaccine protection. The resulting contraction in activity contributed, despite frequent public subsidies, to sharp reductions in income, perpetuating the cumulative process inherent in any crisis.

An inflationary crisis has gradually reappeared in Europe since January 2021. In May 2023, annual inflation rates were running at around 12.5% in Slovakia, Poland and Lithuania, and 22% in Hungary, compared with 8% in Italy, 6.3% in Germany and 6% in France. In addition to the sharp rise in raw material costs linked to a disruption in the geopolitical balance, this crisis is the direct consequence of the stimulus policies implemented at all levels (Member States and ECB within the €uro zone, other Member States and their Central Banks for EU countries outside the €uro zone) to emerge from previous crises. These policies have necessarily generated excess liquidity and spending, as well as lower interest rates, without any commensurate increase in corporate supply, driven by a need for profitability boosted by short-term speculative rather than long-term productive investments.

2 Origins of the Return of Inflation in Europe, in Particular in Its Eastern Part—Including Slovakia

2.1 *The Return of Inflation*

As our economies recover from the health crisis and then the war in Ukraine, inflation, which the European Union and more specifically the €uro zone had kept under control for over twenty years, has suddenly reappeared, starting in 2021, sometimes at very high rates.

2.2 *Monetary Causes*

Inflation, i.e. the irreversible rise in the average price level, has not only monetary origins, but also many other, so-called structural, causes (Achúcarro et al. 2022). The ones most often mentioned are monetary. They correspond to a more rapid increase in the money supply (all means of payment and investments that can rapidly become so) than in Gross Domestic Product (GDP). For the so-called classical authors, based on the quantitative theory of money, the relationship between the evolution of the money supply and that of the price level is strictly proportional and unequivocal. If, for a given GDP, the quantity of money in circulation increases by 6%, this automatically leads to an increase in prices of the same amount. This approach, which has been adapted and reformulated many times (cf. in particular the work of Irving Fisher, Alfred Marshall, Milton Friedman…), has not, however, called into question the strict proportionality of the relationship established by quantitative theory.

While reality confirms a relationship between money supply and price levels, it belies the sense of causality insofar as an exogenous increase in prices can justify an increase in money supply, as well as the principle of strict proportionality. This is illustrated by the European Central Bank's policy of regulating the evolution of the money supply (M3) in line with an inflation rate of around 2%, a desired GDP growth rate and an anticipated reduction in the velocity of money circulation. The resulting rate of increase in the money supply is then adjusted in one direction or another to take account of the need to stimulate or slow down the economy, depending on the economic situation (Buzelay 2022).

Keynesian and post-Keynesian authors allow us to analyze at least five other causes of inflation, which are no longer monetary but structural.

2.3 *Structural Causes*

The first is the "inflationary gap" which, in line with Keynesian analysis, is defined by analogy with the recessionary gap, where aggregate supply exceeds aggregate demand and, consequently, savings exceed investment. In a situation of full employment, with all factors of production fully employed, no expansion of real income is possible, while excess aggregate demand inevitably leads to a rise in prices, with national income rising only in nominal terms.

Then there's the rise in wage costs, which post-Keynesian thinking analyzes as responsible for inflation. Whereas for Keynes, lower interest rates encourage higher investment, national income and employment, for Alban William Phillips (1914–1975), referring to his statistical analysis applied to England for the period 1861–1957, the lower the unemployment rate, the higher the rise in nominal wages, and the higher the rise in prices parallel to that of wages (cf. Phillips curve). While experience requires us to put these relationships into perspective, depending on the period and the direction of causality, it does not allow us to reject them. Let's add that the rise in prices linked to the rise in wages—and more broadly to wage costs—can result from an exogenous increase in payroll taxes.

Another factor is the rise in the cost of goods needed for production, particularly raw materials, which is a function of their growing scarcity, a voluntary reduction in supply in response to geopolitical conflicts, and variations in import prices induced by exchange rate fluctuations.

Finally, there are structural causes linked to the behavior of the various players. Consumer behavior is conducive to inflation. Many of them are victims of a social phenomenon that encourages them to buy at higher prices to satisfy their "need to exist", to assert themselves, to identify with buyers whom they socially rank above them. The same is true of producers, who in an inflationary context can afford to raise their prices above the average increase observed—for various reasons that will be explained in Part Two. It should also be noted that behaviors unfavorable to achieving a social consensus in the sharing of value added between wages and profits can lead to a surge in wage increases, triggering a catch-up on the part of profits, and so on. Adjustments were initially planned in real terms, but rapidly reduced in nominal terms, leading to an inflationary spiral.

2.4 Application to Slovakia

It's worth noting that inflation has risen much faster in the Visegrad states than in other European countries. In May 2023, annual inflation rates showed that Hungary (21.9%), the Czech Republic (12.5%), Slovakia (12.3%) and Poland (12.5%) were the countries most affected by rising prices (see Table 1). The reasons for this very high inflation (Fig. 1) may be questioned, while other European Union countries such as France record much lower rates (6%) (Fig. 2).

Slovakia is one of the countries with inflation well above the Euro-zone average (Fig. 2).

Diagnostics aimed at identifying the causes of inflation may have varied from one country to another, and these differences have had consequences for the instruments put in place to tackle them (see for example Budová et al. (2023).

Table 1 EU inflation rate between May 2022 and May 2023 (date provided by Eurostat)

Country	Yearly inflation rate (in %, May 2022)	Yearly inflation rate (in %, May 2023)
Austria	7.7	8.8
Belgium	9.9	2.7
Bulgaria	13.4	8.6
Cyprus	8.8	3.6
Croatia	10.7	8.3
Czechia	15.2	12.5
Denmark	8.2	2.9
Estonia	20.1	11.2
Finland	7.1	5.0
France	5.8	6.0
Germany	8.7	6.3
Greece	10.5	4.1
Holland	10.2	6.8
Hungary	10.8	21.9
Ireland	8.3	5.4
Italy	7.3	8.0
Latvia	13.1	12.3
Lithuania	16.8	10.7
Luxembourg	9.1	2.0
Malta	5.8	6.3
Poland	12.8	12.5
Portugal	8.1	5.4
Romania	12.4	9.6
Slovakia	11.8	12.3
Slovenia	8.7	8.1
Spain	8.5	2.9
Sweden	7.5	6.7
European Union	8.8	7.1
Euro Area	8.1	6.1

Source data from Eurostat

In Slovakia, prices for the three main household expenditure items—food, energy and housing—rose by double digits.

- Food prices rose by 14.3% between 2022 and 2023. Price increases of 16.4% for bread and cereals, 29.8% for oils and fats, and 26.2% for vegetables had the biggest impact on inflation. In January 2023, food prices throughout the European Union rose by an average of 18.4% year-on-year. At 48.2%, Hungary is the country with the highest increase, followed by Lithuania (32%) and Slovakia (28.6%) (Fig. 3).

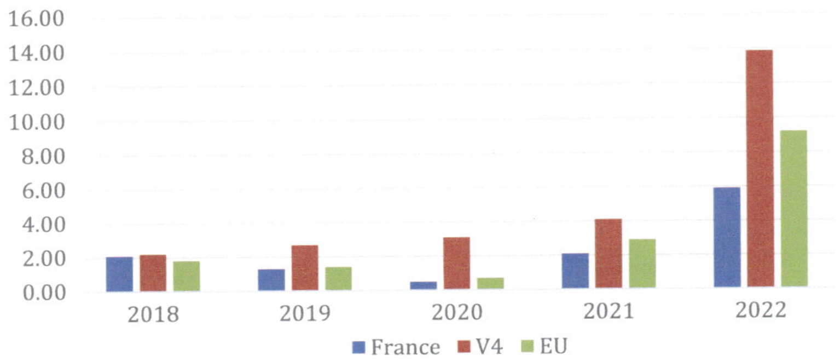

Fig. 1 The inflation rates (in %) in France, Visegrad countries (V4) and in the European Union (EU). *Source* data from Eurostat

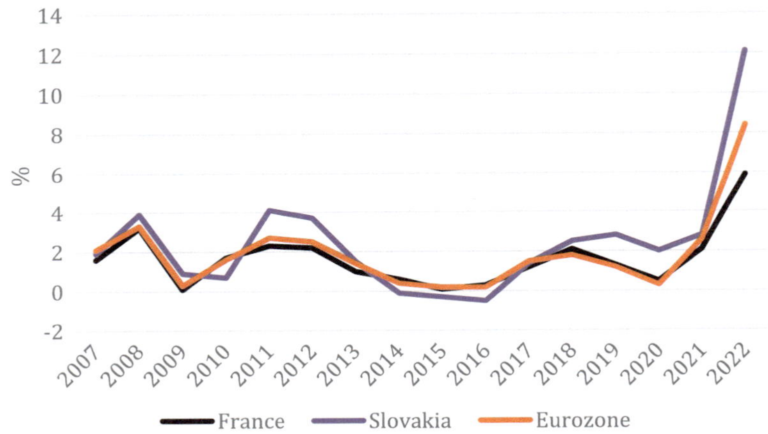

Fig. 2 The inflation rates (in %) in France, Visegrad countries (V4) and in the Eurozone. *Source* Table based on data from Eurostat and Statista. Available at: https://www.statista.com/; https://ec.europa.eu/eurostat

The rise in food prices can be explained by a number of factors. Firstly, soaring food costs are a direct consequence of the war in Ukraine. The Russian invasion has completely disrupted wheat production and exports, with both countries accounting for 30% of the world's harvest. Grain is not the only commodity whose price has soared. Vegetable oil prices, for example, have risen by almost 14% year-on-year, and are now at historic levels. Here again, Ukraine's share of sunflower oil production explains these price rises.

- National measures to combat rising energy prices have differed across Europe: gas price caps, subsidies, pump rebates, tariff shields, reduced VAT rates, etc. Energy policies are essentially a matter for individual states. Even if Article 194 of the Treaty on the Functioning of the EU emphasizes a "spirit of solidarity",

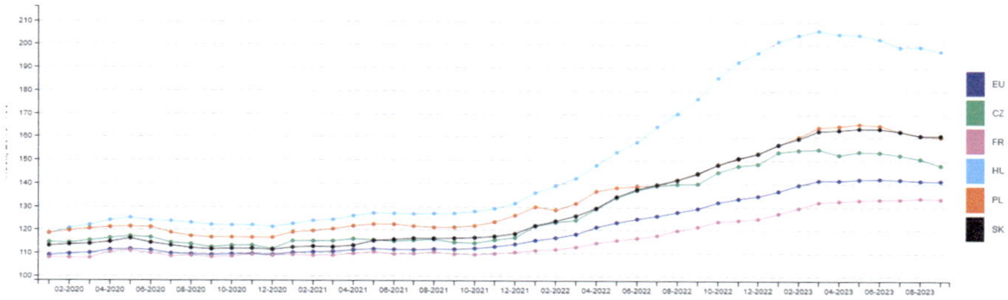

Fig. 3 Food prices development in chosen EU countries. *Source* Eurostat, Harmonized Index of Consumer Prices (HICP), Food, Index 2015 = 100

each State has the right "to determine the conditions for exploiting its energy resources, its choice between different energy sources and the general structure of its energy supply". Depending on national energy mixes and the solutions adopted by governments, prices may therefore have evolved differently. France and Slovakia (but also others such as Spain) have opted for energy price caps. These measures have sometimes been criticized for providing insufficient incentives for protected economic players to reduce their consumption. In Slovakia, the government has opted for price caps on electricity and gas for certain consumers only: households and small businesses. The Office for the Regulation of Network Industries (ÚRSO) has set maximum prices for protected consumers until the end of 2023. In France, the government set up a tariff shield which benefited private individuals, small local authorities and micro-businesses (companies with fewer than 10 employees, annual sales of less than €2 million and subscribed power of 36 kVA or less). This was accompanied by a policy aimed at promoting energy sobriety. One example is the new heating standards for public buildings. This was also the case in Slovakia, where strategies were put in place to optimize heating in universities and ministries.

- At the same time, both Slovakia and France tried to implement supply-side policies by reopening electricity generation capacity. France sought to mobilize its nuclear fleet, but its room for maneuver was limited by the non-availability of certain reactors. Slovakia has also undertaken to develop the nuclear power plant at Mochovce, a town around 150 km east of Bratislava, before the end of 2023. In France and Slovakia, nuclear power plays an important role in the energy mix. Other countries with different energy mixes, such as Germany, have made the environmentally questionable decision to reopen coal-fired power plants.
- Slovakia, like its Central European neighbors belonging to the Visegrad group, remains highly dependent on hydrocarbons from Russia. Slovakia is among the European Union countries most dependent on Russian hydrocarbon imports (Fig. 4).
- At the beginning of 2022, 77.5% of Slovakia's oil was still imported from Russia (46.4% for Hungary). This dependence also explains the Slovak authorities' reluctance to accept the embargo designed to punish the invasion of Ukraine. On May

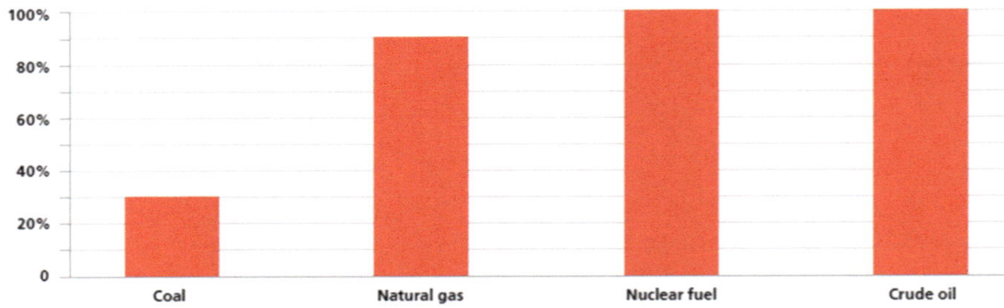

Fig. 4 Imports from Russia in energy supply before the war (2021). *Source* Statistical Office of the Slovak Republic, Eurostat

30, 2022, in Brussels, European leaders granted Slovakia, Hungary and the Czech Republic—landlocked countries with no immediate alternative solutions—a transitional regime on the embargo, enabling them to continue importing Russian hydrocarbons, which will continue to be supplied via the Russian Droujba pipeline until December 2023.

- With regard to housing, the rise in interest rates on real estate loans has had an impact on the increase in household spending on housing. The flexibility of Slovak legislation has enabled many households to borrow to finance the acquisition of property, but the widespread use of flexible rates permitted by the legislation (fixed rates being offered for short terms, 3, 5 or 10 years maximum) has had and will continue to have an impact on household spending on housing. As a result, when the fixed-rate period expires, borrowers are forced to pay interest at the prevailing rates, potentially leading to defaults. As a result of this specific feature of Slovak legislation compared with French legislation, the cost of buying a home is set to rise significantly over the next few years, whereas French buyers who purchased before the rise in interest rates will be protected by legislation that has generalized fixed rates for the entire term of the loan.

We have analyzed the three factors that explain the high level of inflation in Slovakia, despite the measures taken by the government. However, rising energy and commodity prices alone do not explain the galloping inflation in Central Europe and Slovakia in particular (Michálek 2024). While oil and raw material prices have fallen significantly, prices have continued to rise. The practices of companies seeking to protect themselves by anticipating possible further price rises may also have contributed to maintaining an inflationary spiral.

To complete our analysis of the causes of Slovak inflation in the current context, we believe it would be useful to adopt a socio-economic approach, explaining inflation in terms of the structure of household consumption, as well as their subjective and anticipated perception of inflation rates in relation to the official rate measured by the HICP Harmonized Index of Consumer Prices.

Table 2 shows that there are significant differences in consumption practices: for example, households in the Visegrad countries spend more on foodstuffs, to the detriment of spending on transport, leisure and/or culture. These differences may be caused in particular by the different incomes available to households, but also, at least in part, by socio-cultural differences determining individual preferences.

As consumption is often influenced by the rate of perceived inflation (Chava et al. 2022), international comparisons of such rates can help explain inflation differentials. Data on inflation perceptions are available across Europe from consumer surveys. Table 3 shows the differences between countries in perceived inflation rates prior to the 2004–2015 period. Already, during the period of low inflation, households overestimated the real values of inflation rates. A priori, accentuated faults in inflation perception may point to weaknesses in households' financial knowledge. From this point of view, it is not surprising that perceived inflation in 2023 reached up to 35% and Slovaks perceived price growth up to 20 percentage points higher than measured inflation.

Inflation perceptions have the potential to influence inflationary expectations. Inflation expectations represent a risk for future price and wage dynamics. In May 2023, household inflation expectations for the next twelve months were below 25% in Slovakia. By way of comparison, one-year inflation expectations in France in 2023 remained stable at 4% over the same period. As Alligner and Rumler (2023) point out, the difference between actual inflation and declared inflation expectations is considerably higher in the CEECs than in the eurozone countries.

Table 2 Structure of household consumption in selected EU member countries

Consumption categories (COCOIP divisions)	EU-27	France	Slovakia	Czechia	Poland	Hungary
Food products and non-alcoholic beverages	171	143	202	207	247	163
Alcoholic beverages and tobacco	24	25	26	33	26	23
Clothing and footwear	40	40	36	44	37	24
Housing, water, gas, electricity and other fuels	327	287	334	243	327	492
Furnishings, household equipment and routine home maintenance	49	48	51	68	51	29
Health	37	16	26	31	47	34
Transport	108	132	86	96	79	67
Communications	31	24	49	44	44	46
Leisure and culture	67	77	40	95	51	41
Education	9	6	4	10	9	5
Hotels, cafés, restaurants	48	55	54	58	34	24
Other goods and services	89	147	92	71	50	52
TOTAL	1000	1000	1000	1000	1002	1000

Data source Eurostat

Table 3 Perceived inflation rate (average) from January 2004 to July 2015 in selected EU member countries

	EU-27	France	Slovakia	Czechia	Poland	Hungary
Perceived inflation rate (average value)	9.8	6.9	9.1	8.1	13.8	9.1
Inflation rate measured by HICP (average value)	2	1.6	2.7	2.2	2.5	-0.1
Difference between perceived and measured rates	7.8	5.3	6.4	5.9	11.3	9.2

Data source https://www.ecb.europa.eu/pub/pdf/scpops/ecb.op186.en.pdf

In addition to the various reasons for the high level of inflation in Slovakia, we need to mention one that may have to do with the behavior of companies.

3 The Role of the Company in the Inflationary Process

3.1 The Company's Non-neutrality

The company avoids inflation which, in the absence of productivity improvements, cuts into its profits. But it does not necessarily play a neutral role by simply passing on the rate of inflation to its prices, in order to safeguard its margins.

In an inflationary context, producers can more easily raise their selling prices to protect themselves against further anticipated cost increases. Analytically, anticipation occurs when the price increase corresponds to the inflation rate, of which it is a simple extrapolation. This anticipation is said to be unanticipated when the price increases applied are more or less higher than the rate of inflation, due to the need for prevention and anticipated catch-up in the event of an upturn in inflation.

They can also increase their prices more than the rate of inflation demands, in order to boost profits and, above all, the dividends paid to shareholders. These dividends serve to enhance the company's brand image and encourage the inflow of additional equity capital, which is all the more useful when inflationary conditions lead to a rise in interest rates, increasing the cost of borrowing outside capital.

While it is easier for companies to raise their prices in such a context, they can also avoid lowering them by not passing on the drop in their production costs resulting from a fall in the cost of their imports, itself linked to a favorable exchange rate trend, or from an improvement in their productivity. They may also fail to compete by selling on the market at the price of less efficient subcontractors and companies, thus artificially keeping them on the market.

Finally, in an inflationary context, companies may prefer to make short-term speculative investments (cash investments, real estate investments, etc.) rather than long-term productivity investments, which are deemed less profitable and riskier.

This can lead to inter-sectoral and inter-regional differences in productivity, which are themselves inflationary. If high-productivity sector or region A can withstand a pay rise without increasing its prices, this may not be the case for sector or region B.

In Karl Marx's analysis of our market economies, producers are expected to promote the "increasing monopolization" of productive systems, which, along with the growing demand of the highly capitalist classes, are factors in price rises.

3.2 Its Responsibility in Slovakia's Inflationary Process: The Example of Companies Operating in the Food Retail Sector

It makes sense to ask what role companies have played in the recent acceleration in inflation. The dynamics of world commodity and energy prices have played a central role in the inflationary dynamic in Slovakia. However, given the scale of the price rises observed, Slovak manufacturers and distributors seem to have gone beyond simply postponing the increase in production prices, thereby playing a decisive role in the inflationary spiral.

A study carried out by the National Bank of Slovakia and published in January 2023 highlighted that the margin rates of companies operating in the food and beverage manufacturing and retail sectors increased in the first half of 2022, suggesting a decoupling between rising production costs and prices. While estimated margins in relation to production costs have remained relatively stable, margins in food and beverage manufacturing and retailing have shown clear signs of increasing, thus contributing significantly to inflation in Slovakia.

Companies operating in the food distribution sector can easily raise their prices, and they can also avoid lowering them by not passing on the drop in their production costs resulting from a fall in their import costs. Some companies, such as IDC Holding, operating in the food sector saw their profits rise significantly in 2022, even though inflation was at a very high level. Analysts, such as those at the Slovak National Bank, were able to point the finger at "crisis profiteers" who had increased their margins at a time when household purchasing power was at half-mast, even though the issue of retailer margins did not feature as prominently in the public debate as it did in France. Strong post-pandemic demand and forced savings probably enabled vendors to significantly increase food company profits and make up a little for the previous bad period", said Peter Majer from the National Bank of Slovakia.

Slovak companies, particularly those operating in the food and energy sectors, may therefore have played a decisive role in Slovakia's inflationary spiral. By over-anticipating increases in their production costs, they increased their margins to the detriment of household purchasing power and confidence in the economic situation.

4 About Anti-inflation Policies

4.1 From a Long-Standing ECB Stimulus Policy to a Recent Slowdown

While Article 105 of the 1992 Maastricht Treaty, followed by the Lisbon Treaty in 2007, dictates that the ECB should pursue the exclusive objective of price stability (2%), the imperative need for stimulus necessitated by the crises of 2008 and 2020 has forced it to move away from this dogmatic approach in favor of greater pragmatism. Thus, it first resorted to conventional policy instruments by lowering its key rates, imposing a negative rate of return on deposits received from its banks more than the minimum reserve requirement, and introducing a system of long-term loans only to banks committed to lending to businesses (Long-term refinancing operation). Then, from 2015 onwards, the ECB resorted to so-called unconventional measures, chiefly the practice of quantitative easing, which enables it to supply economies abundantly by purchasing numerous private and public assets from their commercial banks and financial institutions.

The sudden return of inflation in 2021 led the ECB to backtrack by gradually raising its key rates and rethinking its quantitative easing strategy, in order to mitigate the sharp disparity in interest rates within the €uro zone, and consequently in national borrowing costs. The instrument designed to protect against fragmentation (of interest rates and borrowing costs) in the €uro zone was thus conceived. This no longer enables the ECB, as before, to buy the same amount of debt securities from each country on an equitable basis, but to do so only for the benefit of those countries that are victims of an exaggerated and speculative rise in their borrowing rates, i.e. those whose vulnerability and fragility may, through a contamination effect, weaken the entire €uro zone.

This austerity policy, redeployed by the ECB, has rapidly reduced inflationary pressures in the €uro zone (see Table 1), but has inevitably slowed activity and growth, with the risk of renewed unemployment if investment prospects worsen. Interest rates that are too low encourage speculative investment, reducing the profitability of specific investments for small savers; interest rates that are too high for too long can slow down growth more than the rise in prices, especially as this rise is not due solely to an abundance of liquidity. Hence the need to accompany monetary policy with other policies that can also promote price stability.

4.2 The Need for Policies to Complement Monetary Action

Among the policies that should complement monetary action to promote price stability, we favour competition policy and industrial policy in the current context. The former is guided by Article 3 of the Treaty of Rome, which requires Member States to establish a system ensuring that intra-Community competition is not distorted within their largest market. And the Maastricht Treaty adds that it must guarantee unity by reference to two groups of prescriptions: the elimination of all tariff and non-tariff barriers, and the maintenance, or even introduction, of rules and standards enabling the desired competitive balance to be respected. Thus, the deregulation sought by the single market does not correspond to the rejection of all rules, but only those deemed to be obstacles to competition. This applies to concerted practices such as price-fixing, supply restrictions, market sharing or market segmentation (cf. Article 8 of the Treaty of Rome or Article 81 of the EU Treaty). This is also the case for abuses of a dominant position which are condemned by the Commission, although it is not always easy to know where the abuse begins and ends (Article 86 of the Treaty of Rome or 86 of the Treaty on the Functioning of the European Union).

It's worth noting that from 2010 onwards, under the influence of an accentuation of the neoliberal trend, competition policy has considerably eclipsed industrial policy, which the recent health crisis of 2020 and the geopolitical tensions arising from the war in Ukraine have just brought back into the limelight.

Industrial policy is the set of guidelines and stimuli coordinated by the public authorities to influence the behavior and decisions of producing agents, with a view to achieving the targeted restructuring and objectives. Within each member state, this policy is more or less pronounced, depending on their willingness and ability to intervene. At European level, following the 1951 Treaty of Paris establishing the European Coal and Steel Community (ECSC), we had to wait until the 1992 Maastricht Treaty to read in Article 3 of its Title XVII that Community action shall include strengthening the competitiveness of Community industry. This competitiveness strategy is intended to be much broader than industrial policy, encompassing research and development, innovation and transport, as well as social and regional action, all of which are more or less factors in improving competitiveness. Nevertheless, this competitiveness strategy, which is intended to support member states' own strategies, remains inadequate in the face of the need to refocus the globalization of our value chains.

The progressive universalization of trade liberalization has led to globalization, i.e. the growing interdependence of economies, following the greater or lesser interpenetration of their production and financial systems. This evolution responds to companies' need to optimize their production by segmenting it geographically, from upstream to downstream, in the territories deemed most appropriate. All these segments form value chains, which are inevitably sources of supply dependency, and which become highly penalizing in the event of disruptions resulting from geopolitical conflicts. Hence the imperative need for the Union to refocus its value chains on geopolitically more secure territories.

5 Conclusion

"Europe will be made in crises, and it will be the sum of the solutions to these crises". This quote from the visionary Jean Monnet is more relevant than ever. We have to admit that the health crisis caused by Covid 19 has enabled political and monetary integration to progress. By accepting the European Commission's indebtedness on behalf of all member states, Germany has agreed to lift the taboo, thus opening the door to a common European debt. This development, which some analysts have described as a Hamiltonian moment (Hall 2020), illustrates that the European Union can sometimes rise to the occasion. The budgetary and monetary measures taken by governments within the framework of the Union, and the action taken by the central bank during the pandemic, helped to protect incomes and limit the impact on European economies. While the ECB's policy may have led to an increase in the money supply in circulation, our Keynesian analysis has enabled us to identify other factors that explain inflation. One of these is the rise in energy costs. The conflict in Ukraine has revealed the energy dependence of certain countries, particularly in Central Europe, on Russian hydrocarbons. These crises have made European elites aware of their dependence on other continents, and of the consequences of supply chain disruptions in a globalized world characterized by the increasing internationalization of the value chain. The contraction of world supply has had a dramatic impact on energy prices, which have exploded in Europe. While some countries, such as France, were able to limit the impact on the various economic players thanks to their proactive policies, others, such as those in Central Europe, were not so lucky. However, even if the action or inaction of the State may have played a decisive role in determining the level of inflation in the States of the Union, which may have varied considerably from one country to another, this does not mean that certain economic players, such as companies, should be exonerated. Producers were able to contribute to price rises by over-anticipating increases in raw material and energy prices, thereby boosting their profits. In Slovakia, various studies have criticized the behavior of certain companies, which are posting substantial profits at a time when the purchasing power of Slovaks is at half-mast. The size of the Slovak market, which is characterized by a limited number of suppliers, has encouraged concerted commercial practices. The limited resources available to the state authorities in charge of monitoring the correct application of European competition law have enabled them to allow cartels to the detriment of consumers, despite the fact that such practices are contrary to EU law. The paradox is that galloping inflation in Central Europe can sometimes fuel sovereignist rhetoric, as illustrated by the results of the recent parliamentary elections in Slovakia. And yet, to cope with these recurring crises, which are increasingly intertwined, the European Union has no alternative but to evolve towards a more integrated European power, with genuine industrial and defense policies, in the face of a threatening Russia which has interpreted the Union's soft power as an admission of weakness. The current situation calls for closer integration of the European powers and the development of the Union's strategic autonomy in an increasingly uncertain multipolar world, where alliances are made and broken according to the interests of the major powers.

Acknowledgements Supported by the grant VEGA 1/0382/23 "Dynamic modeling of nonlinear macroeconomic processes" of the Research of Agency of the Ministry of Education, Research, Development and Youth of the Slovak Republic

References

Achúcarro A, Biagetti M, Braglia M, Cabass G, Caldwell R, Castorina E, Wu WL (2022) Inflation: theory and observations. arXiv preprint arXiv:2203.08128

Budová J, Šuliková V, Siničáková M (2023) When inflation again matters: do domestic and global output gaps determine inflation in the EU? Amfiteatru Econ 25(63):575–592

Buzelay A (2022) Mondialisation économico-financière et sécurité d'approvisionnement en Europe. Revue De L'union Européenne 662:532–535

Chava S, Du W, Shah A, Zeng L. (2022) Measuring firm-level inflation exposure: a deep learning approach. SSRN

Hall B, Fleming S, Chazan G (2020) Is the Franco-German plan Europe's 'Hamiltonian' moment. Financ Times 21

https://fr.euronews.com/2022/04/12/embargo-sur-les-hydrocarbures-russes-les-pays-d-europe-centrale-sont-divises

https://www.ecologie.gouv.fr/bouclier-tarifaire-lelectricite-et-amortisseur-electricite

https://www.euractiv.fr/section/energie-climat/news/rien-ne-change-pour-la-slovaquie-a-lentree-en-vigueur-de-lembargo-petrolier-de-lue/

https://www.euronews.com/next/2022/09/13/slovakia-nuclear

https://www.lesechos.fr/monde/enjeux-internationaux/les-prix-alimentaires-mondiaux-en-repli-mais-toujours-tres-eleves-1894662

Michálek A (2024) Inflácia na Slovensku na národnej a regionálnej úrovni. Acta Geogr Univ Comen 68(1):3–26

Work and Industrial Revolutions: Risks, Opportunities, Policies

Montesi Cristina

Abstract The paper intends to analyze the evolution of work and of work organization across four industrial Revolutions, each linked to waves of technological progress which gave rise to different models of capitalism and of entreprises. More specifically the paper aims to examine generally the benefits and costs of technological innovation and, among the social costs of innovation, the risks relating to work (involution and destruction of work due to technological progress), the opportunities (liberation of work and creation of new professional figures linked to technological progress), the necessary policies to govern ex ante and ex post the processes of technological transition with particular attention to the last frontier, the development of artificial intelligence and of the enabling technologies, with all their multidimensional challenges.

Keywords Industrial revolutions · Great transformation · Artificial intelligence

J.E.L. Classification O33 Technological Change · Choices and Consequences

1 Introduction

The paper intends to analyze the evolution of work and of work organization across four industrial Revolutions, each linked to waves of technological progress which gave rise to different models of capitalism and of entreprises.

More specifically the paper aims to examine generally the benefits and costs of technological innovation and, among the social costs of innovation, the risks relating to work (involution and destruction of work due to technological progress), the opportunities (liberation of work and creation of new professional figures linked to technological progress), the necessary policies to govern ex ante and ex post the processes of technological transition with particular attention to the last frontier, the

M. Cristina (✉)
Department of Economics, University of Perugia, Perugia, Italy
e-mail: montesik@usa.net; cristina.montesi@unipg.it

G. Pellat et al. (eds.), *Data-Centric Business and Applications*, Lecture Notes on Data Engineering and Communications Technologies 258,
https://doi.org/10.1007/978-3-031-95068-1_15

development of artificial intelligence and of the enabling technologies, with all their multidimensional challenges.

2 Methodology

The paper will favor qualitative analysis over quantitative analysis through the study of the works of the "classics" scholars (economists and sociologists) called into question, through the study of the existing scientific literature on the topics (according to a plurality of perspectives), through the consultation of the most up-to-date empirical reports from international research institutes.

3 Technological Innovation and Its Contradictory Effects on Work

This historical phase is a time characterized by profound and rapid economic, social, cultural, anthropological changes due to technological progress associated with artificial intelligence (AI), changes which also concern work and whose consequences are impossible to predict due to the intensity, vastness, speed and heterogeneity of the current transformations. These are "disruptive" technological upheavals that radically and irreversibly change the contexts of work and life, which implement a convergence/combination of different technologies, which simultaneously embrace multiple areas, which occur more rapidly than in the past, which follow one another incessantly in some sectors, which give rise to positive effects, but at the same time can also have negative repercussions on social and economic systems. Some scholars even speak of a "Great Regression" for our Era (Geiselberger 2017), paraphrasing and accentuating, in a negative way, the title of Karl Polanyi's work *The Great Transformation* in which Polanyi critically analyzed the effects of the first industrial Revolution, highlighting, among other aspects, the processes of commodification of work (in addition to the parallel processes of commodification of land and money) that had characterized it (Polanyi 2010).

Today more accentuated pessimism[1] (Roubini 2023), is due to the fact that, on *employment* front, there are two opposing trends both from a *quantitative* and *qualitative* point of view.

From a *quantitative* point of view, technological innovation generates two antithetical effects: the first effect is "destructive" because automation implies *the replacement of labor with capital*, causing, in addition to a lowering of wages, technological unemployment or often forcing workers to invest their skills, once updated, in other sectors different from those in which they were originally active (Rifkin 2015). The second effect is "constructive" because innovation drives the demand for *new goods or services* which, in turn, creates *new jobs*. So the key aspects are the *timing* and the *extent* to which the constructive effect may or may not prevail over the destructive aspect. The fourth industrial Revolution seems to generate, compared to previous Revolutions, fewer *skilled* jobs in *emerging* sectors than it destroys, in a *very quick* way and in a *systemic* way, in the rest of economy (OECD 2023b). In fact: "According to the findings of the *Oxford Martin Program on Technology and Employment*, only 0.5% of American workforce is employed in sectors that did not exist before the turn of the century, a very low percentage compared to 8% of the new jobs which accompanied the creation of new productive sectors in the 1980s and compared to 4.5% of the new jobs created in the 1990s" (Schwab 2016, p. 56). These days only 0.3% of workers are employed in the Artificial Intelligence sector and that they are mainly graduates and male (Green and Lamby 2023).

This trend has even prefigured the hypothesis of the "end of work" and of the advent of "jobless societies" which, according to some scholars (Rifkin 1995; Summer 2014; Crews 2016; Susskind 2022), is going to take the form of a generalized technological unemployment which can only be coped with compensatory monetary transfers (Korinek and Juelfs 2022).

Even from a *qualitative* point of view, technological innovation generates two discordant effects. On the one hand, work, at least in some areas (think of *smart factory*), has been enriched with cultural elements, in an increasingly dense *interweaving between knowledge and work activity*, which began with the ICT Revolution of the 1980s. In third industrial Revolution work assumes unusual forms of autonomy, creativity and responsibility and develops linguistic relational activities. In the fourth industrial Revolution, work translates, even more intensively, into a "linguistic act".

However, there is a difference between *information Revolution* and *digital Revolution*: "There is an essential difference between the *computerized* and *digitalized*

---

[1] Nouriel Roubini in his latest book talks about ten "mega threats" that are outlined against the economic, financial, technological, political, geopolitical, health and environmental fronts and which could cause enormous damage and suffering to humanity in the next two decades. Some of these threats looming over humanity are slower, other are faster, but they are all interrelated, a fact that may raise fears of an unprecedented "confluence of calamities" (Georgieva et al. 2022). According to Roubini, these threats are: debt accumulation, public and private financial crises, aging populations, easy money trap, stagflation, currency collapses, end of globalization, *artificial intelligence*, wars, an uninhabitable planet. As far as work is concerned, Roubini's concern is, above all, for "*a technological revolution that will eliminate more jobs in less time than ever before*" (Roubini 2023, p. 11).

work environment: the first occurs in a context in which the linguistic relations that support cognitive activities are of two types, while in the digital context there are of three types.

In the first case, communications occurs first between man and man and then between man and machine; in the second case, communications between machine and machine are added. In the first case workers talk to each other and then communicate the results of their talks to machines; in the second case men talk to each other and then talk to the machines which, in turn, talk to each other and communicate the results of their interactions to humans, which assume them" (Mari 2020, p. 7).

At the time of Fordist enterprise, however, work activity was silent, production was a silent assembly line and workers were silent. In post-Fordist age, the production process of the "clever" and "open" factory becomes a *complex of performative linguistic acts* because production is based on the strength of "doing things with words", an ability that had already been highlighted by the philosopher of language Austin (1962).

These linguistic acts translate into simultaneous interaction between humans (between workers and between workers and managers), translate into interaction between machines (machine which speak to each other), translate into interaction between machines and men (because machine communicate the results of their interactions to humans). Machines learn continuously thanks to the large amount of data they continuously process and self-produce thanks to artificial intelligence. In this context *production is carried out through communication activities.* Workers are (and should be) loquacious within cyber-physical systems that, thanks to *Internet of Things*, connect people to each other, connect objects to each other, connect people and objects.

On the other hand, in other economic environment, different from smart factory, such as "gig economy" (Crouch 2019), processes of *involution of work* are in progress, characterized by individualization of working relationship, segmentation of work, deskilling of work and "return to slavery" typical of situations in which "the master is an algorithm", processes which characterize the digital capitalism of the platforms (Aloisi and De Stefano 2020).

In this tumultuous and contradictory scenario, the invocations of Pope Francis calling for "work and not income for all" (Papa Francesco 2017) and the need to reconnect the triad *innovation/work/person* do not pass unnoticed, as work does not affect only the dimension, however important, of *having*, but also the dimension of *being* (which refers to safeguarding worker's *dignity* (Lanier 2014), to respecting his desire for *self-realization*, to *giving meaning* to economic activity while we see the proliferation of "meaningless jobs" related to the financialization of economy and related to the spread of ICT (Information and Communication Technologies) (Graeber 2018).

Italian Constitution affirms in several of its articles (art. 1; art. 3, comma 2; art. 4, comma 1 and 2; articles 35, 36, 37, 38, 39) a triple centrality of work: *anthropological*, *ethical*, *economic* (Pennacchi 2010; Di Gaspare 2010; Mari 2019), while in Title IV of EU Charter of fundamental Rights dedicated to "solidarity", social and labor rights find space alongside civil and political rights.

4 Industrial Revolutions, Work and New Technologies

In the epochal transition from industrial to post-industrial society, four industrial Revolutions have occurred. These four industrial Revolutions were each time connected to:

1. a *change in the scientific paradigm* (the "scientific paradigm", according to Thomas Kuhn, is not only the codified knowledge which arises, primarily from basic research, in force in a given Era, but it is the "overall vision of the world" in which a specific scientific theory finds citizenship) (Kuhn 2009);
2. the simultaneous *birth of new sciences* related to the change in scientific paradigm;
3. *technological progress* closely linked to the change in scientific paradigm and the development of new sciences. Georgescu Roegen speaks of *Promethean technologies* (Montesi 2022) linked to *radical innovations* of Schumpeterian memory,[2] which open up to an expansionary phase of the economy and which give rise to a new socio-technical regime with a higher degree of complexity;
4. the *discovery and valorization of one or more strategic "key factors"* for the economy, key factors that are inextricably linked to a "disruptive" technological progress (Clayton Christensen et al. 2015).
5. a *quantitative and qualitative change in work and in its organization*, but also to a parallel transformation of companies and societies.

As regards the paradigm shift related to the fourth industrial Revolution, this change does not only concern "what" to do and "how", but also "who" we are (Schwab 2016, p. 15).

The *radical* innovations, which are the basis of the Industrial Revolutions, bring benefits to a large number of subjects, but can also cause *social costs* of a different nature, of different severity and persistence, to the community, which can fall on different people and which should be carefully identified and quantified before their introduction through a *cost/benefit analysis* (Franzini 2018, pp. 73–76). Neoclassical economic theory is more inclined to exalt the benefits of innovation in its many typologies and to underestimate the social costs of innovation. Neoclassical economic theory is also convinced that market forces are able, on their own, to limit the negative effects of technological progress with their own self-equilibrium mechanisms (the loss of jobs in a sector and/or in a geographical area would be more than compensated

---

[2] According to Joseph Schumpeter, innovations, which consist of those inventions which can be given real commercial application, can be of different kinds (product, process, organizational innovation, expansion of markets, discovery of new sources of supply) and can be *incremental* or *radical*. Other authors have added to Schumpeter's classification *systemic* innovations (which affect, in a correlated way, multiple sectors simultaneously) and innovations which even imply a change in the technical–economic paradigm. Schumpeter went so far as to theorize the figure of the innovator entrepreneur, whose action is based on the "creative destruction" of the existing market equilibrium due to the irruption of *radical* innovation, which is often the basis of an industrial Revolution. Innovation therefore constitutes the "fundamental element of capitalism" (Schumpeter 1942, p. 38).

by the creation of other jobs in other sectors and/or in a geographical-economic elsewhere) (Milanovic 2016).

This vision unrealistically presupposes perfect mobility of labor and perfect substitutability between workers.

But what are the possible benefits of an innovation?

On production side, if we think of a product innovation, the benefits for a firm consist in the formation of a monopoly, although of a temporary nature, on the market which increases profits of the innovative company. Benefits, again on production side, can also fall on firms that have not directly introduced the innovation (with the generation of positive externalities by the innovative company). In fact, if the new product is not protected by a patent, it may happen that other companies can, over time, copy it without having incurred research and development costs faced by the first comer company. On the demand side, benefits arise from the use of a new product by consumers and from lower prices for consumers due to the possible reduction of production costs once the phase of selecting the dominant design has been overcome and the era of retention has been entered by innovating companies.

What are the social costs of an innovation?

Social costs are different and may take the form of:

1. *increase in risks to health and/or to the environment* (think of many of those innovations that were introduced in the past without applying the "precautionary principle", resulting in heavy negative repercussions for habitat and society)[3];
2. *generation of technological unemployment*, with marked gender inequalities, which has historically provoked reactions and revolts (thematized since the third edition of 1817 of David Ricardo's "Principles of Political Economy and Taxation" in the chapter *On Machinery*);
3. *low wages* (think of the wages of the "gig economy", the economy of odd jobs);
4. *increases in income inequalities between individuals, between hyper-skilled work and unskilled work* (the remuneration of those who design/build/update robots and learning machines are stratospheric, as are the remuneration of the owners and administrators of platforms used in platform capitalism. According to Anthony Atkinson[4] platform capitalism would need, together with the adoption

[3] The European Environment Agency has examined the most important lessons resulting from the failure in no having applied the *precautionary principle* when spreading an innovation in European Union and North America over the last hundred years (European Environment Agency 2001). A principle, that of precaution, which should be taken seriously into consideration especially in the context of the current energy transition (OECD 2023a).

[4] Among the 15 proposals that Anthony B. Atkinson puts forward in his book (*Inequality: What Can Be Done?*) to reduce inequalities, proposal 4 expressly calls for the *adoption of a code relating to workers' compensation*: "**Proposal 4**: There must be a national wage policy, based on two elements: a legal minimum wage set at the living wage level and a code of good practice for wages above the minimum, agreed in a "national conversation" involving the Social and Economic Council".

of an ethical code, also a remuneration code that could set a maximum on the remunerations of managers/engineers of platform capitalism)[5] (Atkinson 2016);

5. *increase in functional inequality*, i.e. increase in inequality between wages share and profits share in favor of profits (this is due to the formation, in digital capitalism, of monopoly positions of companies due to the characteristics of both production and demand of their services). The "superstars" companies of digital capitalism control large market shares and earn mega profits that are reinvested in financial securities or in the acquisition of start-up companies[6];
6. *increase in hard skills inequality*, it should be noted that the ongoing technological transformation can exacerbate social disparities not only between workers, but also among citizens, due to digital divide existing between those who are able to access digitalised services/products and those who are unable;
7. *detailed profiling of platform users* and use of their private data as a means of maintaining economic growth and vitality by "surveillance capitalism" companies (Zuboff 2019).

Before analyzing the possible strategies to mitigate inequalities, to combat technological unemployment, to minimize social costs in the event of its incomplete or slow reabsorption, a review of the various industrial Revolutions that took place over time can be traced paying particular attention to the changes that have occurred in work and in work organization and to their impact on employment.

5 First Industrial Revolution

The beginning of first industrial Revolution date back to the last decades of the XVIII century and to the first half of the XIX century and its geographical roots can be found in Great Britain. The first industrial Revolution will then extend to France, Belgium, Germany, Northern Italy, Japan, United States and to some regions of Austro-Hungarian Empire.

The first industrial Revolution is the "mature" applicative fruit of modern science of Copernicus, Kepler, Bacon, Newton, Galileo. It is linked to the discoveries of thermodynamic physics and to the birth of economic science with English Political Economy (whose reference authors are A. Smith, D. Ricardo, C. Marx, J. S. Mill) and Italian Civil Economy (A. Genovesi).

The first Revolution marks the definitive transition from a pre-industrial, agrarian-artisan-commercial society to an *industrial* society. The key factors strategic for the first industrial Revolution were *steam engine* and *other machinery* used, both in

[5] However, it should be noted that the ongoing technological transformation can exacerbate social disparities not only between workers, but also among citizens, due to digital divide existing between those who are able to access digitalised services/products and those who are unable.

[6] The concentration of power of the major companies of digital capitalism is not limited to *economic power*, but extends into other spheres: *technological power, political power, power to control innovative ecosystems*.

agriculture and in industry, relating to the cycle of the first mechanization (such as the hydraulic mechanical loom).

The natural resources most used in the production cycles of the first industrial revolution were *coal, iron, cotton and water*.

The main change that occurred in quality of work is the transition from artisanal work to a more *standardized*, but at the same time to a more *alienated* work. Work is reduced, according to Marx, to a "commodity" and work, including work of minors and of women, is *underpaid* and characterized by *great physical effort* and by *long working hours* (from thirteen to fifteen hours a day).

The most important change in terms of work organization is the application of the principle of *division of labor* between workers who are all concentrated in a single workplace (the famous "pin factory" observed by Adam Smith) with beneficial effects on many fronts: saving time in moving from one job to another, increasing labor productivity, generation of "incremental" product innovation.

The impact on employment, at least at the dawn of the first industrial Revolution, was negative due to the generation of "technological" unemployment and "structural" unemployment.

In this regard can be recalled the emblematic protest reaction of the Luddite movement in 1811 (destruction of mechanical looms because they were a source of unemployment and because they caused wage reduction). Karl Marx in turn denounces the formation of a (permanent) reserve army of labour made by unemployed people due to the specific way of producing of capitalism characterized by the monopoly of ownership of the means of production in the hands of the capitalist class and by the exploitation of the proletariat class due to the asymmetry of power existing between the two classes.

A gradual and partial reabsorption of technological unemployment and an increase in employment occurred, as time passed, thanks to the development of the sectors linked to the leading sectors of the first industrial Revolution (i.e. the sectors linked to the textile, steel, mechanical and extractive industries); thanks to the development of other sectors: transport (locomotives; railways; etc.); electricity (electric machines); communications (telegraph, telephone); thanks to the development of services linked to urbanization processes; thanks to the development of internal and international trade (fueled by the importation of raw materials from the colonies of the British Empire); thanks to the growth in demand for consumer goods due to the gradual improvement in living and wage conditions; thanks to demographic growth (which however remained in many countries lower than GDP growth rate).

6 Second Industrial Revolution

The second industrial Revolution takes place temporally from the 1870s to the 1970s (of particular note are four sub-periods: the *Era of the Great Depression*, the *Era of reconstruction* after the Second World War; the *Era of economic boom* of the 1960s; the *Era of recession* after the two oil crises of 1973 and 1979).

In scientific field the hegemony of neoclassical economics was called into question by Keynesian theory developed to explain and resolve unemployment created by the Great Depression of 1929.

In second industrial Revolution the science/technology/industry combination is strengthened with scientific research that becomes increasingly applied (scientists often become captains of industry). The interpenetration between large companies and banks that finance productive investments is consolidated (in Italy this role is also played by the State).

In second industrial Revolution mass production is the driving force (mainly based on serialized and durable products absorbed by mass consumption influenced by advertising) which has its strong point in the economies of scale achievable thanks to large company size. Precisely for this reason there is a tendency towards corporate concentrations through mergers, trusts, holdings and cartels.

The second Industrial Revolution is marked, in many developed countries, by the advent of heavy industry (steel, chemicals, petrochemicals, electromechanics) and of automotive industry and by the realization of large infrastructure investments by the State (railways, roads, motorways).

The key factor for the economy is *oil* and the natural resources that characterize the production cycles as inputs are predominantly *fossil fuels* and *nuclear power*.

In this productive framework work becomes a *fragmented, merely executive, heterodirected activity*. The organization of work sees the affirmation of the *scientific division of labor* (taylorism) which will find its fulfillment in Fordism which reduces times and unit costs of production and increases the productivity of work. The essential element of Fordism is the introduction of the *assembly line* in production process along which all the different phases of production are interconnected and synchronized and the adoption of other innovations aimed at facilitating the flow of production in the assembly line (conveyor belts, goods lifts, etc.).

The impact of the second industrial Revolution on employment was different depending on the four sub-periods considered. As regards Italy, in the period of the "thirty glorious years" (1945–1975), high GDP growth rates (unimaginable today) corresponded to an increase in employment rate (also favored by migratory flows heading towards the industrial areas of the country). In the period following the two oil crises (the first oil shock occurred in 1973–74 and the second shock in 1979–80) corporate restructuring processes caused in Italy an increase in unemployment, but at the same time gave rise to small and medium-sized enterprises (through hot spin-offs; birth of industrial districts) which partly counterbalanced, thanks to self-employment, the loss of employed jobs.

According to economist Robert Gordon, the period 1870–1970 was the "golden age", the incomparable period of highest development of all humanity (and also of United States): "no other era in human history, neither before nor since, has brought together so many elements in which living standards have improved so rapidly and in which the human condition has been transformed so completely" (Gordon 2016, p. 287) thanks to inventions such as electricity, internal combustion engine, running and drinkable water in homes (Gordon 2016, p. 522).

7 Third Industrial Revolution

The third industrial Revolution is temporally located in the period 1990–2010, giving rise to New Economy or Net Economy (referring to its ability to build, thanks to Information and Communication Technologies, virtual communities and to connect web operators).

This Revolution marks the transition from an industrial society to a *post-industrial* society focused on advanced services (business consultancy, finance, insurance, healthcare, education/education, research and development, culture, entertainment, mass media, databases) rather than based on the centrality of traditional services sector (commerce, tourism, transport, public administration).

Within manufacturing production new sectors emerge: sectors linked to ICT (computers, mobile phones, operating systems and software), fine chemistry and green chemistry, new materials, space activities, automation, industrial robotics, photonics (optical fibers, laser technologies).

From an epistemological point of view the third industrial Revolution insinuates itself into the mature phase of information technology, telematics and microelectronics, into the consolidation phase of the new complexity sciences and of the new life sciences, into the auroral phase of new heterodox trends in economic science (civil economics, ecological economics, bioeconomics, generativity economics, evolutionary economics, experimental economics, behavioral economics).

The key factors of the third industrial revolution are *information* and *microchip* (the microprocessor) which drive the advent of ICT (Information and Communication Technologies).

The natural resources most used in production cycles are always *fossil fuels* alongside *renewable energies*.

The change in the quality of work is summarized in the fact that, thanks to the use of ICT, it becomes a *qualified* activity (it needs continuous investments in training), it becomes a more *clever* activity (and not just an executive and routine activity), becomes a more *autonomous* and *self-managed* activity (linked more to results than to presence in the workplace), it becomes a *group activity* and a *collaborative, participatory, creative, full of meaning activity*. Teamwork attenuates hierarchical relationships and encourages the ability to "think outside the box". It blurs the clear separation between place of work and home and the clear dichotomy between life times and work times.

The assembly line gives way to *lean production*, also known as Toyotaism, named after the Japanese multinational Toyota, which introduced it in the early 1980s. The new system works through *production islands*, made up of technicians and workers, who contribute to the complete creation of the product assigned to them, from the material execution to quality controls and final revisions. The prerequisites of this new organization of work and production are *just in time* (which consists in producing according to orders, avoiding accumulating warehouse stocks) and *total quality*, i.e. the possibility of making continuous improvements to the product throughout the value chain of the business system. The three systems (lean production, total quality, just in time) were subsequently integrated and surpassed in *World Class Manufacturing*, also of Japanese origin, which prevents any type of waste and loss. Company management is further rationalized by reducing important costs such as those for data processing (computing costs), data transmission and communications.

There is also an increase in collaborative relationships between companies because of the reduction of transaction and communication costs thanks to the use of ICT. There is also an improvement in the innovative capacity of companies due to the reduction in times and costs of access to knowledge, to its greater capacity for circulation and replication and to the expansion of the territorial scale for its access.

Company's communication with the customer also changes giving centrality to a consumer who becomes more and more critical and aware, with the development of relational marketing, with the birth of e-commerce and of online consumers' communities. Markets become global thanks to network economies and consumers benefit from a multiplicity of options and greater competition between products.

Net Economy has raised labor productivity albeit not with the same intensity as the second industrial Revolution, except for the period 1995–2004 in which productivity grew significantly thanks to the advances recorded, in this time interval, especially in information technology (Gordon 2016), but Net Economy also destroyed jobs[7] and the fruits of the innovation activity have been no longer shared equally among the population.[8]

The jobs lost in factories as a result of technological innovation have only been partially replaced by jobs created in advanced or traditional services (think of the whirlwind development of logistics linked to electronic commerce) or in other activities with a high content of information and knowledge (Brynjolfsson and McAfee 2011).[9] Carl Benedikt Frey and Michael Osborne, professors at Oxford University, in their 2013 essay entitled *The Future of Employment*: *how susceptible are jobs to*

[7] The third industrial Revolution stumbled upon other important economic-financial events which, in turn, created unemployment or opened up new job opportunities beyond the contribution of technological progress: the bursting on 10th March 2000 of the speculative bubble on the financial markets generated by the tumultuous flourishing of New Economy companies (the Dot-com Bubble); globalization (unregulated); the great economic-financial crisis of 2008; the sovereign debt crisis of Greece, Ireland, Italy, Portugal, Spain, Cyprus; the growing environmental degradation of the Planet which identifies, as the main source of concern above all, the tragedy of global commons (climate change, loss of large forests, loss of biodiversity, water crisis, ocean degradation, etc.) which inspired the design and creation of *Green Jobs* (see the Green New Deal of United States and of European Union).

[8] In many industrialized countries the increasing income disparity between individuals and between classes of income earners (with the growing erosion of the share of wages in the national income) has been the result of the progressive reduction of the weight of trade unions, of the erosion of minimum wage, of the increase of unregulated globalization, of the financialization of economy, of the aging of the population, of the dynamics of public debt, of the progressive reduction of public intervention in the economy until pandemic event.

[9] Here are mentioned some of the new professions that New Economy has given rise to: the *Knowledge Manager* is a specialist in the collection, processing and interpretation of information and knowledge; the *Web portal designer* is a network-professional who manages the creation and operation of Internet portals; the *Web designer* deals with the creation and management of websites; the *Web marketing officer* is the one who deals with online marketing. However there are activities or

computerisation? had calculated that 47% of jobs in United States were at risk of automation (Frey and Osborne 2013).

A study conducted by McKinsey in 2017 on more than 2.000 work activities across 800 occupations in 46 countries showed that 5% of all occupations could be automated *entirely* using "demonstrated technologies" and 60% of all occupations had at least 30% of constituent activities that could be automated (McKinsey & Company 2017). The 78% of *predictable physical works* (physical activities conducted in highly structured and predictable environments as well as the collection and processing of data which exist across a large spectrum of sectors) were the activities most susceptible to automation. In United States these predictable physical works made up the 51% of activities in economy in 2017. The research indicated that sectors like *manufacturing, accommodation and food service, retail trade* and some *middle-skill jobs* had bigger probability to be automated earlier. According to McKinsey study, 73% of the work activities carried out in the *food service sector* could have been automated (food cleaning, food preparation and service, collection and cleaning of dishes), 59% of the *manufacturing activities* (packaging, welding, maintenance, etc.), 53% of *retail trade* (warehousing, packaging, customer and product information management, logistics) (McKinsey & Company 2017). According to McKinsey study the pace and scope of automation would have depended on technical feasibility, on the cost of technology, on competition with labor including skills and supply and demand dynamics, on benefits beyond labor cost savings, on social and regulatory acceptance.

In third industrial Revolution there is a decoupling between productivity and employment (the former increases, the latter decreases). The technologies used in companies make firms more efficient and productive,[10] but technologies save human labor and are renewed too quickly (Acemoglu and Restrepo 2017). Not only blue collars, but also clerks are no longer able to acquire quickly the skills to use technologies (Brynjolfsson and McAfee 2014). Technological evolution marches faster than biological evolution, creating maladaptation and/or rejection in human beings and growing inequalities between workers.

Technologies are beginning to erode, after having affected a range of blue collars' routine physical work activities, not only bureaucratic clerks' work (work based on simplified procedures which do not require particular richness of exchanges with

professions that the third industrial Revolution has put particularly at risk: local commerce, bookshops, translators, tailors, small hoteliers, print journalists, photographers, travel agency employees, mechanics who do not know electronics, postmen, traditional banking, but the list could to continue. Some professions have shown to be more resilient to the impact of technology: these are jobs that require physical contact with the person (such as jobs linked to personal services or the production of relational goods), jobs which take care of public goods or of common goods, jobs aimed at compensating for the lack of free time available to people (domestic work, pets' care, etc.). There is a high demand for these jobs which still remains unsatisfied today and which should be organized and channeled appropriately with the strengthening of *civil economy* (Bruni and Zamagni 2004; Bruni and Zamagni 2009; Montesi 2016).

[10] Automation can enable productivity growth performance through reducing errors, improving quality and speed, increasing throughput, achieving outcomes that often go beyond human capabilities.

customers), but even cognitive and creative work (algorithms nowadays are able to create written texts indistinguishable from those created by humans).

8 Forth Industrial Revolution

It is the industrial Revolution which we are experiencing in the present driven by Artificial Intelligence (AI)[11] and by other technologies (the *enabling technologies*).

Note that technological change is now whirlwind: it is *intragenerational* and no longer intergenerational, a fact which does not allow for adequate discernment on the ethical implications of this hyper fast technological development.

In scientific field there is the synergistic combination of information technologies (in their most advanced version, that of the *enabling technologies*) with nanotechnologies, biotechnologies and cognitive sciences (Nodari 2018, pp. 5–6).

The key factor of this latest industrial Revolution, in addition to Artificial Intelligence, are the *enabling technologies*, which have been divided into 9 categories by a Boston Consulting study: Internet of Things (smart product); Big Data; Cloud Computing; Cyber Security; Simulation; Horizontal and Vertical Integration; Augmented Reality; Additive manufacturing (3D printer); Advanced manufacturing (collaborative robots).

The resources in the limelight for their use in production cycles in forth industrial Revolution are *renewable energies* and *secondary raw materials* thanks to circular economy and blue economy.

As regards the change in quality of work in forth industrial Revolution, what has already been said about the transformations occurred in third industrial Revolution applies with some additional comments. Technology will act on the so-called three Ds of work: "dirty, dangerous, dull". Human work, thanks to the help of robots, will be even less physically tiring and safer from the point of view of the health and well-being of workers (think of the use of exoskeletons) and more stimulating. Work will be increasingly directed towards solving problems, co-designing of products and services, co-decisioning of the development of the factory (with an expansion of industrial democracy).

The new risks are instead represented by:

1. an excessive invasiveness of these technologies in the private sphere which can imply a disorderly management of spaces and times of life outside of the work place (a concrete risk present in smart working if not regulated);

[11] The term Artificial Intelligence was coined back in 1954 by John McCarthy to explain the long history of the attempts to mechanise human intelligence. Its more recent development is linked to progress in computing technologies and in machine learning and predictive processes. AI includes various areas of research, but its core components are machine learning, deep learning, NLP (natural language processing) platforms, predictive APIs (application programming interface), image recognition and speech recognition. Now there is an other typology of AI: *generative* AI which can create music, videos, writings, etc.

2. pathological dependence on technology;
3. a hybridization between humans and technologies which can engulf and assimilate what they penetrate (hypothesis of transhumanism) (the *Internet of beings* has remained the last frontier to cross after the Internet of things) (Zamagni 2018, p. 32).

Digital and specialistic/hard skills are no longer enough, *soft skills* are also necessary to mitigate technocracy: social intelligence, emotional intelligence, ecological intelligence, learning agility, self-management, listening skills, critical thinking skills, persuasion skills, aptitude for learning from experiences (with a strategic role played for their acquisition by *Humanities*).[12]

The *enabling technologies* make it possible to better connect the different production phases and the different areas of the factory and permit to fluidify company's relationships with providers, subcontractors and customers. A better connection allows greater work productivity, greater production flexibility (with increasingly "tailor-made" production, also carried out in small batches), predictive maintenance, shorter reaction times to events and response to market demands, less waste, greater ability to anticipate design errors.

The effects of the *new enabling technologies* and of Artificial Intelligence on employment are controversial. The dilemma of technological unemployment returns, a dilemma which J. M. Keynes had already focused in his essay *Economic possibilities for our grandchildren* in 1930. J. M. Keynes gave this definition of technological unemployment: "an unemployment due to our discovery of the means to economize the use of our arms, faster than the pace at which we can find new uses of workforce. But it is only a temporary misalignment" (Keynes 1963). Keynes considered technological unemployment a transitory "disease" that would have disappeared within a hundred years thanks to the creation of employment linked to the development of new sectors emerging from the Industrial Revolution of his time (the second industrial Revolution).

[12] A recent research based on the analysis of approximately 2.5 million job advertisements downloaded from the web shows that *professional skills* represent the majority of requests in job market with an average of 60% followed by *soft skills* which represent 24% of requests and by *digital skills* which account for 15% (Crisp 2023).

Deep fears of losing one's job or worsening quality of job are arising[13] because of the introduction of AI and other technologies (Confartigianato 2023; Lane et al. 2023), which distress particularly the middle class (even qualified white-collar professions with an intellectual, creative and administrative content are now at risk, not only white-collar jobs doing repetitive tasks and using standardized procedures). A recent study of ILO on the effects of AI generative (in the form of Generative Pre-Trained Transformers, GPTs) on job quantity and quality states that: "we find that only the broad occupation of *clerical work* is highly exposed to the technology *with 24% of clerical tasks considered highly exposed and an additional 58% with medium-level exposure*". As clerical jobs are an important source of female employment, the effects are highly gendered (Gmyrek et al. 2023, p. 1).

These fears arouse regressive feelings of escape from the Future to take refuge in a nostalgic return to the past (Bauman talks of *Retrotopia*) and fuel populisms that know how to exploit these fears artfully (Bauman 2017).

Three schools of thought on the dilemma of technological unemployment in the Era of the fourth industrial Revolution can schematically be identified.

The "apocalyptics" foresee, due to "robocracy", the advent of a "predominantly jobless society", the advent of a society that will have many unemployed people who will have to be assisted by Welfare State (Fondazione Giangiacomo Feltrinelli 2016). This jobless society will guarantee only low-skilled jobs, intermittent and precarious for the majority of people (Colamedici and Gancitano 2023), while a minority will instead have safe and well paid jobs (work polarization hypothesis). This minority will be the technocratic élite, dedicated to the programming of algorithms, to the design/management/control of digital platforms and of other digital tools, to the continuous innovation of technologies (Cowen 2013; Ford 2017).

The "optimists" instead foresee a replacement of lost jobs with others; the liberation of workers' time, some of which will likely be allocated to new and more important tasks in their workplace; the "end of physical fatigue and alienated work"; the reduction of working hours (Keynes speaks in the cited essay of a 15-h working week: 3 h a day for 5 days); the increase in free time which will become the true Kingdom of freedom/affirmation of the person (a free time in which, according to Keynes, one can experience "the arts of life"). This is a reading that exalts *otium* and *liberation from work* (rather than fighting for the *liberation of work*, meant as

---

[13] The *Third Censis-Eudaimon Report on Corporate Welfare*, which conducted a research on a sample of 165 Italian companies, recorded a polarization between the opinions of techno-enthusiastic company management and the opinions of workers and employees who declare themselves techno-resigned. Thanks to the arrival of new enabling technologies, for 97.6% of company managers there will be an increase in productivity, efficiency and competitiveness, for 97% there will be an improvement in working conditions and quality of life in the company, for 77.6% there will be no replacement of workers in many jobs, nor a loss of jobs. For 50.4% of workers, new technologies and automation processes will impose higher work paces and greater work intensity, for 43% they will extend working times, even beyond normal hours, for 42.2% they will destroy work and cause job losses, for 32.7% they will not make people work better or improve the quality of life in the company, for 28.2% they will not make jobs less risky and workers less exposed to the risk of suffering accidents or damages in the exercise of one's profession (Censis-Eudaimon 2020, pp. 14–16).

the transformation of work into a quality, freer and more creative activity) (De Masi 2002, 2016).

The "cautiously optimistic" believe that the fear of job losses should not be exaggerated, because new jobs will take gradually the place of others[14] (Autor 2015) and old jobs will be revisited in their tasks (creating a complementarity between human work and machines), provided however that the technological transition is wisely managed with retraining of workers victims of technological unemployment, with continuous training of workers, with adaptation of school and university systems to technological progress.

The fear of robots, which has to do with their anthropomorphism,[15] should not be exaggerated because, in the current state of technology, what robots cannot do yet is to replicate man's creativity, to have his ability to solve problems and take decisions in unpredictable contexts and in situations of uncertainty resorting to "lateral" thinking, to have the capability to manage the emotional and relational aspects of working life (World Economic Forum 2023).

Despite these differences the gap between organic intelligence and artificial intelligence is progressively narrowing, becoming AI more and more *generative* of content (texts, videos, audios, music, images) being able, like ChatGPT and Bard, to write lyrics and even movie scripts, compose music and create digital art.

According to a recent study of McKinsey (McKinsey & Company 2023) the new capabilities of *generative* AI (*increased ability to understand natural language*) have increased the share of time spent on work activities that theoretically could be automated today (from 50% of the previous study to 70%) (McKinsey & Company 2017).

Unlike most automation in the past *the impact of generative AI will fall most heavily on knowledge work and on occupations requiring higher levels of education and commanding higher wages.*

Workers will see some jobs disappear (fully or partly for some tasks) in categories and sectors that until recently looked safe (entertainment, media creation, education, health care, engineering, design, etc.). Using data on the task content of over 900 occupations in the US Goldman Sachs economists Joseph Briggs and Devesh Kodnani estimate that roughly two-thirds of current jobs are exposed to some degree of AI automation and show that inside those occupations which are exposed, most have a significant-but partial-share of their workload (25–50%) that can be replaced (Briggs and Kodnani 2023).

[14] Digital Economy has already given rise to new professions like: *Digital Innovation Officer*, *Technology Innovation Manager*, *Data Protection Officer*, *Data Scientist*, *App Developer*, *Cloud Architect*, *Digital Brand Manager*, *Graphic Designer*.

[15] "The obsession with, or the fear of, robots has to do, I believe, with our fascination with their anthropomorphism" (Milanovic 2016).

9 Strategies to Limit the Social Costs of Technological Innovation

Strategies aimed at limiting the social costs of technological innovation can be divided into two main types: those to be adopted AFTER the innovation has been introduced and those to be adopted BEFORE the innovation is introduced (Franzini 2018, pp. 80–83). Both ex-ante and ex-post strategies need adequate institutions and policies. Some institutions (the State) are already at work, other institutions should be created ex novo, but there is a problem: institutional evolution is slower than technological evolution, failing to keep pace with technological development.

Among the possible strategies to be adopted AFTER the innovation has been introduced to limit its social costs can be included:

1. *monetary compensation for those who are damaged* (the damage consists in the loss of a job or in the impoverishment due to technology). These monetary compensations can be found within the measures of social assistance of classic Welfare State (among the possible instruments to be used: unemployment benefits; subsidies to help working poor; subsidies provided for the fight against poverty; a minimum wage which should be foreseen for the "jobs" of gig economy);
2. *interventions to contain technological unemployment*. These interventions fall within the borders of the Welfare State of the promotion of people capabilities and may take the form of the retraining of workers (reskilling). To prevent technological unemployment the following actions should instead be adopted: continuous training of workers (upskilling); education in hard and soft skills that keep pace with innovation. All these tasks should be carried out by the following institutions: schools, universities, training system.
3. *other more radical measures* such as: giving workers the ownership of robots and technologies (as suggested by R.B. Freeman) (Freeman 2015); divert part of the dividends of digital capitalism companies into a public mutual fund for Welfare State with the institution of a universal basic dividend (UBD) financed from the returns on capital (as hypothesized by Y. Varoufakis) (Varoufakis 2017); reduce working hours for the same salary (Autonomy 2023).

Among the possible strategies to be adopted BEFORE the innovation is introduced to limit its social costs can be included:

1. *a regulatory activity of technological change carried out by the State*. According to some scholars, the direction of technological change should be oriented by the State. In which direction? Should be encouraged that innovation which is complementary to human work and which is not a substitute for human work and which responds to needs of people that are still unsatisfied (needs which are so pressing in the fields of aging, health, immigration management, management of common goods, governance of climate change) (Atkinson 2016; Franzini 2018; Pennacchi 2018).

The regulation by the State of technological change, formulated by Anthony Atkinson in his book *Inequalities. What can be done*?, figures in first place among his 15 proposals to contrast inequalities:

Proposal 1: The direction of technological change must be an explicit policy concern; innovation must be encouraged in a form that increases employment, highlighting the human dimension of service provision.

Atkinson refutes the reading that the neoclassical economist R. Solow gave of technological progress as a "manna that fell from heaven". According to Atkinson, technology is neither exogenous nor neutral and the power relations existing in capitalism dictate its development and shape its form. The trajectory that innovation follows therefore depends primarily on companies, but also on consumers and *governments who must act as guides for technological progress.*

According to Atkinson, a public authority should be established both for defining the criteria that a "socially good" technological innovation should satisfy and for the evaluation ex ante of the net social benefits of an innovation, even if this process of evaluation is difficult and not free from possible errors. Further regulatory difficulties could arise in the case of the innovative activity of global companies where public authorities should be supranational.

Criticisms have been formulated against this proposal by Atkinson because it is considered too dirigistic and because State's regulation of technological progress could slow down innovative activity which, according to some scholars, should remain free and autonomous. A self-regulating function could be exercised, in the case of *open innovation*, by all the actors who contribute collectively to it or are indirectly engaged in it.

However, even the *Future Life Initiative*, an Institute that includes researchers, managers, entrepreneurs, has declared itself concerned about the spontaneous evolution of AI which can cause dangerous outcomes for humanity, asking, through a public letter signed in 22th March 2023 by some of its members (including Elon Musk, CEO of Tesla and Space X), a six-month moratorium on developments of AI systems more powerful than Gpt-4 to make such systems "safe, interpretable, transparent, robust, aligned, reliable and fair".

At the same time the letter also calls for *government action to regulate and control the development of AI* and avoid "disruptive economic and political upheavals (especially for democracy) that will be caused by AI".

Some scholars suggest putting filters to moderate technological evolution: *ethical filter* (it is not necessary to do everything that is possible from a technical point of view); *social filter* (innovation can be implemented gradually, staggered over time); *filter relating to corporate governance* (this means giving listening and voice to multiple stakeholders in the company, not just to shareholders); *filter connected to productivity* (a washing machine equipped with devices similar to those of Apollo spacecraft should not be designed to take us to the Moon, but to help us only in washing the clothes) (Meyer 2016).

Regarding algorithms which can become real "weapons of mathematical destruction", some scholars suggest answering the following questions before applying them: is the algorithm opaque or invisible to the people involved in its operation? Is the algorithm fair to those involved in its operation? Does the algorithm harm or destroy lives? Can the algorithm expand its effects beyond its intended task?

A second strategy to be adopted BEFORE the innovation is introduced to limit its social costs is that of:

1. *entrepreneurial State*, which means to boost up public research along with its commercial use, orienting it towards innovations that are not enemies of "human work" and aimed at environmental sustainability and social equity. Mariana Mazzucato, in her book *The Entrepreneurial State*, has demonstrated how the State in USA, through DARPA (Defense Advanced Research Projects Agency) together with other public research institutes, contributed to giving rise to some of the most important innovations of XX century (nuclear energy, jet, computers, lasers, internet, nanotechnology, biotechnology, drugs for rare diseases, iPhone components) (Mazzuccato 2014).

This solution is also not immune to criticism for different reasons. Mariana Mazzucato highlighted how the profits from the commercialization of innovations created by the State are often appropriated by the private sector, suggesting as possible remedies the *creation of investment banks controlled by the State*, or the *imposition of royalties on the application by private companies of technologies created by the State*, or *shareholdings* of the State in innovative private companies that the State itself finances. Furthermore public financial resources are not enough for AI investments.

"Until 2014 in United States most significant machine learning models were released by academia. Since then, industry has taken over. In 2022, there were 32 significant industry-produced machine learning models compared to just three produced by academia. Building state-of-the-art AI systems increasingly requires large amounts of data, computer power, and money-resources that industry actors inherently possess in greater amounts compared to non profits and academia" (AA.VV 2023, p. 23). However, the strategy of the "entrepreneurial State" does not protect against the fact that private research can continue to generate high social costs that can only be mitigated by public regulation.

10 Conclusion

In parallel with the innovations characterizing the four industrial Revolutions, many changes have occurred in quality of work and in employment, but also in companies which employ labor. Many "transformations of capitalism" have occurred, because capitalism is truly protean, but always vital seeking incessantly new territories to conquer in order to extract profit. During the four industrial Revolutions different forms of capitalism have alternated over time together with different forms of enterprises that have characterized them according to the following scheme.

First Industrial Revolution

- *Manufacturing capitalism* (factory)

Second Industrial Revolution

- *Oligopolistic capitalism* (big corporations, multinational companies)
- *Flexible capitalism* (the network company and business networks)

Third Industrial Revolution

- *Cognitive capitalism* characterized by centrality of knowledge (company is a "cognitive system" in continuous learning and its competitive advantage derives from "knowledge management", from "information" which is a new productive input, from "interactive communication" thanks to ITC)
- *Financial and patrimonial capitalism* characterized by hypertrophic and self-referential finance and by the increase in income and wealth inequalities between individuals and between classes (this capitalism finds its counterpart, on the production side, in the global company without borders) (Chancel et al. 2022).

Fourth Industrial Revolution

- *Digital capitalism* founded on *enabling technologies* and on technological *mix* of automation, artificial intelligence and internet connection (smart factories and smart industries/Industry 4.0).
- *Surveillance capitalism* (Zuboff 2019) based on various types of platforms (platform of advertising, cloud, industrial, product, lean) (Smicek 2017), on the control of the data of platform-users, on the triad algorithm-platform-people that supplants the triad idea-product-company (McAfee and Brynjolfsson 2017).
- *Sharing capitalism* (the sharing economy, always facilitated by Internet and platforms, but which moves on the register of gratuitousness).

From all analysis conducted it is clear how important is the role of various institutions (State, public regulatory Agency, educational and training institutions) upstream and downstream of technological innovation process, so that a median path can be found between the apocalyptic scenarios and the salvific conceptions of technology.

It is a path of improvement of living and working conditions to be walked, on the one hand through the best use of technologies mediated by some filters and, on the other hand, through their control both at a political and community level (OECD 2023c). Only in this way could we move towards a digital neo-humanism and a model of capitalism with a human face (Pennacchi 2018, pp. 49–55).

The most pressing problem today is to regulate, through the right institutions, the activities of the platforms and of the monopolistic "giant companies" of digital/surveillance capitalism not only on the side of the different social costs generated by their technological progress in order to prevent or contain them (paying special attention to the negative effect of innovation on labour market), but also on the side of the protection of competition, copyright, privacy, taxation, economic democracy, political democracy (fake news and misinformation, hate speech, etc.) and others rights.[16]

The ownership of the platforms does not necessarily have to be private, but platforms could be managed by the community as common goods (see the case of Wikipedia) or could be managed by cooperatives or could be publicly owned with positive effect on work (quality, employment levels, labor organization).

References

AA.VV. (2023) The AI index 2023 annual report. Institute for Human-Centered AI, Stanford University, Stanford, AI Index Steering Committee

Acemoglu D, Restrepo P (2017) Robots and jobs: evidence from labour markets. Boston University, MIT

Aloisi A, De Stefano V (2020) Il tuo capo è un algoritmo. Contro il lavoro disumano, Laterza, Roma-Bari

Atkinson AB (2016) Disuguaglianza. Che cosa si può fare? Raffaello Cortina, Milano

Austin JL (1962) How to do things with words, Oxford U.P., Oxford, II ed. riv. 1975, tr it. Come fare cose con le parole, Marietti, Genova, 1987

Autonomy (2023) The results are in: the UK's four-day week pilot, pp 1–69

Autor DH (2015) Why are there still so many jobs? the history and future of workplace automation. J Econ Perspect 29(3):3–30

Bauman Z (2017) Retrotopia, Laterza, Roma-Bari

Briggs J, Kodnani D (2023) The potentially large effects of artificial intelligence on economic growth. https://www.gspublishing.com/content/research/en/reports/2023/03/27/d64e052b-0f6e-45d7-967b-d7be35fabd16.html

Bruni L, Zamagni S (2004) Economia civile. Efficienza, Equità, Felicità pubblica, Il Mulino, Bologna

Brynjolfsson E, McAfee A (2011) Race against the machine: How the digital revolution is accelerating innovation, driving productivity, and irreversibly transforming employment and the economy. Digital Frontier Press, Lexington, MA

Brynjolfsson E, McAfee A (2014) The second machine age: work, progress, and prosperity in a time of brilliant technologies. W.W.Norton, New York

Bruni L, Zamagni S (eds) (2009) Dizionario di Economia Civile, Città Nuova, Roma

Censis-Eudaimon (2020) Terzo Rapporto Censis-Eudaimon sul Welfare aziendale, Roma, 5 febbraio 2020, pp 2–75

Chancel L, Piketty T, Saez E, Zucman G et al (2022) World inequality report 2022, World Inequality Lab wir2022.wid.world

Clayton CM, Raynor ME, McDonald R (2015) What is disurptive innovation? Harv Bus Rev 93(12):44–53

Colamedici A, Gancitano M (2023) Ma chi me lo fa fare? Come il lavoro ci ha illuso: la fine dell'incantesimo, Harper Collins, Milano

Confartigianato (2023) Intelligenza artificiale, lavoro e imprese, report presented on 12th June 2023

---

[16] European Union has already imposed rule on companies offering digital services in the common space. See: *General Data Protection Act*, *Digital Markert Act*, *Digital Service Act*, *Artificial Intelligence Act*. See also USA President Biden's *Executive Order on Safe, Secure, and Trstworthy Artificial Intelligence* which has recently been revocated by Donald Trump, the new president of United States.

Cowen T (2013) Average is over: powering America beyond the age of the great stagnation. Dutton, New York
Crews J (2016) Robonomics: prepare today for the jobless economy of tomorrow. CreateSpace Independent Publishing Platform
Crisp (2023) Un mercato del lavoro che cambia. Le professioni low-level skill, Fondazione per la sussidiarietà, Milano
Crouch C (2019) Se il lavoro si fa gig. Il Mulino, Bologna
European Environment Agency (2001) Late lessons from early warnings: the precautionary principle 1896–2000. In: Environmental issue reportn, n 22, pp 3–211
Fondazione Giangiacomo Feltrinelli (2016) Decalogo della jobless society. https://fondazionefeltrinelli.it
Ford M (2017) Il futuro senza lavoro. Accelerazione tecnologica e macchine intelligenti. Come prepararsi alla rivoluzione economica in arrivo Il Saggiatore, Milano
Francesco P (2017) Discorso del Santo Padre tenuto nello stabilimento Ilva di Cornigliano, 27 maggio 2017 nel quadro della visita pastorale a Genova
Franzini M (2018) La direzione del cambiamento tecnologico come problema politico: riflessioni su una proposta di Tony Atkinson. In: Pennacchi L, Sanna R (2018) Lavoro ed innovazione per riformare il capitalism. Ediesse, Roma
Freeman RB (2015) Who owns the Robots rule the World. IZA World of labor
Frey CB, Osborne MA (2013) The future of employment: how susceptable are jobs to computerisation? Oxford Martin School, Programme on the Impacts of Future Technology. University of Oxford 17:1–72
Di Gaspare G (2010) Lavoro ed economia di mercato nella Costituzione della Repubblica. In: Pinelli C, Treu T (2010) La Costituzione economica: Italia, Europa, Il Mulino, Bologna, pp 57–89
Geiselberger H (2017) (a cura di) La grande regressione. Quindici intellettuali da tutto il mondo spiegano la crisi del nostro tempo, Feltrinelli, Milano
Georgieva K, Gopinath G, Pazarbasioglu C (2022) Why we must resist geoeconomic fragmentation-and how. International Monetary Fund
Gmyrek P, Berg J, Bescond D (2023) Generative AI and Jobs: a global analysis of potential effects on job quantity and quality. In: ILO Working paper, n 96, (Geneva, ILO), pp 1–55
Gordon RJ (2016) The rise and fall of american growth. Princeton University Press, Princeton
Graeber D (2018) Bull shit jobs, Garzanti, Milano
Green A, Lamby L (2023) The supply, demand and characteristics of the AI workforce across OECD countries. OECD Soc Employ Migr Work Pap 1–56
Keynes JM (1963) Economic possibilities for our grandchildren. In Essays in persuasion. W.W. Norton, New York, pp 358–373
Korinek A, Juelfs M (2022) Preparing for the (non-existent?) future of work. In: NBER working paper, no 30172, pp 1–42
Kuhn TS (2009) La struttura delle rivoluzioni scientifiche. Einaudi, Torino
Lane M, Williams M, Broecke S (2023) The impact of AI on the workplace: Main findings from the OECD AI surveys of employers and workers. In: OECD social, employment and migration working papers, N 288. OECD Publishing, Paris, pp 1–156
Lanier J (2014) La dignità ai tempi di Internet. Per un'economia digitale equa, Il Saggiatore, Milano
Mari G (2019) Libertà nel lavoro. La sfida della rivoluzione digitale, Il Mulino, Bologna
Mari G (2020) La "fine del lavoro" e il lavoro come atto linguistico. Lavoro Diritti Europa 2:2–9
De Masi D (2002) Ozio creativo. Conversazioni con Maria Serena Palieri, Bur, Milano
De Masi D (2016) Una semplice rivoluzione. Lavoro, ozio, creatività: nuove rotte per una società smarrita, Rizzoli, Milano
Mazzucato M (2014) Lo Stato innovatore, Laterza, Roma-Bar
McAfee A, Brynjolfsson E (2017) Machine platform crow: harnessing our digital future. Norton & Company, New York
McKinsey & Company (2017) A future that works: automation, employment and productivity. McKinsey Global Institute, pp 1–137

McKinsey & Company (2023) The economic potential of AI. The next productivity frontier, McKinsey Global Institute, pp 1–68
Meyer H (2016) Five filters moderate the technological revolution. Soc Eur
Milanovic B (2016) Robotics or fascination with anthropomorphism Soc Eur
Montesi C (2016) Il Paradigma dell'economia civile. Radici storiche e nuovi orizzonti, Umbria Volontariato Edizioni, Terni
Montesi C (2022) The promethean fate of economy: will hydrogen really be prometheus III's new gift to humanity? In: Pellat G, Jovan Zafiroski J, Šuplata M (eds) (2022) Cooperation and enlargement: two challenges to be addressed in the european projects, Springer, pp 325–346
Nodari F (2018) "Introduzione" a Zamagni S (2018) Come e quanto la quarta rivoluzione industriale ci sta "toccando", Mimesis, Sesto San Giovanni, pp 5–23
OECD (2023a) Undestanding and applying the precautionary principle in the energy transition. OECD Publishing, Paris
OECD (2023b) The supply, demand and characteristics of the IA workforce across OECD countries. In: OECD social, employment and migration working papers, n 287
OECD (2023c) Artificial intelligence in science: challenges. OECD Publishing, Paris, Opportunities and the Future of Research
Pennacchi L (2010) Lavoro, costituzionalizzazione della persona, istituzioni economiche nella Costituzione italiana. In: Pinelli C, Treu T (2010) La Costituzione economica: Italia, Europa, Il Mulino, Bologna, pp 39–55
Pennacchi L (2018) Lavoro e innovazione per un nuovo umanesimo. In: Pennacchi L, Sanna R (2018) Lavoro ed innovazione per riformare il capitalismo, Ediesse, Roma, pp 29–69
Polanyi K (2010) La grande trasformazione. Le origini economiche e politiche della nostra epoca, Einaudi, Torino (ed.or. 1944)
Rifkin J (2015) The zero marginal cost society: the internet of things, the collaborative commons, and the eclipse of capitalism. St. Martin's Griffin, New York
Rifkin J (1995) La fine del lavoro, il declino della forza lavoro globale e l'avvento dell'era post-mercato, Baldini & Castoldi, Milano
Roubini N (2023) La grande catastrofe. Dieci minacce per il nostro futuro e le strategie per sopravvivere, Feltrinelli, Milano
Schumpeter JA (1942) Capitalism, socialism, and democracy. Harper & Brothers, New York (tr. it.: Schumpeter JA (1955) Capitalismo, socialismo, democrazia, Edizioni di comunità, Milano)
Schwab K (2016) La quarta rivoluzione industriale, Franco Angeli, Milano
Srnicek N (2017) Capitalismo digitale. Luiss University Press, Roma
Summer LH (2014) Lawrence H. Summers on the economic challenge of the future: jobs. online.wallstreet journal.com. 7th July 2014
Susskind D (2022) Un mondo senza lavoro. Come rispondere alla disoccupazione tecnologica, Bompiani, Milano
Varoufakis Y (2017) Taxing Robots won't work. www.weforum.org
World Economic Forum (2023) Future of jobs report, Geneva
Zamagni S (2018) Come e quanto la quarta rivoluzione industriale ci sta "toccando". Mimesis, Sesto San Giovanni (Milano)
Zuboff S (2019) Il capitalismo della sorveglianza. Luiss University Press, Il futuro dell'umanità nell'era dei nuovi poteri

Human Resources Management During and After the Pandemic Crisis

Ivana Kuráková, Ružena Lukášová, and Jana Marasová

Abstract This communication extends and deepens our article presented at the ERECO PGV conference in Porto in 2021. The object of our study consists of changes in human resources management during and after the pandemic crisis. The problem is addressed on the one hand from the point of view of the changes made in management itself and on the other hand, from the perspective of the difficulties that companies have had in switching to digital mode of operation and remote working. The analysis in question is based on our own survey of companies in the two European countries, as well as other existing and available surveys on this subject. Apart from the level of digital capabilities of employees which are of primary interest to us, and which have been linked to threats of a digital divide, we also examine the impact that remote work can have on employee performance and consequently, on those of companies. The content of the last part of our communication is devoted to the lessons that companies have learned from the pandemic crisis in terms of human resources management, based on the challenges that these represent in the current disrupted and uncertain context.

Keywords Crisis · Czech and Slovak companies · Human resources · Mangers · Digital skills · Adapting employees

I. Kuráková · J. Marasová (✉)
Faculty of Economics, Matej Bel University, Banská Bystrica, Slovakia
e-mail: jana.marasova@umb.sk

I. Kuráková
e-mail: ivana.kurakova@umb.sk

R. Lukášová
Faculty of Economics and Administration, Masaryk University, Brno, Czechia
e-mail: Ruzena.Lukasova@econ.muni.cz

G. Pellat et al. (eds.), *Data-Centric Business and Applications*, Lecture Notes on Data Engineering and Communications Technologies 258,
https://doi.org/10.1007/978-3-031-95068-1_16

1 Introduction

After a long period of relatively stable economic development, the economy of the first decades of the new millennium was marked by several crises. Their consequences were, and still are, reflected in the macroeconomic indicators of nation-states and their integration groups, but also in the corporate environment. Business operations and management were primarily affected by the consequences of the pandemic crisis in 2020–2021, as it directly affected the key production factor—human resources, which set all other factors in motion. The Covid-19 virus affected the health of many employees and managers, and excluded them from their normal working lives for some time. In addition, as part of the preventive measures taken by governments against the spread of the virus, companies were forced to stop their activities, in several cases causing long-term problems in supply chains, some of which persist to this day. From the outset, the pandemic crisis also presented a challenge in terms of increasing employees' digital capabilities and skills, as well as the necessary technological equipment. The content of our article focuses on how human resource management in companies has addressed the following questions: how the content of people management activities and processes has changed on the one hand, and to what extent these changes have helped employees adapt to new conditions and requirements in the area of digital skills on the other hand.

2 Nature, Causes and Consequences of the Crisis

The theoretical level reveals a number of difficulties in all scientific disciplines concerning the definition of crisis, the multiplicity of its causes and characteristics, and its consequences on the environment in which it manifests itself. According to Tardy (2009), the term crisis represents a situation of anomie caused by a certain change, as opposed to what is considered normal. A given situation becomes a crisis if it lasts a certain length of time, and if the standard regulatory tools put in place prove insufficient. In other words, it's when the relevant organizations fail to restore normality. Such a definition of crisis is automatically linked to the existence of a regulatory mechanism, or even to the more or less obvious assumption of intervention. Borraz (2020), relying on the Greek term krisis, perceives crisis as an unpredictable turning point, but the social sciences attribute to it broader characteristics such as a loss of meaning, a disappearance of organizational boundaries or a complex, urgent and dynamic character, as the crisis situation evolves and transforms.

In principle, a crisis thus evokes an undesirable and unfavorable situation in the personal or professional life of an individual or certain groups of individuals, in organizations, in the life of society or even of a country as a whole, but also in the functioning of global organizations. Although the definitions of crisis from the point of view of the various scientific disciplines differ in terms of causes, manifestations and consequences, what they have in common is that the negative nature of the crisis generally also means a new beginning in the sense of taking on the necessary changes, innovative procedures, evolution of practices, methods and tools used. Humanity has always overcome crises, whether religious, moral, political or economic. Since the advent of the capitalist system of production and consumption, crises have been a natural part of this (Attali 2009).

2.1 *The Economic and Managerial Perspective of the Crisis*

It is generally known that economic activity in a free-market system develops in cycles that result from shocks to supply or demand, such as an increase in the prices of production factors, variability in investments, a global demand for purchasing power that is insufficient in relation to the existing global supply (endogenous causes), political influences, war conflicts or climatic changes, or changes in government economic policy (exogenous causes). According to economists, an economic crisis is a state of imbalance in the economy (Muchová and Lisý 2009; Denape 2021), caused by a certain event or sequence of events.

In the economic sense, a crisis is therefore defined as a turning point in the business cycle, when there is a mismatch between the production of goods and services, i.e. their supply and the demand for them, which manifests itself in a slowdown in economic activity and, as a general rule, an increase in unemployment, and which simultaneously reveals certain malfunctions in production and demand. Explanations for the latter traditionally depend on existing schools of economic thought. Liberals believe that crisis is a natural part of economic life, and should be respected. Its consequences can be offset by the reserve resources the economy has created during periods of growth. Representatives of the interventionist school of economic thought recommend mitigating the crisis through state economic policy, even at the cost of excessive indebtedness, which becomes dangerous (Attali 2009; Delaneuville and Sakáčová 2024). This approach is applied today by most developed economies, leading to a disproportionate increase in the quantity of money in the economy and becoming the main cause of inflationary pressures.

The cyclical development of the economy is largely associated with technological progress, which renders certain sectors inefficient, meaning that they stop making a profit. As a result, they lose their appeal to investors, who stop investing further capital in them, and it becomes necessary to replace them with other sectors. Supporters of Schumpeter's theory believe that the cause of crises lies precisely in the economy's difficulty in adapting to technological development. According to Barta (2009), crisis may be a painful consequence of certain government decisions in the past, but it nevertheless remains a natural part of our lives. In this context, the author wonders whether it is really necessary to fight against this aspect of the natural economic cycle, and whether its consequences cannot be seen as positive. Whether the ideas of classical economists on the importance of private property, of the right to decide on the use of one's own resources, of the belief in the creativity of the human spirit, which should not be unnecessarily limited by government regulation, whether these principles, the application of which has historically led to periods of economic prosperity, should not be restored even if today they are described as outdated, or frozen, dogmas.

Combalbert (2012) characterizes the managerial vision of crisis as a situation that goes beyond the usual framework of known incidents, with the need to make strategic decisions quickly. The challenges associated with the situation seem vast and multifaceted, and generally only become apparent after some time has elapsed. Roux-Dufort (2003) sees a crisis as a process that begins with a certain triggering event, and then reveals a whole series of dysfunctions. Boumrar (2010) also sees a crisis as a process, in a global context, whose phases can be examined separately, whereas it is necessary to preserve the whole or a global vision of the issue. According to the author, the process in question may have a low probability but serious consequences, and may lead to significant organizational changes, positive or negative, depending on the context, as well as on how the crisis is managed, since the interpretation of the actors managing the crisis is closely related to both these factors. As an example of the positive consequences of the crisis during the Covid-19 pandemic, on which our research focuses, we mention the findings of PWC (Press Release 2021), according to which up to 40% of employees out of 32,500 in 19 countries declared that their digital skills had improved during the crisis.

The causes of the crisis are manifold, from both economic and managerial point of view. Among the most common are managers' persistence in maintaining tried-and-tested practices and their reluctance to change them, or unexpected events threatening businesses and the associated need for managers to take urgent decisions, or an accumulation of sudden events that disrupt the standard course of processes and activities, and create an atmosphere of great uncertainty in companies. From a managerial point of view, the crisis situation in the company is often directly linked to the general economic crisis, but this is not always the case. According to Černaj (2015), a crisis is a stage in a company's life cycle, during which unfavorable developments in performance potential, market value, liquidity and other indicators manifest themselves, resulting in the realization of losses and long-term insolvency. Preventive measures include the development and ongoing updating of a crisis plan, while repressive crisis management tools are aimed at identifying potential risks as possible sources of a crisis, and ensuring that managers react appropriately in the event of a crisis, so that it can be brought under control and its consequences minimized.

2.2 The Impact of the Crisis on Human Resources Management

Every economic crisis, including the current one triggered by the Covid-19 pandemic and subsequently amplified by the war of invasion in Ukraine, has sooner or later affected many people in their social and professional lives. Thus, on the one hand, the crisis directly affects the management of people in the work process, whether in terms of employment policies, work organization, employee training, remuneration, etc., and on the other hand, it reveals the accumulation of unmet expectations as well as employee frustration. Based on the well-known definitions of human resource management (HRM), according to which its mission is to ensure a match between the need for the workforce in terms of the desired structure of qualifications in relation to the purpose of the company's activities, Thévenet (2009) argues that in times of crisis, this mission can be summarized in three fundamental facts. These are:

- interruption, which must be managed in terms of employment, e.g. by laying off temporary staff, consolidating operations and activities, re-evaluating budgetary procedures or the way in which established GRM processes are secured;
- maintenance, consisting of sufficient internal cohesion in companies weakened by a depressed context of crisis, mistrust and uncertainty on the part of employees and customers, where the responsibility of HR management is to maintain a favorable collective climate and the necessary motivation of individuals;
- preparation, which applies to the post-crisis period, even if we don't yet know when and how it will manifest itself: according to the author quoted, this preparation should be guided by two principles—involvement, i.e. the cooperation of all members of the company on the basis of cohesion and reciprocity, and innovation, focusing mainly on ways and methods of managing people.

Referring to the current evolution of the human resources function, several authors (Frimousse and Peretti 2020; Galambaud 2014; Lelarge 2014; Peretti 2017) claim that it has been in crisis for a long time. The latest stage of its evolution, in the form of HRM, lasts around 40 years, whereas the duration of the previous stages (personnel management and personnel management under the influence of the human relations school) was from 15 to 20 years. Since the concept of HRM has not yet been transformed into a new stage in which this function would acquire a new formal identity and a new professional ambition, it would seem that it is no longer in a position to respond adequately to people management needs in the context of changes in the economic and social environments, which have been particularly significant since the start of the new millennium. This fact is also pointed out by the entrepreneurial community itself (FoxRH 2016). The criticisms that need to be taken seriously relate to several facts including: societal change and employee expectations are not sufficiently taken into account in companies, HR managers are subject to fashion trends, imitations or even snobbery, says Storhaye (In Barabel 2017), and often act against their own ethics and values. Managers also struggle to think proactively, autonomously and creatively about the future of the HR function.

Réale (2018) argues that the existing problems of HR management are linked to the metamorphosis of the business context, which has entered the VUTCA phase: it has become volatile—V, uncertain—U, turbulent—T, complex—C and ambiguous—A. Nowadays, contradiction and uncertainty are more the rule than the exception. Digital technologies have affected all sectors of economic activity, and the recent pandemic crisis has made this impact even stronger. Increasingly, work is carried out in virtual and project-based teams, new technologies enable tasks to be performed more or less outside the standard workplace, the abstract nature of work is reinforced, and employees' expectations of their own professional and personal development are heightened. Against this backdrop, the question arises as to whether the universal mode of HR management, whose roots go back to the 1980s, can still accompany, anticipate or react appropriately to these changes, and whether it is capable of ensuring that companies have enough flexible and adaptable employees, even under crisis conditions. According to the cited author, the new mission of HRM will probably be to create and develop a portfolio of talents involved in joint projects. Perhaps the search for and management of talent will become part of HRM's new identity and professional ambition, and their realization will restore the lost dynamism of the HR function, and the impact of the pandemic crisis will also contribute to this.

HRM in companies, which is essentially about managing people in their working lives, has been directly affected by the crisis in question, as it has considerably destabilized the already unstable business environment of recent years, as a result of the process of increasing globalization, the development of the knowledge economy, the management of change, but also as a consequence of unfavorable demographic trends in terms of skilled labor requirements. In direct proportion to the emergence of the pandemic, the need for companies to take swift and effective action has increased, particularly in the areas of work organization, health and safety, and employee training in digital skills. Several authors have already defined this term at the start of the Internet era. According to De Haan (2004) or Hargittai (2002), it refers to the individual's ability to use ICT and to use it to obtain information quickly and efficiently. Since the onset of the pandemic crisis, but also in its aftermath, companies are, among other things, increasingly confronted with a mismatch between their employees' existing digital skills and those they would actually need to implement digital transformation.

3 Impact of the Pandemic Crisis on Human Resources Management Activities

First in this section, we present the methodology of our study, which was carried out as part of VEGA project no. 1/0668/20 "Digital inequality and digital exclusion as a challenge for human resources management in the enterprise". We then discuss the most relevant findings of the survey.

3.1 Research Methodology

The aim of our research was to identify the consequences of the coronavirus pandemic crisis on HRM in companies, from two points of view: on the one hand, in terms of the changes that had to be made in the work of human resources departments themselves, and on the other hand, in terms of employees' adaptation to new ways

Table 1 Sample of companies by size

Companies	Slovakia (SK)		Czech Republic (CZ)	
	Number	%	Number	%
Averages	132	67	101	58
Large	66	33	73	42
Total	198	100	174	100

of working, developing in connection with employers' increased demand for digital skills.

The main information resource for our survey was our own data collection via an online questionnaire, where we formulated questions based on several available research studies. Our research was carried out on a sample of medium-sized and large companies in the Slovak Republic (SK) and the Czech Republic (CZ), two European countries with a similar historical and cultural background. During the research period, the years 2021–2023, according to information obtained from the statistical offices of both countries, there were 2,944 companies in the Slovak Republic, of which 2,326 were medium-sized and 619 large. In the Czech Republic, there were 9,498 companies, of which 7,587 were medium-sized and 1,911 large. As for the list of Slovak companies, we sent the questionnaire to all of them, while for Czech companies, we first generated the 5,540 companies (4,440 medium-sized and 1,100 large) by a simple random selection, and then contacted them. We contacted all these companies by e-mail to ask them to take part in the study. The e-mails were sent to the heads of HRM departments, or to the managers responsible for managing them in companies where there was no HRM department.

Only 198 of the Slovak companies (out of 2,944 contacted) returned a completed questionnaire, representing a return rate of 6.7%. Of the Czech companies contacted (5,540), we received 174 completed questionnaires, so the return rate was only 3.1%. Our total sample thus comprised 372 respondents. After processing the questionnaire responses using mathematical and statistical methods, we used some of the data and information obtained to calculate Pearson's correlation coefficient, using the SPSS program, which then enabled us to detect some relevant correlations, or the absence of correlations. The results are presented in several tables and graphs. Table 1 presents the composition of our sample in terms of company size, and Table 2 according to the NACE nomenclature of economic activities.

3.2 The Crisis and Post-crisis Period in the Work of HR Managers

To the first question Did the content of HR department activities have to be adapted to the new situation linked to the crisis caused by the Covid-19 pandemic? 177 Slovak companies (89%) and 149 Czech companies (86%) answered "Yes". The responses

Table 2 Company structure by NACE

Activity	Slovakia		Czech Republic	
	Number	%	Number	%
Agriculture. forestry and fishing	7	3.54	9	5.17
Mining and quarrying	3	1.52	2	1.15
Manufacturing industry	66	33.33	52	29.89
Electricity, gas, steam and air conditioning production and distribution	4	2.02	5	2.87
Water production and distribution, wastewater treatment, waste management and pollution control	4	2.02		0.00
Construction	12	6.06	14	8.05
Trade: repair of automobiles and motorcycles	30	15.15	8	4.60
Transport and warehousing	13	6.57	13	7.47
Accommodation and catering		0.00	1	0.57
Information and communication	7	3.54	2	1.15
Financial and insurance activities	5	2.53	2	1.15
Real estate activities		0.00	1	0.57
Specialized, scientific and technical activities	3	1.52	5	2.87
Administrative and support service activities	6	3.03	1	0.57
Public administration	8	4.04	14	8.05
Arts, entertainment and recreation	2	1.01	3	1.72
Other service activities	28	14.14	42	24.14
Total	198	100.00	174	100.00

therefore show that in more than 4/5 of the sample companies in both countries (198 SK and 174 CZ), the Covid pandemic had an impact on the content of HRM services. Managers from companies that answered "yes" to this first question were then asked to respond to this one: In your opinion, what are the most significant changes in the content of HR department activities? They could indicate several of the answer variants presented. Table 3 documents the number of "yes" responses for each change in HR department work content listed in the questionnaires.

Table 3 shows that in companies in both countries, the most frequently cited change in HR department activities associated with the Covid-19 pandemic was the implementation of activities related to the need to ensure employee health protection. 88% of Slovak companies and up to 97% of Czech companies reported this change.

From 2nd to 5th place, the order in terms of frequency of occurrence of changes implemented during the pandemic in HRM department activities is different. In

Table 3 Frequency of occurrence of changes in the content of HR department activities in surveyed companies (in absolute terms and as a percentage of the number of companies (SK—177, CZ—149)

Changes in HRM activities	Slovak Republic			Czech Republic		
	n°	%	Order	n°	%	Order
Provide additional health protection for employees (supply of protective equipment, organization of work so that employees come into contact with each other as little as possible)	155	88	1	144	97	1
Create new guidelines, work procedures and job descriptions in connection with the implementation of remote working	120	68	2	86	58	3
Crisis management (the need to create crisis teams, crisis plans, their reflection and adaptation to new conditions)	117	66	3	66	44	5
Address labour shortages related to employees' inability to work or care for family members	105	59	4	96	64	2
Equip employees with the right hardware and software for remote working	102	58	5	72	48	4
Coaching and mentoring online management in the new situation	43	24	6	34	23	6
Organization of training/workshops for employees in the field of psycho-hygiene and mental health	28	16	7	19	13	7
Organize digital skills training for employees	23	13	8	14	9	8
Major dismissal	20	11	9	11	7	9

Slovakia, the managers surveyed indicated the creation of new guidelines, work procedures and job descriptions as the 2nd most frequent change, this being linked to the implementation of teleworking (68%), and the 3rd was the implementation of crisis management (66%). In Czech companies, crisis management was the 5th most frequently mentioned activity (44%) requiring time and attention from HR departments. In the 2nd position in the Czech ranking was the search for solutions to the shortage of workforce caused by covid disease (64%), and in the 3rd position was the development of new guidelines, work procedures and job descriptions relating to remote working (58%). In Slovakia, the problem of workforce shortages due to illness of employees or their family members, which had to be solved by HRM, came only in the 4th position (59%). The need to equip employees with appropriate hardware and software to enable them to work remotely, or even from home, came in the 5th (58%) in the sample of Slovak companies, and the 4th (48%) in the sample of Czech companies.

As for the other changes in the content of HRM department activities indicated by respondents, their order was identical in terms of frequency of appearance in the Slovak and Czech Republics—in 6th place is coaching and mentoring of middle managers (e-managers), in 7th place is the organization of training (workshops) focusing on psycho-hygiene and mental health, aimed at employees. In 8th place is the provision of training in digital skills, and in last place is the major dismissal of employees. We regard the latter result as positive, in that neither the Slovak nor the Czech companies surveyed had to make any major redundancies as a result of the pandemic. This suggests that managers were looking for every possible way to maintain employment levels during the pandemic.

A closer comparison of the frequency rate of different changes in the activities of companies' HR departments clearly shows that in the vast majority of Slovak companies, this rate was slightly higher than in Czech companies. The only exception is the change listed in Table 3 as the first, concerning additional employee health protection, where the frequency of occurrence was higher in the Czech Republic (97% vs. 88% in Slovakia). We then identified the most significant differences in the frequency of occurrence of the changes examined in the following three cases—the implementation of crisis management (22% more in Slovakia), the creation of new guidelines, work procedures, job descriptions (10%) and the provision of technological equipment enabling employees to work remotely (10%). Other differences were minimal.

As for the case of crisis management, where the need for its implementation proved to be up to 22% higher in Slovak than in Czech companies, we can deduce that the standard way of managing human resources in the Czech business environment is more stable, or more resistant to economic turbulence than in Slovakia. However, we must be cautious in generalizing the results of our survey, as the two company samples were not sufficiently representative in relation to the number of companies in the two countries. Nevertheless, they are revealing, all the more so as no other similar comparative study between Slovak and Czech companies has so far been carried out on the subject in question.

The Covid-19 pandemic forced managers and employees in many companies to work from home, insofar as the nature of their work allowed. In order to obtain information on the rate of activities carried out digitally, i.e. remotely, we asked our respondents the question What HR activities do you currently consider to be carried out in your company digitally (remotely), including by telephone? They had a choice of five activities listed in Table 4, and could tick more than one. The latter shows the results of responses to the question asked, expressed as a percentage of the company samples in both countries (198 SK and 174 CZ).

Although the percentage of different activities differs to some extent in the responses of Slovak and Czech managers, the order of activities is the same. The largest number of HR managers identified employee search and recruitment as the activity they carry out and probably continue to carry out remotely (in SK 69%

Table 4 HRM activities carried out in digital companies

	Share of Slovak companies (%)	Share of Czech companies (%)
Employee search and recruitment	69	75
Employee training	66	52
Employee selection	47	27
Employee retention activities	36	25
Employee evaluation	29	21

and in CZ 75% of companies). On the other hand, the lowest representation, as an activity carried out digitally, had in both countries employee evaluation (29% SK and 21% CZ). In connection with the above responses, we consider it important to emphasize that it is not possible to state unequivocally that the percentages of companies performing the indicated activities digitally are entirely the result of the pandemic crisis. It is possible that some of these activities were already being carried out digitally prior to the crisis. But in view of the message accompanying our request to complete the questionnaire, as well as the introductory explanations in the questionnaire itself, we conclude that managers answered this question in relation to the consequences of the crisis. What's more, there is no comparable research on the digital way of carrying out HRM activities prior to the pandemic crisis to verify this.

3.3 *Company Difficulties in Adapting Employees to Increased Digital Skills Requirements*

The Covid-19 pandemic crisis, as is already generally known, has considerably accelerated the pace of implementation and use of new technologies and the digitization of work processes. The growing level of digitization means new demands on the digital skills of employees in companies, and thus new challenges for HRM. To find out whether and to what extent employees have been able to meet the expected requirements, we asked our respondents to estimate What percentage of employees in your company were unable to adapt to the increased digital skills requirements during the crisis? Table 5 shows the results of Slovak and Czech company managers' answers to this question, both in absolute and relative terms. We have selected the most significant results.

Despite our expectations that the percentage of employees who were unable to adapt to the crisis conditions in terms of digital skills would be high, the survey showed that, according to the estimates of Slovak (30 companies) and Czech (34 companies) managers, the highest rate of non-adaptable employees was just 5%. The second largest group of managers (21 Slovak and 12 Czech companies) estimated that 10% of their employees were unable to adapt to the new conditions. The share

Table 5 Share of companies in the Slovak (198) and Czech (174) samples and number of employees not adaptable to digital working methods

Number of employees who have been unable to adapt to increased digital skills requirements (%)	Number of managers expressing the estimate			
	Slovak Republic		Czech Republic	
	n°	%	n°	%
1	14	7	11	6
2	10	5	6	3
3	1	1	3	2
5	30	15	34	20
10	21	11	12	7
15	3	2	4	2
20	7	4	11	6
30	3	2	4	2
35	0	0	1	1
50	3	2	1	1
60	0	0	1	1
80	0	0	1	1
90	1	1	0	0

of employees unable to adapt to increased demand for digital skills, at over 30%, appears only exceptionally in our respondents' answers. However, it is not possible to deduce unequivocally that this low percentage of employees with insufficient digital skills could mean a fully satisfactory level of this type of skill in companies. The responses of the managers questioned only show how they themselves perceived the adaptation of employees to the increased demands for digital capabilities during the pandemic crisis. The total number of Slovak companies in which managers felt some degree of employee inadequacy in terms of necessary digital skills was 93 (out of 198), and the total number of Czech companies was 89 (out of 174). The results show that in 105 companies in Slovakia and 75 companies in the Czech Republic, adapting employees to digital skills was not a problem.

We then wanted to know whether, under the influence of the pandemic crisis, companies had started to train their employees in digital skills development. Management responses showed that 32 Slovak and 24 Czech companies had done so. Compared with the number of companies reporting a certain percentage of non-adaptable employees (93 SK and 89 CZ), we can see that employees have been trained in around a third of companies in Slovakia and in around a quarter in the Czech Republic. However, this does not mean that the companies that trained their employees were those that, according to our respondents, had employees unable to adapt to the growing demand for digital skills. In this context, we found that Slovak companies in which the estimate of unadaptable employees at the level of

5% (30 companies) and 10% (21 companies) implemented training for them to a small extent. There were only 4 companies out of 30 (5% level of unsuitability) and 6 companies out of 21 (10% level). The same was true of Czech companies, where for 5% of employees unable to adapt to new conditions (34 companies), training was organized in only 5 of them, and for 10% of unadaptable employees (12 companies), this was the case in only 4 of them. Overall, we can conclude that, when it comes to employee training in digital skills, we expected a higher proportion of companies to be involved, even a year or two after the start of the Covid-19 pandemic. Despite the fact that our survey in the Czech Republic took place a year later than in Slovakia, this was not reflected in the intensity of training at Czech companies.

We were also interested in a potential correlation between ICT companies and the provision of e-skills training. Their number in our sample was minimal—there were only 7 Slovak and 2 Czech companies, which could not significantly affect the results from the point of view of training provision. We did, however, check their situation: none of these 9 companies needed to organize digital skills training, which is what we expected.

In relation to the insufficient level of digital capabilities of their employees, we also looked at what obstacles to the development of digital capabilities our manager respondents perceived. They could indicate several of these obstacles. It seems to us that analysing these factors is important, as they are directly linked to the pandemic crisis and to managers' decisions regarding the following training of employees. As Table 6 shows, in the company samples from both countries, digital illiteracy was considered the main obstacle to the development of digital skills (30% in Slovakia and 39% in the Czech Republic). We explained in the questionnaire that this illiteracy represents a lack of basic digital experience, often linked to an unwillingness to acquire a computer and overcome this barrier by working with it.

Table 6 also shows that, based on the frequency of responses from surveyed managers, mental barriers, the root of which lies in the individual's lack of awareness of the need for digital skills today, are the second most serious obstacle to the development of digital skills. This result is rather surprising, given that the problem of mental barriers concerns up to around a quarter of the Slovak and Czech companies

Table 6 Barriers to the development of digital skills

Types of barriers	Slovak Republic		Czech Republic	
	n°	%	n°	%
Digital illiteracy	60	30	67	39
Mental barriers	53	27	44	25
Motivation barriers	45	23	38	22
No barriers	23	11	13	8
Material barriers	12	6	9	5
No answer	6	3	2	1

surveyed. On the contrary, a positive result in this direction is the fact that motivational barriers, consisting in the lack of motivation of individuals to develop their digital skills, only come in third place (23% out of 198 SK and 22% out of 174 CZ). This partially confirms what other available studies show (PwC 2021) that the pandemic crisis has had a positive impact on the one hand on stimulating company management to digitize as many processes and activities as possible, and on the other hand on the motivation and even willingness of individuals to improve their digital skills, insofar as the increasing level of the latter increases their own employability on the labour market. Another positive result of our research findings concerns material obstacles to the development of digital skills, as these are only a minor problem for Slovak and Czech companies (6% in SK and 5% in CZ).

Barriers to the development of employees' digital skills are more or less connected to the level of these skills of job seekers on the labor market. Therefore, we asked our managers who, in their opinion, are job seekers who are disadvantaged from this perspective. Their answers confirmed the results of several studies, the authors of which (Ertl et al. 2022; Hargittai 2002) claim that an individual's age is a key factor in their knowledge, or even their digital skills. As shown in Table 7, up to 79% of Slovak managers and 69% of their Czech counterparts are convinced that the most disadvantaged job seekers due to their insufficient digital skills are those aged 50 and over. The importance of this factor is obvious for this category of people.

The responses of the interviewed managers also showed that gender, which was considered an important factor influencing the digital skills of individuals until the advent of ITC at the beginning of the new millennium (Ono and Zavodny 2002; Campos-Castillo 2014), has almost no influence. Only in 8 cases from Slovak companies and 5 cases from Czech companies did the respondents identify gender as a determinant of the unsatisfactory digital skills of job seekers.

Our respondents, both Slovaks (57%) and Czechs (46%), consider long-term unemployment of job applicants to be the second most important factor influencing the level of their digital skills.

Table 7 also shows that according to both groups of managers, the third factor of a disadvantageous position of job seekers is their language barriers, but unlike the two previous factors, in the eyes of Czech managers these are more of an obstacle to the development of digital skills (34%) than in the eyes of Slovak managers (24%). Both groups of respondents consider language difficulties to be serious, especially

Table 7 Jobseekers are disadvantaged in the labour market due to inadequate digital skills

Types of disadvantages	Slovak Republic		Czech Republic	
	n°	%	n°	%
People over 50 years of age	60	79	67	69
Long-term unemployed	53	57	44	46
People with language barriers	45	24	38	34
High school graduates	23	10	13	3
University graduates	12	2	9	1

Table 8 HR changes made in companies during the pandemic crisis due to employees' lack of digital skills

Types of changes	Slovak Republic		Czech Republic	
	n°	%	n°	%
Dismissal	15	8	14	8
Reassigning employees to other positions	26	13	24	14
No change	162	82	133	76

for foreigners applying for jobs in our countries, but also for members of national minorities. In fourth place on the list is the fact that even secondary education can be a cause of insufficient digital skills.

The aim of the question What personnel changes were you forced to make during the coronavirus crisis due to the lack of digital skills among your company's employees? was to find out whether and to what extent the problem of employees' insufficient level of digital skills forced companies to lay them off or transfer them to other positions. Respondents were allowed to tick more than one answer. The processed results showed no significant dependency between the variables mentioned (Pearson correlation had a value of 0.0615 for Slovak companies and 0.3153 for Czech companies). The responses to the question showed that, according to the vast majority of managers surveyed, no changes had been made to human resources due to insufficient digital skills (Table 8). In Slovakia, 162 out of a total of 198 respondents answered this way, and in the Czech Republic there were 133 out of 174.

HR managers had to transfer employees to another position due to insufficient digital skills in 26 companies in Slovakia and in 24 companies in the Czech Republic. On a positive note, only in 15 companies in Slovakia and 14 companies in the Czech Republic did HR managers have to dismiss employees due to insufficient digital skills. The result means that company managers were able to solve the problem of employees' insufficient adaptability to the demands of digital skills in a way other than by reassigning or dismissing them.

4 Conclusion

Every crisis, to one extent or another, destabilizes economic activity and stalls its driving force—companies. It is above all the brain of the companies, human resources, that is affected, which justifies the attention paid to their management in crisis situations. By having a direct impact on the human factor in companies, both in terms of people's physical and mental health, the pandemic crisis has represented a considerable challenge for managers from the outset, including human resources managers. Based on the many existing analyses on this subject, we have analysed in this paper how Slovak and Czech companies have faced to the challenge. In line with the defined objective, the first part of our study focused on identifying the changes in the work of human resources departments that were imposed on them by the crisis, and then we concentrated on the problem of adapting employees to new ways of working, relating mainly to employees' digital skills. With regard to the first part, the results show that HR managers had to deal with issues such as employee health protection, a regulatory framework for teleworking, and labour shortages. In terms of digital skills, according to manager respondents, the highest occurrence rates of employees not adaptable to the demands of this type of skill were 5 and 10%, which can be considered a standard level. This findings are consistent with those of Horehájová and Lacová (2022). Only few companies have had to lay people off due to their inability to adapt to the new digital environment. However, digital illiteracy and mental barriers represent a serious problem in the companies studied. Our analysis has confirmed that the pandemic crisis has affected the way companies operate, as well as the content and working methods of their HRM departments, in a number of ways. Remote working has strengthened its position in companies, it helps to motivate employees to invest in improving their digital knowledge and skills.

Acknowledgements Supported by the grant VEGA 1/0668/20 "Digital Inequality and Digital Exclusion as a Challenge for Human Resources Management" and VEGA1/0029/25 "Specifics of human resource management in Slovak companies related to Industry 4.0" of the Research Agency of the Ministry of Education, Research, Development and Youth of the Slovak Republic

References

Attali J (2009) La crise et après? Fayard, Paris
Barabel M (2017) Pour une fonction RH inspirante: Une réponse au RH Bashing. Entreprise&Carrières, Paris
Barta J (2009) Na čo je kríza dobrá? Týždeň, Február, 28
Borraz O (2020) What is a crisis? SciencesPo, Centre de sociologie des organisations. Dossier Les sciences sociales en temps de crise: le Covid 19. https://www.sciencespo.fr/cso/fr/content/qu-est-ce-qu-une-crise.html. Accessed 30 Aug 2023
Boumrar J (2010) La crise: levier stratégique d'apprentissage organisationnel. Vie & Sciences De L'entreprise 3:13–26. https://doi.org/10.3917/vse.185.0013
Boustil A (2016) Climate in the HR function: issues and paradoxes. FoxRH. https://www.slideshare.net/AzizBoustil/le-climat-dans-la-fonction-rh-enjeux-et-paradoxes. Accessed 26 June 2024
Campos-Castillo C (2014) Revisiting the first-level digital divide in the United States: gender and race/ethnicity patterns, 2007–2012. Soc Sci Comput Rev 33(4):423–439. https://doi.org/10.1177/0894439314547617
Černaj T (2015) Krízový manažment. https://www.euroekonom.sk/krizovy-manazment/. Accessed 12 Dec 2022
Combalbert L (2012) Le management des situations de crise. ESF Editeur, Paris
Delaneuville F, Sakáčová A (2024) Building bridges to employment: EU actions to facilitate the professional integration of graduates. CESGED. https://cesged.com/. Accessed 12 Dec 2024
Denape T (2021) Comment les économistes expliquent-ils les crises? Economie Politique 802(803):67–71
De Haan J (2004) A multifaced dynamic model of the digital divide. IT Soc 1:66–88

Ertl MM, Trapp SK, Alzueta E, Baker FC, Perrin PB, Caffarra S, Yüksel D, Ramos-Usuga D, Arango-Lasprilla JC (2022) Trauma-related distress during the COVID-19 pandemic in 59 countries. Counsel Psychol 50(3):306–334. https://doi.org/10.1177/00110000211068112
Frimousse S, Peretti JM (2020) Les répercussions durables de la crise sur le management. Quest Manag 28:159–243. https://doi.org/10.3917/qdm.202.0159
Galambaud B (2014) Réinventer le management des ressources humaines. Editions Liaisons, Paris
Hargittai E (2002) Second level digital divide: differences in people's online skills. First Monday 7.https://doi.org/10.5210/fm.v7i4.942
Horehájová M, Lacová Ž (2022) Mapping the skills needed for advanced technologies use in Europe. In: Proceedings of international scientific conference Economic and societal challenges of the European economy. Silesian University, School of Business Administration, vyd. – Karviná, pp 116–125. ISBN 978-80-7510-527-1
Lelarge G (2014) Gérer les ressources humaines: finalités actions outils. Arnaud Franel, Paris
Muchová E, Lisý P (2009) Fiškálna politika v hospodárskej a menovej únii. Iura Edition, Bratislava
Ono H, Zavodny M (2002) Race, internet usage, and E-commerce. Rev Black Polit Econ 30(3):7–22. https://doi.org/10.1007/BF02717319 (Original work published 2002)
Peretti J-M (2017) Ressources humaines. Vuibert, Paris
Pwc (2021) Enquête sur les espoirs et les craintes 2021. Pwc—Press release. https://www.pwc.fr/fr/espace-presse/communiques-de-presse/2021/mars/hopes-and-fears-report.html. Accessed 24 Sept 2024
Réale Y (2018) De la GRH au management stratégique des RH. Groupe Eyrolles, Paris
Roux-Dufor C (2003) Gérer et décider en situation de crise. Outils de diagnostic, de prévention et de décision. Dunod, Paris
Tardy T (2009) Introduction à la gestion de crise. In: Gestion de crise, maintien et consolidation de la paix: Acteurs, activités, défis. De Boeck Supérieur, Louvain-la-Neuve, pp 13–34
Thévenet M (2009) Crises et GRH. Revue française de gestion 193:37–41. https://www.cairn.info/revue--2009-3-page-37.htm

AI Adoption by the European Companies: Drivers and Barriers in the EU

Ján Huňady, Žaneta Lacová, and Anna Vallušová

Abstract The use of artificial intelligence (AI) by companies seems to be inevitable to get additional gains in productivity, efficiency, and in market shares. However, the differences among the EU Member States are important. Their gradual reduction is desired and supported by the European Commission Initiatives. In our paper, we try to identify the specific features of companies (size, sector, strategic goals, approach to digital transformation, approach to business model innovation, approach to organization and cultural change, approach to financial and economic leverage, etc.) correlated with the AI adoption in different EU Member States. For this purpose, we use the micro-level data coming from Advanced Technologies for Industry Surveys (ATIS). Our ambition is to identify the drivers and the barriers eventually contributing to the AI adoption across companies in the EU Member States.

Keywords Artificial Intelligence · Companies · European Union

1 Introduction

The crisis usually generates innovation and transformation. As being pointed out by Fuhrmann and Mooney (2021), the COVID-19 pandemic appears to have accelerated the pace of AI adoption by prompting businesses to seek innovations that reduce costs and circumvent the need for co-located employees and physically present customers. In addition, the current crisis of climate change is also catalyzing the digital transformation of our economies. One of the main driving forces of the ongoing digital

J. Huňady · Ž. Lacová · A. Vallušová (✉)
Faculty of Economics, Matej Bel University, Banská Bystrica, Slovakia
e-mail: anna.vallusova@umb.sk

J. Huňady
e-mail: jan.hunady@umb.sk

Ž. Lacová
e-mail: zaneta.lacova@umb.sk

G. Pellat et al. (eds.), *Data-Centric Business and Applications*, Lecture Notes on Data Engineering and Communications Technologies 258,
https://doi.org/10.1007/978-3-031-95068-1_17

transformation comes from the massive adoption of the AI worldwide, while other advances technologies (such as nanotechnology, quantum computing, biotechnology, artificial intelligence, robotics, the Internet of Things, fifth-generation wireless technology, fully autonomous vehicles, 3D printing and so on) are also playing their roles in this transformation. The AI technology is an irreversible trend now. It finds its suitability for different economic sectors (mainly automotive, health care, financial services, transportation and logistics, technology, communication and entertainment, retail, energy, or manufacturing), in different styles of companies management and different business models, in different parts of companies everyday operational functioning and processes (production, marketing and communication, finance etc.). However, the technology is not adopted to the same extent by all companies and in all countries. From the global perspective, the European companies are not leading this AI adoption process, rather it is dominated by the North America and Asian countries (China, Singapore, India etc.) according to data published in the Artificial Intelligence Index Report 2023, as well as IBM Global AI Adoption Index 2022. From the European perspective, the differences among the EU Member States could be identified. The goal of our paper is to identify the gaps in the AI adoption realized by different EU Member States. In the limits of data availability, we focus on different factors which are rising or reducing the probability of the AI use as well as the effects of this use for the companies in the EU.

2 Literature Review

There is an expansionary growth of the scholars who are focusing their research on different aspects of the AI adoption. Besides the stream of research who are concentrating their research on the technology itself, a growing number of authors is now publishing papers about the socio-economic prerequisites as well as consequences of the technology adoption. The latter is underlying the human-centric approach to the AI, and are contributing to the discussion about necessary guidelines and (not exclusively) ethical principles for the AI adoption in our societies—like the White Paper on Artificial Intelligence: A European approach to excellence and trust (Kilian 2020).

Radhakrishnan and Chattopadhyay (2020) propose a comprehensive literature review on the determinants and barriers of the AI adoption. A solid literature overview is needed because different theories, models and frameworks have been used to study technology adoption, some explaining the determinants of adoption at the individual level, some at the organizational level and some at both. The authors have reviewed journal articles and other reports pertaining to AI adoption and understand the adoption theories used and the factors that facilitate and those that hinder AI adoption. Factors influencing adoption at the individual level were related to trust, security, purchase price, intrinsic motivation, social influence, utilitarian benefit whereas at the organizational level, it was related to the technical competencies, strategic road mapping for AI, top management support and the digital maturity of the organization.

Chatterjee et al. (2021) identify the socioenvironmental and technological factors that influence the adoption of artificial intelligence embedded technology by digital manufacturing and production organizations. The results from Indian companies highlight that all the relationships, except organizational readiness, organizational compatibility and partner support on perceived ease of use, were found to be significant in the context of digital manufacturing and production organizations. The results further indicated that leadership support acts as a countable factor to moderate such an adoption.

Van Nguyen and Viet (2021) explore critical factors impacting AI adoption in Vietnam. The findings indicate that technical compatibility, relative advantage, technical complexity, technical capability, managerial capability, organizational readiness, government involvement, market uncertainty, and vendor partnership are significantly related to AI application adoption. Interestingly, the study results indicated that the relationship between organization size and AI adoption is not statistically significant.

From the European perspective, Mihai et al. (2023) investigate the influence of emerging digital technologies, such as artificial intelligence (AI), the Internet of Things (IoT), and cloud computing, on the digital intensity index (DII). The results demonstrate that the DII is positively influenced by emerging IoT and cloud computing digital technologies, as well as the use of AI technologies based on machine learning and AI-based robotic process automation (RPA) software. Furthermore, the same positive influence was identified in human resource management and recruitment processes compared to the intensity with which these technologies are used in other economic processes. Based on these findings, this study offers persuasive arguments for implementing emerging digital technologies at the EU organizational level to achieve significant increases in digitalization levels.

Moreover, the findings of the study of Neumann et al. (2024) based on Swiss public organizations research, suggest that the importance of technological and organizational factors varies depending on the organization's stage in the adoption process. Surprisingly, the environmental factors are generally less critical. In addition, the study advances our theoretical understanding of the specificities of AI adoption in public organizations throughout the different adoption stages.

3 Methodology

The analysis is based on microdata collected for Advanced Technologies for Industry' (ATI) project. The project is the joint effort of European Commission, Directorate General for Internal Market, Industry, Entrepreneurship and SMEs and the European Innovation Council and Small and Medium-sized Enterprises Executive Agency. The main aim of the project was to analyze "the maturity level and adoption rate of advanced technologies, the trends in key enabling factors such as skills, investment or entrepreneurship and comparison of the EU27 performance to key competing economies" (Izsak et al. 2020). Partial activities of the project were two waves of

Table 1 Research sample

Country	#	Industry		Size band	#
France	311	Financial services	171	10–249	464
Germany	303	Government/education	167	250–499	334
Italy	279	Healthcare	130	500–999	345
Poland	165	Manufacturing discrete	120	1000+	404
Spain	281	Manufacturing process	154		
Denmark	80	Professional services	187		
Sweden	128	Retail, wholesale	161		
		Telecom, media	116		
		Transport, logistics	130		
		Utilities, oil, gas	111		
		Agriculture	100		

Source Advanced technologies for industry—methodological report (Izsak et al. 2020)

primary data collection in 2019 and 2020, where executives of firms with more than 10 employees described the situation of the company regarding implementations of advanced technologies.

In this paper, we present the analyses of the data from second (2020) wave of data collection, and we focus solely on implementation of artificial intelligence by the companies (the questionnaire covers altogether 16 types of advanced technologies). Table 1 describes the sample of the companies covered by the research.

To describe the usage of AI according to country, the most frequent usage purposes declared by firms, business goals, that drives the adoption of advanced technologies and sources of financing the advanced technology adoption, frequency tables were used. To identify the types of the companies, which has the highest probability to adopt artificial intelligence, we run the logistic regression with AI usage as the dependent variable and firm characteristics as the independent variables.

4 Results

Table 2 shows the level of AI adoption in researched countries—both frequency and percentual share of the respondents (within the country) which declare actual usage, planned usage, evaluation or not usage of the technology. In this realm, we can see Denmark and Sweden as a leaders of AI adoption, where 36.2% and 35.9% of respondents respectively declare actual implementation of AI in their operation. For Sweden, if we take to account also planned usage (26.6% of respondents), there are 62.5% of the firms which already use or plan to use artificial intelligence in their operation. France, Italy and Spain are catching up with around 30% portion of respondent with actual usage and 30% portion of planned use. In Poland, 40% of the respondents declare that they do not use and do not plan to use artificial intelligence, and only 16.4% is already using the technology.

To account for different corporate structure in the individual countries, we run a logistic regression to identify characteristics of the companies which correlates with higher probability of AI adoption. As dependent variable we calculated new dichotomous variable, which was equal to 1 if the company was already using or planning to use artificial intelligence and 0 otherwise. The independent variables were country of origin, company size, industrial segment, and multinational status. The regression results are shown in Table 3.

The regression showed statistically significant lower probability of AI usage in Poland (p-value < 0.05) even after controlling for other characteristics. Highest probability of AI usage is correlated with the seat of the company in Sweden. As to the other company characteristics, the size of company increases the probability of the AI adoption. Lower rates of AI usage were identified in the following segments: Government/Education, Retail and wholesale, Transport and logistics and Agriculture. Higher rates of AI usage could be identified in the companies with multinational status.

Table 2 Usage of AI by firms according to country

	Denmark		France		Germany		Italy	
	#	%	#	%	#	%	#	%
Not using and no plans	15	18.8	53	17.0	59	19.5	50	17.9
Already using	29	36.2	91	29.3	89	29.4	72	25.8
Plan to start using in the next 12 months	14	17.5	95	30.5	62	20.5	91	32.6
Evaluating, but no plans to adopt yet	22	27.5	60	19.3	70	23.1	59	21.1
Not aware of, or never heard of this technology	0	0.0	8	2.6	15	5.0	3	1.1
Don't know	0	0.0	4	1.3	8	2.6	4	1.4

	Poland		Spain		Sweden	
	#	%	#	%	#	%
Not using and no plans	66	40.0	37	13.2	21	16.4
Already using	27	16.4	82	29.2	46	35.9
Plan to start using in the next 12 months	29	17.6	84	29.9	34	26.6
Evaluating, but no plans to adopt yet	35	21.2	73	26.0	26	20.3
Not aware of, or never heard of this technology	2	1.2	1	0.4	0	0.0
Don't know	6	3.6	4	1.4	1	0.8

Source Own calculation

Table 3 Logistic regressions results (probability of using AI based on firms' characteristics)

	Coeff.	Std. er.	z	*p*-value	95% CI
Country (base Denmark)					
France	0.39	0.27	1.46	0.1432	[−0.13, 0.91]
Germany	−0.12	0.26	−0.44	0.6577	[−0.64, 0.40]
Italy	0.25	0.27	0.94	0.3461	[−0.27, 0.78]
Poland	−0.73	0.29	−2.52	0.0119	[−1.31, −0.16]
Spain	0.31	0.27	1.17	0.2426	[−0.21, 0.84]
Sweden	0.50	0.30	1.66	0.0970	[−0.09, 1.10]
Company size (base 10–249)					
250–499 empl	0.65	0.15	4.25	0.0000	[0.35, 0.95]
500–999 empl	0.88	0.16	5.68	0.0000	[0.58, 1.19]
1,000 + empl	1.06	0.15	6.97	0.0000	[0.76, 1.36]
Industry segment (base financial services)					
Gov/Edu	−0.65	0.23	−2.78	0.0054	[−1.11, −0.19]
Healthcare	−0.13	0.25	−0.53	0.5972	[−0.62, 0.36]
Manufacturing discrete	0.10	0.26	0.39	0.6990	[−0.42, 0.62]
Manufacturing process	−0.27	0.24	−1.13	0.2565	[−0.74, 0.20]
Professional services	−0.08	0.23	−0.37	0.7141	[−0.54, 0.37]
Retail, wholesale	−0.66	0.24	−2.78	0.0054	[−1.12, −0.19]
Telecom, media	0.33	0.27	1.22	0.2238	[−0.20, 0.85]
Transport, logistics	−0.69	0.25	−2.76	0.0058	[−1.18, −0.20]
Utilities, oil, gas	−0.10	0.26	−0.39	0.6995	[−0.62, 0.41]
Agriculture	−1.07	0.28	−3.85	0.0001	[−1.62, −0.53]
Multinational status (base national)					
Multinational	0.64	0.14	4.74	0.0000	[0.38, 0.91]
Intercept	−0.41	0.30	−1.36	0.1727	[−1.00, 0.18]

Source Own calculation

Table 4 shows the most frequent purposes of AI implementation declared by the respondents. The purposes are obviously segment specific; therefore the answers reflect also AI use by segments. In the first places there are task connected to automation of business operation. The next most frequent usage purposes are various intelligent assistance. AI is also widely used in marketing.

Table 5 shows the frequencies and proportion of the respondents who identified individual goals of advanced technology implementation. The table was created for the subsample of the companies, that claim to use or plan to use artificial intelligence. For respondents from all the countries, it was among the most important goals to

Table 4 Purpose which firms declare to use AI for

Purpose	#
IT automation	134
AI-powered robotic process automation software to support business	132
Intelligent assistants for internal decision support (providing information to business decision makers in a natural way)	126
Intelligent assistants for customer interactions (including chatbots and speech recognition)	118
Next best action for supply chain, operations and maintenance	112
Regulatory/compliance intelligence	107
Next best action for sales and marketing	105
New digital channels (voice banking, webchat, chatbots, virtual assistant)	92
Automated threat intelligence and prevention systems	91
Cognitive intelligence embedded in the final product	90
Automated threat intelligence and prevention systems	88
Automation of manual repetitive tasks (robotic process automation)	87
Accounts payable/accounts receivable automation	84
Governance, risk and compliance (e.g., fraud analysis and investigation, cybersecurity management/automated threat intelligence and prevention systems, compliance management)	83

Source Own calculation

implement the technology to reduce or optimize business processes and to improve or innovate product, service or program. What stands out as to the most important business goals which was pursued by advanced technology implementation, is the fact that in Poland, the companies stressed most the marketing goals (Attracting and retaining customers, expanding into new markets, segments or geographies, and improving reputation and brand awareness). Also, the least important goals distinguished the companies' approaches in the individual countries. In Denmark, the goal which was named least frequently was Improving reputation and brand awareness. French companies were least motivated by driving operational performance. In Germany, respondents were not motivated by acquiring, integrating, spinning off business. The same was true for respondents from Spain. In Poland, the least popular goal was Empowerment, development and acquisition of talent.

Table 6 describes sources of financing the implementation of advanced technologies. Again, the table covers only the subpopulation of the companies which declared present or future usage of artificial intelligence in their operation. Unsurprisingly, the most companies throughout all the countries declares financing form internal resources, what is typical for European companies. Interesting is, however, the fact that half of the Polish respondents declares government and EC investment as the source of financing.

Table 5 Business goals, that are driving adoption of advanced technologies according to countries (in % for companies that are using AI)

	Denmark	France	Germany	Italy	Poland	Spain	Sweden
Driving operational performance (EBITDA, revenues)	11.6	8.6	11.9	14.1	17.9	14.5	12.5
Attracting and retaining customers	23.3	23.1	14.6	19.6	30.4	19.9	18.8
Reducing operational and/or product costs, optimising business processes	44.2	34.4	41.1	34.4	25.0	36.1	36.2
Product, services or programme improvement and innovation	32.6	25.3	25.2	28.8	25.0	25.3	31.2
Expanding into new markets, segments or geographies	32.6	17.2	23.8	19.6	30.4	21.1	18.8
Managing regulatory compliance	18.6	23.1	21.2	20.9	17.9	14.5	21.2
Acquiring, integrating, spinning off business	7.0	11.8	6.6	12.9	21.4	9.6	17.5
Strengthening detection and resilience capabilities to guarantee security of people, facilities and resources	16.3	16.7	17.2	19.0	8.9	16.9	20.0
Improving detection and resilience capabilities against digital attacks	16.3	21.5	15.2	16.0	19.6	15.7	15.0
Empowerment, development and acquisition of talent	25.6	14.0	25.2	14.7	7.1	19.9	18.8
Improving reputation and brand awareness	4.7	16.1	13.2	16.0	28.6	20.5	20.0
Development of a broader, connected (partner) ecosystem	20.9	15.1	23.2	17.2	23.2	21.7	11.2
Commitment to sustainability and social welfare	9.3	17.7	12.6	19.6	25.0	18.7	20.0

Table 6 Source of funds for financing implementation of advanced technologies according to countries (for companies, which are using AI)

	Denmark	France	Germany	Italy	Poland	Spain	Sweden
Internal IT budget	46.5	44.6	55.0	44.8	57.1	50.0	45.0
Internal line-of-business budget	37.2	35.5	38.4	39.9	55.4	47.0	36.2
External investment through banks	20.9	21.5	14.6	20.9	46.4	24.7	11.2
External investment from venture capitalists	16.3	22.6	26.5	24.5	19.6	20.5	16.2
Government and EC investment in technology	27.9	20.4	13.9	20.2	50.0	16.3	21.2
Collaborative projects with organisations in the same value chain	23.3	25.3	18.5	22.1	16.1	17.5	25.0

5 Conclusion

The first goal of our paper was to identify the gaps in the AI adoption realized by different EU Member States. Based on our data sample, we could see Nordic countries (Denmark and Sweden) as the leaders in AI adoption. The South-Western countries (France, Italy and Spain) are catching the gap and the Eastern and Central European countries (Poland) are the least advanced in the AI adoption process. The sample do not allow to formulate solid conclusion about differences in levels of AI use in the EU countries (only data on 7 countries available), however the result of our comparison in line with other literature which identifies Nordic countries as technological leaders, and CEE countries as laggers (CEE countries are represented only by Poland) in the EU.

The second goal of our paper was to identify some specific drivers and barriers to AI adoption in the context of EU (manufacturing) firms. To identify the types of the companies, which has the highest probability to adopt AI, we run the logistic regression with AI usage as the dependent variable and firm characteristics as the independent variables. The larger the company is, the higher probability that it will use artificial intelligence in its operation. The higher probability is also correlated with multinational status. Companies are using AI mostly to reduce their cost and improve their products and services. The implementation of advanced technologies (including AI) is financed mainly from internal resources, in Poland, there is high proportion of government and EC investments.

To sum up, we can confirm that while the AI adoption advances in Europe, the foundational barriers between EU Member States persist to exist. Thus, the strategy of the EU institutions to booster the digital transformation in the EU (by the national recovery and resilience plans for example) has a potential to indoctrinate the existing country-specific differences. However, the further research on the impact of the EU Digital Decade initiatives is needed to assess the concrete results of the EU strategical decisions.

Acknowledgements Supported by the grant VEGA 1/0411/24 "Impact of public policy economic tools on digital readiness of companies" of the Research of Agency of the Ministry of Education, Research, Development and Youth of the Slovak Republic

References

Chatterjee S, Rana NP, Dwivedi YK, Baabdullah AM (2021) Understanding AI adoption in manufacturing and production firms using an integrated TAM-TOE model. Technol Forecast Soc Change 170:120880. https://doi.org/10.1016/j.techfore.2021.120880

Fuhrman P, Mooney JG (2021) Business adoption of artificial intelligence: an analysis of scope, intent and realized business benefit. Graziadio Bus Rev 24:1–18

Izsak K, Markianidou P, van de Velde E, Frietsch R, Kroll H, Wydra S, Glennon M, Garcia JM (2020) Advanced technologies for industry—methodological report. European Commission, European Union, Brussel

Kilian G (2020) White paper on artificial intelligence—a European approach to excellence and trust. European Border and Coast Guard Agency, Frontex

Mihai F, Aleca OE, Gheorghe M (2023) Digital transformation based on AI technologies in European Union organizations. Electronics 12:2386. https://doi.org/10.3390/electronics12112386

Neumann O, Guirguis K, Steiner R (2024) Exploring artificial intelligence adoption in public organizations: a comparative case study. Public Manag Rev 26:114–141. https://doi.org/10.1080/14719037.2022.2048685

Radhakrishnan J, Chattopadhyay M (2020) Determinants and barriers of artificial intelligence adoption—a literature review. In: Sharma SK, Dwivedi YK, Metri B, Rana NP (eds) Re-imagining diffusion and adoption of information technology and systems: a continuing conversation, vol 617. IFIP advances in information and communication technology. Springer International Publishing, Cham, pp 89–99. https://doi.org/10.1007/978-3-030-64849-7_9

The artificial intelligence index report 2023. Stanford University. https://aiindex.stanford.edu/report/

The global AI index. https://www.tortoisemedia.com/intelligence/global-ai/

The IBM global AI adoption index 2022. https://www.ibm.com/watson/resources/ai-adoption

The PWC's global artificial intelligence study: exploiting the AI revolution. https://www.pwc.com/gx/en/issues/data-and-analytics/publications/artificial-intelligence-study.html

Van Nguyen P, Viet DDT (2021) The retail industry's impact of artificial intelligence chatbots and augmented-reality applications: an examination of trust and satisfaction of the youth Vietnamese online shopping. Int J Supply Oper Manag 8

Conclusion: Learning from Crisis, Rethinking Europe

Ghislaine Pellat, Žaneta Lacová, and Marian Šuplata

This volume offers a wide-ranging reflection on the ways Europe is being transformed by a series of overlapping crises. From the first to the last chapter, the reader is invited to consider how individuals, firms, public institutions, and territories are navigating a time marked by economic fragility, political shifts, environmental urgency, and technological disruption.

The chapters collectively support the idea that today's crises, whether social, ecological, or geopolitical, cannot be treated as isolated shocks. They form part of a deeper condition of instability, often referred to as "polycrisis," which challenges traditional ways of managing public policy, economic development, and social cohesion. In this context, the contributions in the book provide valuable insights into how Europe is adapting, or failing to adapt, to this new reality.

A central message that emerges is the changing role of the individual and especially the expectations and actions of young people in the face of social and political uncertainty. Whether it is through civic engagement, educational innovation, or cultural exchange, youth appears as both vulnerable and active agents in the process of European reconstruction. Their desire for more meaningful and cooperative forms of belonging reflects broader shifts in the relationship between citizens and institutions. The evolution of work and labor relations is a recurring theme. This book explored

G. Pellat (✉)
Université Grenoble Alpes, Grenoble, France
e-mail: ghislaine.pellat@univ-grenoble-alpes.fr

Ž. Lacová
Matej Bel University, Banská Bystrica, Slovakia
e-mail: zaneta.lacova@umb.sk

M. Šuplata
Faculty of Management, Comenius University, Bratislava, Slovakia
e-mail: Marian.Suplata@uniba.sk

G. Pellat et al. (eds.), *Data-Centric Business and Applications*, Lecture Notes on Data Engineering and Communications Technologies 258,
https://doi.org/10.1007/978-3-031-95068-1_18

how technological change, especially digital tools and artificial intelligence, is transforming not only how people work, but also what they expect from work. Flexibility, purpose, and continuous skill development have become central to how younger generations engage with the job market. At the same time, companies must rethink how they attract and retain talent in a context where organizational culture and human values are as important as salary and benefits.

The role of businesses themselves is also redefined. Firms are no longer judged only by their economic performance, but also by their contribution to social and environmental sustainability. New reporting standards frameworks are helping shift corporate strategies toward transparency, responsibility, and long-term thinking. In the future, the companies which would embrace this evolution are better positioned to face future shocks.

A similar transformation is visible at the level of public institutions and territories pointing to the importance of local and regional initiatives in managing crises. After reading this book, the lecturer can understand that innovation is not limited to large cities or high-tech sectors but the resilience often depends on the ability of communities to adapt, cooperate, and act with limited resources.

To summarize this book, it invites readers to rethink Europe's place in the world. Whether through the lens of cross-border cooperation, digital sovereignty, or climate diplomacy, the European Union is shown to be at a crossroads. On the one hand, it faces internal divisions and structural weaknesses. On the other, it has the potential to offer a unique model of integrated, democratic, and sustainable development, if it can build the right tools and frameworks to support it. This volume offers not only a diagnosis of the present but also the outlines of a possible future. They encourage us to see crisis not just as breakdown, but as a space for learning and transformation. If there is one conclusion to draw, it is that Europe's strength lies in its ability to adapt while remaining committed to cooperation, equity, and democratic values. In this sense, the work presented here contributes to a wider conversation about how to make Europe not only more resilient, but also more just and forward-looking.

MIX
Papier aus verantwortungsvollen Quellen
Paper from responsible sources
FSC® C105338

FSC
www.fsc.org

If you have any concerns about our products,
you can contact us on
ProductSafety@springernature.com

In case Publisher is established outside the EU,
the EU authorized representative is:
Springer Nature Customer Service Center GmbH
Europaplatz 3, 69115 Heidelberg, Germany

Printed by Libri Plureos GmbH
in Hamburg, Germany